ETHICAL VEGETARIANISM AND VEGANISM

The protest against meat eating may turn out to be one of the most significant movements of our age. In terms of our relations with animals, it is difficult to think of a more urgent moral problem than the fate of billions of animals killed every year for human consumption.

This book argues that vegetarians and vegans are not only protestors, but also moral pioneers. It provides 25 chapters which stimulate further thought, exchange, and reflection on the morality of eating meat. A rich array of philosophical, religious, historical, cultural, and practical approaches challenge our assumptions about animals and how we should relate to them. This book provides global perspectives with insights from 11 countries: US, UK, Germany, France, Belgium, Israel, Austria, the Netherlands, Canada, South Africa, and Sweden. Focusing on food consumption practices, it critically foregrounds and unpacks key ethical rationales that underpin vegetarian and vegan lifestyles. It invites us to revisit our relations with animals as food, and as subjects of exploitation, suggesting that there are substantial moral, economic, and environmental reasons for changing our habits.

This timely contribution, edited by two of the leading experts within the field, offers a rich array of interdisciplinary insights on what ethical vegetarianism and veganism means. It will be of great interest to those studying and researching in the fields of animal geography and animal-studies, sociology, food studies and consumption, environmental studies, and cultural studies. This book will be of great appeal to animal protectionists, environmentalists, and humanitarians.

Andrew Linzey is director of the Oxford Centre for Animal Ethics, and a member of the Faculty of Theology in the University of Oxford. He is a visiting professor of animal theology at the University of Winchester, and the first professor of animal ethics at the Graduate Theological Foundation, Indiana.

Clair Linzey is the deputy director of the Oxford Centre for Animal Ethics. She holds an MA in theological studies from the University of St Andrews, and an MTS from Harvard Divinity School. She is currently pursuing a doctorate at the University of St Andrews on animal theology and Leonardo Boff.

ETHICAL VEGETARIANISM AND VEGANISM

Edited by Andrew Linzey and Clair Linzey

Routledge
Taylor & Francis Group
LONDON AND NEW YORK

First published 2019
by Routledge
2 Park Square, Milton Park, Abingdon, Oxon OX14 4RN

and by Routledge
711 Third Avenue, New York, NY 10017

Routledge is an imprint of the Taylor & Francis Group, an informa business

British Library Cataloguing in Publication Data
A catalogue record for this book is available from the British Library

Library of Congress Cataloging-in-Publication Data
A catalog record has been requested for this book

ISBN: 978-1-138-59096-0 (hbk)
ISBN: 978-1-138-59099-1 (pbk)
ISBN: 978-0-429-49074-3 (ebk)

Typeset in Bembo
by Taylor & Francis Books

For
MURIAL ARNAL
President of One Voice
and a courageous pioneer of animal protection

CONTENTS

ILLUSTRATIONS

Figures

Tables

CONTRIBUTORS

Marjorie Corbman is a doctoral candidate in systematic theology at Fordham University in New York City. She also holds a master's degree in the study of religions from the School of Oriental and African Studies in the University of London. Her interests include: theological interpretations of violence and suffering, the intersections of race and religion in the United States, and religious narratives shared between Judaism, Christianity, and Islam. Her dissertation research focuses on the influence of Black Muslim thought on the development of black liberation theology in the US.

Max Elder is an associate fellow of the Oxford Centre for Animal Ethics. His publications in books and academic journals range in topics from animal ethics to metaphysics to the philosophy of comedy. His research interests include animal ethics, fish ethics, and food ethics.

Andrew Fisher is a master's candidate in philosophy at the University of Cape Town. He holds a law degree and a bachelor's degree in social sciences (politics, philosophy and economics) with first-class honours in philosophy, and is an admitted attorney of the High Court of South Africa. His main areas of research lie at the intersection of law and practical ethics, including biomedical law and ethics, and animal law and ethics.

Bob Fischer is an assistant professor of philosophy at Texas State University and he co-directs the Society for the Study of Ethics and Animals. He is the editor of *The Moral Complexities of Eating Meat* (Oxford University Press, 2015) and *College Ethics* (Oxford University Press, 2016), and he is the author of *Modal Justification via Theories* (Springer, 2017).

Michael J. Gilmour is associate professor of New Testament and English literature at Providence University College in Manitoba, Canada, and a fellow of the

Oxford Centre for Animal Ethics. He gained his doctorate from McGill University. His most recent book is *Eden's Other Residents: The Bible and Animals* (Cascade Books, 2014).

Robyn Hederman is an associate fellow of the Oxford Centre for Animal Ethics. She is the Principle Law Clerk for a Supreme Court Justice of the State of New (Kings County). She has worked on various publications on behalf of the Animal Law Committee of the New York City Bar Association. She contributed to the search and seizure section of the committee's legal manual on animal fighting entitled "Prosecuting Animal Fighting and Live Depictions: Legal Issues under New York and Federal Law." She is also a member of the Animal Law Committee of the American Bar Association where she sits on the Animals in Science and Technology Subcommittee. Her research interests include, the study of the commonalities between animal advocacy and other nineteenth-century reform movements in the United States.

Jonas House is a lecturer in the sociology of consumption and households group at Wageningen University in the Netherlands. His doctoral research from the University of Sheffield investigated public acceptance of insects as food in the Netherlands, with a focus on how insect-based foods were integrating into established practices of food provisioning and consumption.

Monique R. E. Janssens is a doctoral candidate of the Rotterdam School for Management of Rotterdam University in the Netherlands. She is an independent communications consultant for ethical issues. Her publications include, *Dieren en Wij* ("Animals and us," A3boeken, 2008). Her research interests include, business ethics, animal ethics, and communications.

Rebecca Jenkins is a graduate of Trinity College Dublin School of Law, and of Lewis and Clark Law School's Animal Law LLM degree, the first program of its kind in the world. She has been published in Lewis and Clark's *Animal Law* Journal. Her research interests include, intersectional approaches to critical animal studies, critical race studies, and feminism.

Jeff Johnson is an associate professor of philosophy at St Catherine University in Saint Paul, Minnesota, where he regularly teaches courses on food ethics, with a special focus on the ethics of eating animals.

Robert Patrick Stone Lazo is an associate fellow of the Oxford Centre for Animal Ethics. His research interests include, complexity theory and emergence, philosophy of biology, American Pragmatism, and the animal ethics of the Roman poet Lucretius.

Andrew Linzey (*editor*) is director of the Oxford Centre for Animal Ethics, and a member of the Faculty of Theology in the University of Oxford. He is a visiting professor of animal theology at the University of Winchester, and the first professor of animal ethics at the Graduate Theological Foundation, Indiana. He is the author or editor of more than 20 books, including *Animal Theology* (SCM Press/University of Illinois Press, 1994); *Why Animal Suffering Matters* (Oxford University Press, 2009); and *The Global Guide to Animal Protection* (University of Illinois Press, 2013), and co-editor of the *Palgrave Macmillan Handbook of Practical Animal Ethics* (2018).

Clair Linzey (*editor*) is the deputy director of the Oxford Centre for Animal Ethics. She holds an MA in theological studies from the University of St Andrews, and an MTS from Harvard Divinity School. She is currently pursuing a doctorate at the University of St Andrews on animal theology and Leonardo Boff. She is associate editor of the *Journal of Animal Ethics* and assistant editor of the Palgrave Macmillan Animal Ethics Series. She is also director of the Annual Oxford Animal Ethics Summer School. She is co-editor with Andrew Linzey of *Animal Ethics for Veterinarians* (University of Illinois Press, 2017) and *The Ethical Case Against Animal Experiments* (University of Illinois Press, 2018).

Hadas Marcus is an associate fellow of the Oxford Centre for Animal Ethics, and a member of the Human-Animal Bond research forum in Israel. She teaches English for academic purposes at Tel Aviv University and Oranim Academic College of Education. She has published scholarly articles and spoken at numerous international conferences on animal welfare, the arts, ecocriticism and environmental justice.

Patricia McEachern is the Dorothy Jo Barker Endowed Professor of Animal Rights at Drury University, and the Director of the Drury University Forum on Animal Rights. In 2012, she established an innovative minor in Animal Studies at the undergraduate level. She directs and participates in a one-of-a kind course on Animal Ethics taught by a team of eight professors, all experts in their prospective fields. In addition, she teaches courses on Animals in Literature and on Animals and Contemporary Issues. Her publications include, *Deprivation and Power: Eating Disorders in Nineteenth-Century French Literature* (Greenwood, 1998) and *A Holy Life: The Writings of Saint Bernadette of Lourdes* (Ignatius, 2005).

Patrick Meyer-Glitza is a doctoral candidate at the Humboldt University of Berlin in the Faculty of Life Sciences, Department of Agricultural Economics. His areas of research include, non-killing cattle husbandry, agriculture, and art.

Corine Pelluchon is a full professor in philosophy at the University of Paris-Est-Marne-La-Vallée. She is the author of several essays and articles in moral and political philosophy (especially on Leo Strauss and on Emmanuel Levinas) and in applied ethics (medical and biomedical ethics, animal ethics, and environmental ethics). Her books include, *L'autonomie Brisée. Bioéthique et Philosophie,* (PUF, 2009); *Eléments pour une Éthique de la Vulnérabilité: Les Hommes, les Animaux, la*

Nature (Le Cerf, 2011). She is the winner of the Moron Prize of the French Academy for her *Les Nourritures: Philosophie du Corps Politique* (Le Seuil, 2015). She is also the winner of the Prize Edouard Bonnefous of the French Academy for Moral and Political Science and the Prize Paris-Liège 2016 for her *Éthique de la Consideration* (Le Seuil, 2018).

Simon Pulleyn is an independent scholar. He studied Classics at Balliol College, Oxford, in the 1980s and completed his doctorate at Merton College, where he taught until 1999. He has an LLM in Canon Law from the University of Cardiff and has a particular interest in the medieval period. He has published two books with Oxford University Press: *Prayer in Greek Religion* (1997) and *Homer: Iliad I with Introduction, Translation and Commentary* (2000). His *Homer: Odyssey I edited with Introduction, Translation, Commentary and Glossary* was published in 2018. He is a fellow of the Royal Historical Society and a fellow of the Oxford Centre for Animal Ethics. His interest in animals is interdisciplinary and crosses boundaries between literature, history and religion.

Kurt Remele is an associate professor of ethics and Catholic social thought at the University of Graz in Austria. He is a fellow of the Oxford Centre for Animal Ethics. His most recent book is *Die Würde des Tieres ist unantastbar. Eine neue christliche Tierethik* (Butzon and Bercker, 2016).

Jim Robinson is a doctoral candidate in systematic theology at Fordham University, in New York. His research unfolds at the intersection of ecotheology and religious pluralism.

Philip Sampson is a fellow of the Oxford Centre for Animal Ethics, and formerly a research fellow at the University of Southampton. His publications include, *Six Modern Myths* (IVP, 2001); "Humans, Animals and Others" in C. Falke (ed.) *Intersections in Christianity and Critical Theory* (Palgrave, 2010); and "Lord of Creation or Animal Among Animals?" in A. Linzey and C. Linzey eds., *Routledge Handbook of Religion and Animal Ethics* (2018). Research interests include, discourse and social change, the discourses of animal ethics and theology.

Jeanette Thelander is a student of veterinary medicine at the University of Copenhagen and the president of Animal Welfare Northeast Scania. She is a Swedish author who gained her MA degree from the University of Uppsala. Her master thesis in the ethnological field was entitled "Cannibals and Vegans: Distances and Borders in the World of Meat". She combines writing with freelance journalism.

Lucille Claire Thibodeau is writer-in-residence at Rivier University in Nashua, New Hampshire. A professor of English, she was president of Rivier University from 1997–2001. She earned her PhD from Harvard University in Comparative Literature and is a fellow of the American Council on Education. She has published a biography of the founder of Rivier University, *Crucible and Charism* (Hollis Publishing, 2009), and has presented widely on the poetry of Peter Abelard as well as on biblical literature. A member of the New Hampshire Animal Rights League and the newly formed New Hampshire Wildlife Coalition, her research interests include factory farming, animal ethics, animal law, and animal rights. Her current work focuses on ethical, political, and legal issues concerning the treatment of free living animals.

Tobias Thornes is a physicist, environmental scientist, and writer. He is currently nearing completion of his doctorate of philosophy in climate physics at the Atmospheric, Oceanic and Planetary Physics department of the University of Oxford. He obtained a master's degree in physics and astronomy from the University of Durham in 2014. His research interests include astrophysics, atmospheric physics, mathematics, general philosophy, history, environmental history, and climatic change. He has written a number of research articles on these and related topics published in *Weather*, the *Quarterly Journal of the Royal Meteorological Society*, and the *Journal of Animal Ethics*.

Kenneth Valpey (Krishna Kshetra Swami) received his doctorate at University of Oxford in 2004, with a dissertation on Vaiṣṇava-Hindu temple ritual. In 2006, his dissertation was published in revised form with the Routledge/OCHS Hindu Studies Series as a monograph entitled *Attending Krsna's Image: Caitanya Vaisnava Murti-seva as Devotional Truth*. He has taught courses in Indian and Asian religions at the University of Florida and at the Chinese University of Hong Kong. He is currently dean of studies at Bhaktivedanta College, Septon, Belgium, and works with Dr Ravi M. Gupta on the OCHS Bhāgavata Purāṇa Research Project, out of which they have produced two volumes published by Columbia University Press. Dr Valpey is also a fellow of the Oxford Centre for Animal Ethics.

Floryt van Wesel is an assistant professor at the Department of Methodology and Statistics, Faculty of Social and Behavioural Sciences at Utrecht University in the Netherlands. She is a consultant in qualitative and mixed-methods methodology. Her publications include, "Priors and Prejudice: Using existing knowledge in social science research" (Dissertation, 2011). Her research interests include, mixed methods methodology and systematic review methodology.

Stephen Patrick Kieran Walsh is a research fellow at Vrije Universiteit Brussel, and chair of The Vegan Society. He is the author of *Plant Based Nutrition and Health* (The Vegan Society, 2003). His articles include "Dietary Patterns in Plant-Based, Vegetarian, and Omnivorous Diets," in *Vegetarian and Plant-Based Diets in Health and Disease Prevention* (2017). His areas of research are vegan nutrition and health, vitamin B12, and numbers of vegans.

Joe Wills is a lecturer in law at the University of Leicester. His publications include: *Contesting World Order? Socioeconomic Rights and Global Justice Movements* (Cambridge University Press, 2017).

INTRODUCTION

Vegetarianism as ethical protest

Andrew Linzey and Clair Linzey

I

Theologian Karl Barth wrote against vegetarianism on the ground that it represented "a wanton anticipation of … the aeon for which we [i.e. Christian believers] hope". And yet even Barth admitted that meat eating was so similar to homicide that it was only proper and necessary that it should be accompanied by a "radical protest" against the whole possibility.[1]

Although most people in the world have no moral qualms about eating meat, there is a growing and determined minority worldwide who choose vegetarianism or veganism as a moral protest. Of course, there are longstanding traditions of vegetarianism in China and particularly India, but it is largely in the West that we have seen a recent upsurge of campaigning non-meat eaters. Estimates indicate that there are already 3.5 million vegans in the UK alone.[2] According to a recent report 6 per cent of US consumers identify as vegan.[3]

Some definitions may be helpful. *Ethical* vegetarianism or veganism should, of course, be distinguished from non-ethical forms, for example vegetarianism motivated by economic considerations or for purely health reasons. It is often supposed that vegetarians are people who simply abstain from meat, whereas the official definition supplied by the Vegetarian Society defines a vegetarian as someone who "lives on a diet of grains, pulses, legumes, nuts, seeds, vegetables, fruits, fungi, algae, yeast and/or some other non-animal-based foods (e.g. salt) with, or without, dairy products, honey and/or eggs".[4] Notice how flesh foods (fish, flesh, and fowl) are all excluded, but consuming dairy produce, honey and eggs are permitted. Contrary to common belief, the eating of fish, but not meat (sometimes called pescetarianism), is not officially recognized as a vegetarian diet.

The subsequent line of the definition, however, makes the foregoing rather problematic. It reads:

> A vegetarian does not eat foods that consist of, or have been produced with the aid of products consisting of or created from, any part of the body of a living or dead animal. This includes meat, poultry, fish, shellfish, insects, by-products of slaughter or any food made with processing aids created from these.[5]

This is problematic because, as defined, all milk products (including butter, cream and cheese) are necessarily excluded, since it is difficult to see how milk can be produced without "the aid of products consisting of or created from, any part of the body of a living or dead animal".

Perhaps better expressed: There are different kind of vegetarians variously described as lacto-ovo-vegetarians who eat both dairy products and eggs, lacto-vegetarians who eat dairy products but avoid eggs, and ovo-vegetarians who eat eggs but not dairy products.

The Vegan Society defines veganism as "a way of living which seeks to exclude, as far as is possible and practicable, all forms of exploitation of, and cruelty to, animals for food, clothing or any other purpose".[6] Vegans are well known for their rejection of dairy produce and eggs, but the above definition does add a critical qualifying sub-clause "as far as is possible and practicable", which presumably allows for situations of scarcity or conflict, though these are undefined. It is not difficult of course to imagine a variety of circumstances in which strict veganism can be compromised.

Again, however, as with the Vegetarian Society, the subsequent line of the definition makes the foregoing rather problematic. It reads: "One thing all vegans have in common is a plant-based diet avoiding all animal foods such as meat (including fish, shellfish and insects), dairy, eggs and honey – as well as products like leather and any tested on animals." It is the last line, namely "products … tested on animals" that gives cause for concern. The reason is simple: It is impossible to live without utilizing products that have been tested on animals, since testing has at one time or another included virtually all vegetables, and even salt or water. As has been noted elsewhere:

> it would be impossible to stop using each and every product that has, at some time or another, been tested on animals for the simple reason that every commercial (and not even only commercial) product has been tested, at some time or another, on animals. The range of usage of animals is so extensive – including, but not limited to, fire extinguisher substances, dyes, paints, hair sprays, weaponry, poisons, radiation, plastics, agrichemicals, and even vegetable and herbal products – that no one can live entirely free of products tested on animals. The notorious LD50 poisoning test – designed to ascertain the dose at which 50 percent of the animals to whom the substance is given die – has also been carried out using water, the very stuff of life.[7]

We can see immediately, then, that the official definitions are not entirely watertight, consistent, or indeed even practicable. The difficulty of establishing clear, unambiguous definitions is not entirely surprising. The vegan ideal of living without any recourse to animal products is pretty much unattainable in modern society. Animal products can be found everywhere, not only in food, but also in dyes, glues, paints, film making equipment, clothing, footwear, furniture, household items, most drinks (including wine and beer), ceramics, toiletries, and even UK five pound notes. Elsewhere, one of us has indicated how parts of a pig's body can be found in more than one hundred items we may come in contact with,[8] and the pig is only one of the hundreds of animals that we daily exploit throughout the world. Living free of such exploitation is an immense task and can, if pursued rigorously, consume one's whole waking life. Our exploitation of animals is relentless and unremitting: We hunt, ride, shoot, fish, wear, eat, cage, exhibit, factory farm and experiment on billions every year. None of us are completely free of this exploitation, directly or indirectly, either through the food we eat, the products we buy, or the taxes we pay.

Quite apart from practical difficulties, there is a social cost as well. Humans are social creatures who live in communities and institutions that help sustain our daily life. Since eating is one of the most important forms of social interaction, the adoption of a different diet can provoke among observers resentment, feelings of guilt or defensiveness. More specifically, vegetarians or "veggies" – as they are now often called – suffer from ridicule or downright hostility. Families or close-knit social groups can be the most uncomprehending of all. Individuals may have difficulty in surmounting these forms of social ostracism, let alone maintaining their own commitment to consistency. Vegetarianism is the only moral issue that requires at least three practical choices every day.

Moreover, while vegetarian food can be outstandingly delicious, the public provision of vegetarian meals in restaurants, at social gatherings and in institutions often falls below acceptable standards. One example may suffice. Following many complaints from students concerning the provision (or rather the lack of it) of vegetarian and vegan food at Oxford University, a survey showed that students had many experiences of substandard, uninteresting, or even no vegetarian and vegan food options. Comments from students (who are after all customers who have to pay for their food in colleges and halls) include the following:

> Vegetarian/vegan meals at most colleges are only pretty average. They have a long way to go provide nutritionally adequate and tasty food for veggies.
>
> Incredibly expensive food and compulsory catering charges, and never a vegan option in hall anyway so I could never eat there.
>
> All colleges need to improve. Vegetarians need protein. Vegetarians don't all want cheese!
>
> When I tell the kitchen staff that I've ordered a vegan meal at a formal [hall], they usually react negatively and seem annoyed.[9]

In the light of all this, the question might not unnaturally be asked: What is it then that drives individuals to tolerate such hardships? To protest when simply conforming would make life so much easier? Even running the risk of marginalizing oneself from full social life, which so often centres around the consumption of animal products?

To understand this protest, we need to examine the three overlapping but separate underlying concerns, which are reflected in the structure of this book.

II

The first is the moral concern about killing sentient beings when it is not strictly necessary. Whether eating meat was necessary at one point or another in the history of humankind is a moot point. We just don't know for certain. Neither is it clear whether some humans living in remote and inaccessible places nowadays require animal protein in order to survive. But what we can be clear about is that in almost all countries, especially in the Western world, it is not now necessary to kill animals in order to sustain a healthy diet. Indeed, as Stephen Patrick Kieran Walsh ("Why food derived from animals are not necessary for human health") points out, a diet completely free of animal products has now been declared a healthy option by competent authorities – a fact that assumes massive moral significance. The position of the Academy of Nutrition and Dietetics is that:

> appropriately planned vegetarian, including vegan, diets are healthful, nutritionally adequate, and may provide health benefits for the prevention and treatment of certain diseases. These diets are appropriate for all stages of the life cycle, including pregnancy, lactation, infancy, childhood, adolescence, older adulthood, and for athletes. Plant-based diets are more environmentally sustainable than diets rich in animal products because they use fewer natural resources and are associated with much less environmental damage. Vegetarians and vegans are at reduced risk of certain health conditions, including ischemic heart disease, type 2 diabetes, hypertension, certain types of cancer, and obesity … Vegans need reliable sources of vitamin B-12, such as fortified foods or supplements.[10]

Of course, the need for vitamin B12 has long been recognized as a weakness in strict vegan diets, but with the addition of fortitude foods or supplements it no longer poses the issue it once did.

The moral question then is straightforward: Whatever may have been true in the past and may yet be true for some living in remote and inaccessible parts of the world, why should one kill sentient beings for food when it isn't necessary? The stress on sentience here requires elaboration. Some people suppose that vegetarians oppose the killing of any form of life, hence the rather silly comment sometimes heard that "I don't know what vegetarians have against carrots". But the vegetarian position needs to be sharply distinguished from a general reverence for life position. Although reverence for all living things is clearly an important sensitivity, the case for vegetarianism most clearly focusses on killing sentient beings. By "sentient", we mean philosophically speaking the fact that some animals are capable of pleasure and pain. There is now a great deal of scientific literature that shows that mammals, birds, reptiles and some fish are capable of suffering. The importance of this capacity for suffering should not be underestimated. To suffer, in this context, does not just mean the experience of adverse physical stimuli, but also mental and emotional suffering, such as fear, distress, foreboding, trauma, terror, anticipation, stress – only to a greater or lesser extent than humans themselves experience.[11] Moreover, to be a being that can suffer presupposes not only feelings, but a unique experience of self, cognitive awareness, and subjectivity. Such a being has not only a biology but also a biography.

It follows that killing a sentient being when one has no good reason to do so cannot be morally right. The act of killing extinguishes a unique form of life that has value to the individual concerned. It takes away a form of individual awareness and subjectivity that cannot be replaced. While not perhaps the ultimate injury, since prolonged and un-relievable suffering may sometimes be worse, death is still an injury and, what is more, in many circumstances, an avoidable one.

Part I further details the philosophical and religiously inspired arguments. Isn't it justifiable to kill animals so long as they have a happy life and are killed humanely? Andrew Fisher ("Against killing 'happy' animals") explains why it is wrong to kill, even if the animals concerned had a happy life (accepting of course that many don't). On balance isn't it better that they existed, even if they are eventually killed? Since they are sentient beings that are self-aware, their needless destruction cannot be justified by reference to the so-called "replaceability" argument. Fisher is clear:

> If animals count morally, we should not sacrifice their vital interests to promote the non-vital interests of ourselves. Raising and killing animals for food (even if raised happily and killed painlessly) overrides vital interests of animals to remain alive through their natural lifespan.[12]

Other philosophical voices variously amplify and support this position. Corine Pelluchon ("Food ethics and justice toward animals") argues that an ethics of consideration must logically include consideration for other sentient beings. Eating reflects who we are and indicates the limits or otherwise of our moral solicitude. Bob Fischer ("Animals as honorary humans") rejects Cora Diamond's view that relating to animals as "fellow creatures" is sufficient ethically. Rather he proposes describing sentients as "honorary humans". He defends the approach of strict veganism, "namely, that it's *pro tanto* wrong to eat *any* animal products, whatever their origin".[13] But since animals may not have a concept of death how can they be harmed by it? Robert Patrick Stone Lazo ("Nonhuman animals' desires and their moral relevance") reacts against the argument that animals cannot conceive of themselves existing over time and do not therefore have future orientated desires. He argues that "Death can be an evil for nonhuman animals, since it can frustrate their desires for the future that we understand as temporally extended, whether they are aware of that extension or not".[14]

It is sometimes supposed that the early Greeks were animal-friendly, and it is true that some, like Pythagoras, advocated a vegetarian diet. Indeed, vegetarianism was known for centuries as the Pythagorean diet. But Simon Pulleyn ("Why vegetarianism wasn't on the menu in early Greece") shows how inconceivable it was for many Greeks "to think of animals as deserving of protection and a life of their own". And the reason it seems was the prevalence of animal sacrifice:

> the early Greeks were not unsophisticated. They knew that animals had breath and mental apparatus. To some extent, this made them kin with humans. But their sense of a soul had not developed to a point where it caused the eating of animals to seem problematic. And for all that Orphics, Pythagoreans, and neo-Platonists put the contrary case from time to time, the fact is that early notions of sacrifice being normative were so deeply imbedded by Homer's own day that nothing was likely to change it and it persisted until the end of antiquity and beyond.[15]

It is often supposed that meat eating is divinely sanctioned, and the Bible in particular is frequently castigated in this regard. But, as Pulleyn notes, the notion of an original Golden Age of peaceableness between creatures found in the thought of Hesiod and other Greek philosophers is paralleled in the Hebrew Bible itself. In Genesis 1.29–30 both humans and animals are given a vegan diet, a position only changed after the fall and the flood symbolizing the descent of humans into violence. Vegetarians and vegans then can claim some canonical authority for their position and argue that God's original will was for a peaceful and non-violent creation. Why is it then that some parts of Christianity, and its evangelical form in particular, appear hostile to vegetarianism? Philip Sampson ("The ethics of eating in 'evangelical' discourse: 1600–1876") traces the history and theology of evangelicals and their Puritan and Calvinist forebears. It becomes clear that there was a strong sub-tradition of care for animals within evangelical discourse, one that saw flesh eating as a justifiable result of human sin yet strongly abhorred cruelty. And some certainly adopted vegetarianism, including Methodist founder John Wesley (although perhaps only intermittently), celebrated Baptist preacher C. H. Spurgeon, and the founders of the Salvation Army William and Catherine Booth. But the strange thing is that Christian blindness to Genesis 1.29–30 (the divine command to be vegan) continues even in writers that were otherwise known for their progressive views on animals. Michael J. Gilmour ("Myth and meat: C. S. Lewis sidesteps Genesis.1.29–30") cites the example of C. S. Lewis whose own myth making was inspired by the prelapsarian theology of Genesis, yet utterly failed to grasp the practical import of Genesis 1.29–30.

For some, going the whole hog (an unfortunate speciesist expression) and going vegan is beyond them, and so they variously seek middle positions, like having some meat-free meals (demi-vegetarianism), eating only white meat (usually on health grounds), or eating fish but not meat. Max Elder ("The moral poverty of pescetarianism") takes issue with those who believe that pescetarianism is a moral improvement over omnivory. He points to the latest scientific work that indicates that fish feel pain, echoing the delightfully understated view of the Cambridge Declaration of Consciousness that "the absence of a neocortex does not appear to preclude an organism from experiencing affective states".[16] Elder concludes that "The morally motivated pescetarian appears to be an oxymoron".[17] Kurt Remele ("There is something fishy about eating fish, even on Fridays: on Christian abstinence from meat, piscine sentience, and a fish Called Jesus") continues the theme by exploring the way in which the Christian tradition has required fasting and abstinence, especially of fish on Fridays. Remele shows how the discoveries of piscine sentience should modify even this restrictive tradition and argues for Christian vegetarianism in honor of a fish called Jesus.

III

The second concern relates to the harms or cruelty involved in routinized and institutionalized killing of sentient beings. Not all individuals are predisposed to vegetarianism or veganism because of a principled opposition to the unnecessary killing of sentients, rather it is the harm or suffering involved that is their primary motivation. Collected in Part II are some of the voices that most reflect this concern.

Robyn Hederman ("'The cost of cruelty': Henry Bergh and the abattoirs") begins by reminding us of the close historical connection between abattoirs and cruelty to animals. Bergh was the lauded founder of the American Society for the Prevention of Cruelty to Animals – the first animal protection society in the United States. In addition, he pioneered the first US anti-animal cruelty legislation. While not a vegetarian, Hederman shows how he was "repulsed by how animals were transported, housed and slaughtered".[18] Although he "exposed the slaughterhouses to the public",[19] it has to be admitted that this only drove such slaughter underground, eventually making public slaughter unacceptable. In this, animal protectionists were victims of their own success. While public slaughter was a practice that lent itself to the brutalization of bystanders, private abattoirs hid the awfulness of killing from the public gaze. Thus, the link between cruelty and meat eating became opaque. The task now for animal advocates is to expose the realities of slaughter, as Bergh himself did.

The next three chapters seek to lift the lid on cruelty in contemporary animal farming. Lucille Claire Thibodeau ("'All creation groans': The lives of factory farmed animals in the United States") and Patricia McEachern ("L'Enfer, c'est nous autres: Institutionalized cruelty as standard industry practice in animal agriculture in the United States") both focus on the cruel practices that have now become commonplace in the biggest meat-eating country in the world. Both make for grim, but essential, reading. McEachern, who has the distinction of holding the first chair of animal rights in the world, also shows how difficult it is to open up rational discussion of these issues even in universities. When she sought to establish an animal rights course at Drury University, she found that "cattlemen attempted to muscle and intimidate Drury's president into squashing the class".[20] Likewise "ag-gag" laws now passed in some states "essentially [punish and] gag journalists, activists, undercover investigators, and even animal agriculture employees" who dare to speak out.[21] Yet, the need for people to know precisely what transpires in animal farming (which, after all, is mostly subsidized by the tax payer) could not be greater.

Jeff Johnson ("Welfare and productivity in animal agriculture") argues that when questions are raised about standard practices in animal agriculture, we are often told that we ought not to worry too much, since producers who engage in these practices have an interest in promoting animal welfare. This sort of claim is designed to offer us a kind of moral cover. But, on closer inspection, it becomes clear that "A conception of animal welfare that's so closely tied to the notion of productivity and, ultimately, profit seems to me to have far more to do with the interests of producers than it does with the interest of the animals themselves".[22]

These empirical critiques are then reinforced by theological ones. Jim Robinson ("Taking on the gaze of Jesus: Perceiving the factory farm in a sacramental world") maintains that mass industrial slaughter is incompatible with a sacramental view of the world, which sees divinity, specifically the Holy Spirit, as suffused within creation. Robinson argues that factory farms are "obscured sites of desecration".[23] Desecration because we fail to see the value of other living creatures. In the case of hogs alone there are,

> Annually, 80 million unique beings, who are loved into being by God, whose bodies are saturated with mystical meaning and revelatory potential, who might communicate to us the divine caress in the textures of their flesh, are ensnared in and eclipsed by the machinery of our factory farms. We cannot see them. We cannot reach them. We cannot sense their suffering. The workers recruited to electrocute, stab, and carve the bodies of these pigs – those who do the dirty work of desecration – wear earplugs in order to muffle the howls.[24]

And Marjorie Corbman ("'A lamb as it had been slain': Mortal (animal) bodies in the Abrahamic traditions") in a moving meditation explains how the narratives from the Abrahamic religions (Judaism, Christianity, and Islam) can help us in sustaining a proper perspective on animals as fellow creatures and nurture compassion.

But does animal farming *have* to involve suffering and death? Patrick Meyer-Glitza ("Cattle husbandry without slaughtering: A lifetime of care is fair") rejects the abolitionist view of some animal advocates that all animal use is immoral, and maintains that a humane and symbiotic relationship, specifically with cattle, is feasible and workable in practice. He enumerates five principles of care that can undergird such a model of "animal husbandry" based on interviews with farmers who keep cattle but do not send them to slaughter at the end of their "productive" lives. This takes us to the heart of the disagreement between vegans and vegetarians. The usual vegan argument against dairy production is that the relationship with cattle is inherently exploitative: In almost all cases, the calf is separated from its mother after just a few hours or days (which causes emotional trauma to both); the male calves are then usually sent for slaughter; excessive milking reduces cattle to little more than commodities and engenders poor health and physical exhaustion, and then, when they can no long produce the quantities of milk that the farmer requires, they are sent for slaughter.[25] These considerations certainly make dairy produce morally problematic. But, if cattle can be kept while avoiding all these objections, it may be morally acceptable. The problem is, of course, that it is always a mistake to argue from some limited occasions to a general rule. Even if *some* (actually tiny) amount of milk can be obtained in this symbiotic way, it still does not justify the general practice in the vast number of occasions when milk is the result of unethical treatment. The matter is further complicated when it is further appreciated that many forms of cheese comprise animal rennet (normally sourced from the stomach of slaughtered calves or sheep)[26] and therefore are not vegetarian at all.

The same, we should say, is also true of egg production. In principle, there may be no objection to eating eggs since they are not sentient beings, and hens naturally produce without male intervention eggs that have no chance of growing into chicks. Hens in their natural state are not forced into egg laying, it is only in industrialized farming that their natural cycle is manipulated. Contemporary industrialized egg production is predicated on two further ugly features. The first is the way in which male chicks are gassed, suffocated or homogenized (fed through chopping machines en masse) within 72 hours of birth. The reason is simple: Male chicks cannot lay eggs. Second, when their productivity declines, all hens are slaughtered usually by gassing or electrical stunning.[27]

Ironically, more animal advocates have become keepers of hens than ever before. Instead of the expense of having battery chickens "cleared out" when they are no longer judged as sufficiently productive, hen farmers sometimes give animal people the chance to save them by taking them away.[28] The result is that many animal protectionists, including vegetarians and vegans, now keep hens and use their eggs for food or give them away to others in order to lessen the demand for commercial egg production.

But, as is sometimes remarked, what about insects? Jonas House ("Are insects animals? The ethical position of insects in Dutch vegetarian diets") examines the position in the Netherlands where "entomophagy" (insect consumption) has become established, even to the point of selling buffalo worms in supermarkets. Despite the Vegan Society's definition, which lists insects as one of the forbidden foods for vegans,[29] the moral (as distinct from the ascetical) case against eating insects is unclear. The unclarity arises from the lack of evidence for sentiency among insect species, such as crickets, grasshoppers, mealworms, and buffalo worms. When it comes to birds, mammals and reptiles, even many fish, we may be clear that a line is being crossed, but when it comes to insects we know so little about them as to be unable to make informed judgements. House is tantalizingly equivocal:

> My primary intention is to demonstrate how the introduction of a novel animal species to the human food system in an industrialized Western country "problematizes" ethically oriented diets, by raising a number of hitherto

> largely unexplored questions about animal ethics, and illuminating a number of taken-for-granted assumptions about the nature of animal life on which ethical diets frequently appear to be based.[30]

Clearly this particular debate has only just begun.

In short, then, while there may be areas of unknowing and even honorable exceptions, it is evident that to consume egg and dairy products is to be complicit in the systemic exploitation of animals. The moral message seems clear: Even if we are only concerned for harm and suffering, rather than death, then ethical vegetarianism and veganism is the only way to opt out the systemic exploitation of animals as represented by contemporary farming.

IV

The third concern variously relates to the human and environmental costs of institutionalized killing. Eating meat and dairy does not just affect animals, of course – it affects the farmers who raise the animals, the slaughterhouse workers who kill them, the commercial companies that propagate animal products, and the consumers who buy and eat them. What impact do these facts have on our lives and the way we live?

Jeanette Thelander ("Our ambivalent relations with animals") begins by highlighting the dichotomy that meat eating poses in human culture. On one hand, empathy for animals abounds and yet, on the other, we routinely hurt and kill billions of animals. Thelander offers an ethnological approach and discovers that such ambivalence is maintained by two strategies – creating distance and drawing lines – that both foster denial and dishonesty. But she concludes that these strategies which create cognitive dissonance are weakening, and the future requires a new strategy of "facing up to reality, meeting eye to eye with the absent referent and stop the eating of meat".[31] From a different perspective, Kenneth Valpey (Krishna Kshetra Swami) ("From devouring to honouring: A Vaishnava-Hindu therapeutic perspective on human culinary choice") suggests that meat eating is nothing less than a "civilizational disease" perpetuated by bad habit.[32] Valpey says that we have, in effect, become desensitized to violence and blind to the reality of mass industrial slaughter. His solution is nothing less than therapy to help kick the habits of a lifetime. He utilizes the resources within the Vaishnava-Hindu tradition – requiring attention to three currents of thought and practice – Dharma, Yoga, and Bhakti – to liberate ourselves as well as the animals.

Rebecca Jenkins ("The other ghosts in our machine: meat processing and slaughter house workers in the United States of America") drives home the human cost of industrialized killing. Slaughterhouses are known, of course, for their killing of animals, but what is not always grasped is the appalling working conditions for slaughterers themselves. Some may think that since Upton Sinclair's classic American novel, *The Jungle* (1906), which famously exposed the harsh conditions in the Chicago stockyards that things would have improved. Apparently not. Jenkins details the human rights abuses of slaughterhouse workers, many of them immigrants, who are subjected to exploitative practices. It is striking that the Human Rights Watch report in 2005 reported that the abuses are directly linked to the vulnerable immigration status of workers.[33] It seems then that we have two exploited subjects in slaughterhouses, not only animals, but also human workers. In addition to the psychological and social problems experience by workers, Jenkins also raises the negative impact that such facilities have on the environment. Just one fact must suffice: In North Carolina there are 7.2 times as many swine CAFOs (concentrated animal feeding operations, and associated slaughtering and marketing facilities) located within the areas of highest poverty as compared within the areas of lowest poverty.[34]

Tobias Thornes ("Animal agriculture and climate change") demonstrates the detrimental impact of animal agriculture in contributing to greenhouse gas emissions. His conclusion is that the widespread adoption of a vegetarian diet provides the best solution to this urgent problem. But is a vegetarian or vegan diet really less harmful to the environment, and animals themselves? Although specifically veganism is often portrayed as a way of saving animals from death, it cannot be denied that growing crops often involves the elimination of competitor species, which means that many so-called "pest" species are routinely killed. Joe Wills ("The intentional killing of field animals and ethical veganism") tries to resolve this conundrum by offering morally salient distinctions between intentional and unintentional killing. He concludes that while the unintentional killing should not incur the same moral opprobrium as intentional killing, "this does not make the unnecessary and disproportionate killing of field animals morally permissible, it merely makes such acts less wrongful in one sense, than the opportunistic killing of farmed animals".[35]

The book concludes with two chapters devoted to the mechanisms of change. Moral exhortation is not sufficient to create social change, let alone one that involves what people put on their plate three times a day. Vegetarians need to devise new strategies in order to highlight the negative aspects of meat eating. Hadas Marcus ("How visual culture can promote ethical dietary choices") stresses the moral significance of seeing, of perceiving animals in a new light. This requires animal people paying special attention to the ways in which animals are portrayed in visual culture and in relation to meat eating in particular. Our insensitivity to animals is part of a much less well understood cultural blindness that needs to be challenged. Monique R. E. Janssens and Floryt van Wesel ("Leadership, partnership and championship as drivers for animal ethics in the Western food industry") seek to locate through sociological analysis the drivers for change in the Western food industry. Such an international, multi-million business might appear impervious to change, yet Janssens and Wesel are adamant that change is possible and can be enabled by companies taking seriously their own declared corporate social responsibility to people, nature, animals, and sustainability.

Ultimately, however, the choice is an individual one, but companies are capable of change and consumers have real power to influence that change.

V

We end where we began. Vegetarians, as Barth rightly observed, are protestors, modern day protestants. The protest against meat eating may turn out to be one of the most significant protest movements of our age, as important as previous campaigns for the rights of women or the rights of same-sex people. In terms of our relations with animals, it is difficult to think of a more urgent moral problem than the fate of literally billions of sentient creatures killed every year for human consumption. In terms of suffering, harm, and death, the issue is compelling for animal protectionists, and also, as we have shown, for environmentalists and humanitarians more generally. Vegetarians and vegans are perhaps not just protestors, but moral pioneers.

All in all, this book provides 25 original chapters, which we hope will stimulate further thought, exchange, and reflection. They offer a rich array of philosophical, religious, historical, cultural and practical approaches many of which challenge our taken for granted assumptions about animals and how we should relate to them. We are indebted to all the contributors who have opened new doors, challenged well-worn assumptions, and offered new points of view. In the words often attributed to Max Weber: "all knowledge comes from a point of view".

As we say in most of our books, we have, as editors, tried to pay special attention to the question of ethical language. So much of our historic language denigrates animals as "beasts", "brutes", "subhumans", or "dumb brutes" or deploys negative metaphors about animals, such as "snake in the grass", "cunning as a fox", "greedy as a pig", and "stupid cow". With these terms we libel animals, and not only animals, of course. Therefore, we have found it essential to pioneer an ethical or at least more objective terminology. We have used "he" or "she" instead of "it" for individual animals. We have used "free-living", "free-roaming", or simply "free" instead of "wild" because wildness has negative connotations. We have also used the term "farmed animals" rather than "livestock" or "farm animals". Where substitution has proved problematic, we have simply placed the objectionable words in quotation marks. Needless to say, exceptions to ethical language have been made in the quotation of texts, particularly historical writings.

This book is a project of the Oxford Centre for Animal Ethics, of which we are the directors. The Centre aims to pioneer ethical perspectives on animals through academic research, teaching, and publication. Through the *Journal of Animal Ethics*, the Palgrave Macmillan book series on animal ethics, and the Annual Oxford Animal Ethics Summer Schools, we are putting animals on the intellectual agenda and helping people to think differently about animals. We cannot change the world for animals without changing our ideas about them. More than 100 academics from the humanities and the sciences are fellows of the Centre, and we are grateful to many who have contributed in various ways to the making of this book.

We would also like to express our gratitude to the editorial team at Routledge, especially Faye Leerink, Kelly Cracknell, and Ruth Anderson for their encouragement and expert assistance. All the royalties from the sale of the book will go to the Oxford Centre.

Oxford Centre for Animal Ethics
www.oxfordanimalethics.com
May 2018

Notes

1 Karl Barth, *Church Dogmatics: The Doctrine of Creation*, Vol. III, pt. 4 (Edinburgh, Scotland: T and T Clark, 1961), n. 355–356.
2 Olivia Petter, "Number of Vegans in the UK Soars to 3.5 Million, Survey Finds," *The Independent*, April 3, 2018, www.independent.co.uk/life-style/food-and-drink/vegans-uk-rise-popularity-plant-based-diets-veganism-figures-survey-compare-the-market-a8286471.html.
3 Top Trends in Prepared Foods 2017, "Exploring Trends in Meat, Fish and Seafood; Pasta, Noodles and Rice; Prepared Meals; Savory Deli Food; Soup; and Meat Substitutes," June 2017, www.reportbuyer.com/product/4959853/top-trends-in-prepared-foods-2017-exploring-trends-in-meat-fish-and-seafood-pasta-noodles-and-rice-prepa red-meals-savory-deli-food-soup-and-meat-substitutes.html.
4 Vegetarian Society, "What is a Vegetarian?" last modified October 2016, www.vegsoc.org/definition.
5 Vegetarian Society, "What is a Vegetarian?" our emphases.
6 The Vegan Society, "Definition of Veganism," accessed May 30, 2018, www.veganso ciety.com/go-vegan/definition-veganism.
7 Working Group of the Oxford Centre for Animal Ethics, "Normalizing the Unthinkable: The Ethics of Using Animals in Research," in *The Ethical Case Against Animal Experiments*, eds. Andrew Linzey and Clair Linzey (Urbana, IL: University of Illinois Press, 2018), 87.
8 Andrew Linzey, "Foreword to the 2016 Reprint Edition," in *Christianity and the Rights of Animals* (Eugene, OR: Wipf and Stock, 2016), xvii.
9 The Vegetarian Norrington Table, "Anonymised Comments," accessed May 30, 2018, www.veggienorringtontable.com/anonymised-comments/.
10 Vesanto Melina, Winston Craig and Susan Levin, "Position of the Academy of Nutrition and Dietetics: Vegetarian Diets," *Journal of the Academy of Nutrition and Dietetics* 116, no. 12 (2016): 1970–1980.
11 For a discussion of these issues, see Bernard E. Rollin, *The Unheeded Cry: Animal Consciousness, Animal Pain and Science* (Oxford, England: Oxford University Press, 1990); and Bernard E. Rollin, "Animal Pain," in *The Global Guide to Animal Protection*, ed. Andrew Linzey (Urbana, IL: University of Illinois Press, 2013), 256–7.
12 See p. 12 in this volume.
13 See p. 56 in this volume.
14 See p. 68 in this volume.
15 See p. 78 in this volume.
16 Philip Low, "Cambridge Declaration of Consciousness," Declaration Presented at the Francis Crick Memorial Conference on Consciousness in Human and Non-Human Animals, at Churchill College, University of Cambridge, July 7, 2012, http://fcmcon ference.org/img/CambridgeDeclarationOnConsciousness.pdf.
17 See p. 109 in this volume.
18 See p. 129 in this volume.
19 See p. 129 in this volume.
20 See p. 154 in this volume.
21 See p. 153 in this volume.
22 See p. 170 in this volume.
23 See p. 173 in this volume.
24 See p. 178 in this volume.

25 See Joyce D'Silva, "The Welfare of Cows," in Linzey, *The Global Guide*, 173–5.
26 Vegetarian Society, "Fact Sheet: Cheese," last modified March 2017, www.vegsoc.org/cheese.
27 See Karen Davis, "Birds Used in Food Production," Linzey, *The Global Guide*, 164–5.
28 See, for example, "Re-home Some Hens," British Hen Welfare Trust, accessed May 30, 2018, www.bhwt.org.uk/rehome-some-hens/.
29 The Vegan Society, "Definition of Veganism."
30 See p. 202 in this volume.
31 See p. 220 in this volume.
32 See p. 229 in this volume.
33 Human Rights Watch, "Immigrant Workers in the United States Meat and Poultry Industry," December 15, 2005, www.hrw.org/legacy/backgrounder/usa/un-sub1005/.
34 Nicole Wendee, "CAFOs and Environmental Justice: The Case of North Carolina," *Environmental Health Perspectives*, 121 (2013): A182-A189, http://ehp.niehs.nih.gov/121-a182/.
35 See p. 261 in this volume.

Bibliography

Barth, Karl. *Church Dogmatics: The Doctrine of Creation*, Vol. III, pt. 4. Edinburgh, Scotland: T and T Clark, 1961.

British Hen Welfare Trust. "Re-home Some Hens." Accessed May 30, 2018. www.bhwt.org.uk/rehome-some-hens/.

Davis, Karen. "Birds Used in Food Production." In *The Global Guide to Animal Protection*, edited by Andrew Linzey, 164–165. Urbana, IL: University of Illinois Press, 2013.

D'Silva, Joyce. "The Welfare of Cows." In *The Global Guide to Animal Protection*, edited by Andrew Linzey, 173–175. Urbana, IL: University of Illinois Press, 2013.

Human Rights Watch. "Immigrant Workers in the United States Meat and Poultry Industry."December 15, 2005. www.hrw.org/legacy/backgrounder/usa/un-sub1005/.

Linzey, Andrew. "Foreword to the 2016 Reprint Edition." In *Christianity and the Rights of Animals*, xi–xxi. Eugene, OR: Wipf and Stock, 2016.

Low, Philip. "Cambridge Declaration of Consciousness." Declaration Presented at the Francis Crick Memorial Conference on Consciousness in Human and Non-Human Animals, at Churchill College, University of Cambridge. July 7, 2012. http://fcmconference.org/img/CambridgeDeclarationOnConsciousness.pdf. http://online.liebertpub.com/doi/abs/10.1089/fpd.2006.0066?url_ver=Z39.88-2003&rfr_id=ori%3Arid%3Acrossref.org&rfr_dat=cr_pub%3Dpubmed&.

Melina, Vesanto, Winston Craig and Susan Levin. "Position of the Academy of Nutrition and Dietetics: Vegetarian Diets." *Journal of the Academy of Nutrition and Dietetics* 116, 12 (2016): 1970–1980.

Petter, Olivia. "Number of Vegans in the UK Soars to 3.5 Million, Survey Finds." *The Independent*. April 3, 2018. www.independent.co.uk/life-style/food-and-drink/vegans-uk-rise-popularity-plant-based-diets-veganism-figures-survey-compare-the-market-a8286471.html.

Rollin, Bernard E. "Animal Pain." In *The Global Guide to Animal Protection*, edited by Andrew Linzey, 256–257. Urbana, IL: University of Illinois Press, 2013.

Rollin, Bernard E. *The Unheeded Cry: Animal Consciousness, Animal Pain and Science*. Oxford, England: Oxford University Press, 1990.

The Vegan Society. "Definition of Veganism." Accessed May 30, 2018. www.vegansociety.com/go-vegan/definition-veganism.

The Vegetarian Norrington Table. "Anonymised Comments." Accessed May 30, 2018. www.veggienorringtontable.com/anonymised-comments/.

Top Trends in Prepared Foods. "Exploring Trends in Meat, Fish and Seafood; Pasta, Noodles and Rice; Prepared Meals; Savory Deli Food; Soup; and Meat Substitutes." June 2017. www.reportbuyer.com/product/4959853/top-trends-in-prepared-foods-2017-exploring-trends-in-meat-fish-and-seafood-pasta-noodles-and-rice-prepared-meals-savory-deli-food-soup-and-meat-substitutes.html.

Vegetarian Society. "Fact Sheet: Cheese."Last modified March 2017. www.vegsoc.org/cheese.

Vegetarian Society. "What is a Vegetarian?" Last modified October 2016. www.vegsoc.org/definition.

Wendee, Nicole. "CAFOs and Environmental Justice: The Case of North Carolina." *Environmental Health Perspectives* 121(2013): A182–A189. http://ehp.niehs.nih.gov/121-a182/.

Working Group of the Oxford Centre for Animal Ethics. "Normalizing the Unthinkable: The Ethics of Using Animals in Research." In *The Ethical Case Against Animal Experiments*, edited by Andrew Linzey and Clair Linzey, 13–100, Urbana, IL: University of Illinois Press, 2018.

PART I

Killing sentient beings

1.1

WHY FOODS DERIVED FROM ANIMALS ARE NOT NECESSARY FOR HUMAN HEALTH

Stephen Patrick Kieran Walsh

Introduction

This chapter sets out the evidence that vegan diets (no animal products at all) can be a healthy choice provided certain known issues are addressed, particularly vitamin B_{12}. Other nutrients are reviewed leading to the conclusion that appropriately planned vegan diets are nutritionally adequate in the light of current, well developed, nutritional science. The Global Burden of Disease[1] model of diet and health is used to give a perspective on the role of animal and plant foods in improving healthy life expectancy, showing that this model emphasizes the quality of plant foods rather than the quantity and provides no support for an irreplaceable benefit from any food derived from animals. Mortality data from long-term studies of vegans is reviewed, confirming that vegans in the USA and UK experience comparable (and possibly slightly favorable) death rates compared with omnivores. The limited literature on failure to thrive in vegans is also reviewed, showing that failure to thrive affects only a minority of former vegetarians/vegans. Based on this combined evidence, it is concluded that "necessity" cannot be generally claimed as an ethical justification for consuming foods derived from animals. The ethical argument should therefore move beyond the question of necessity.

Vitamin B_{12} and the history of vegan diets

The oldest vegan organization still in existence (The Vegan Society) was founded in 1944. This was not the first attempt at veganism, but unlike earlier attempts, including by Mohandas K Gandhi, it has endured.

The path to sustained success was not entirely smooth. In 1955 multiple health problems were reported in UK vegans[2]: Amenorrhea, sore tongue, paresthesia (abnormal sensations such as tingling, tickling, pricking, numbness or burning) and "poker" back were common complaints and one case of subacute combined degeneration of the spinal cord was mentioned. Except for amenorrhea, which may be related to low weight, these are classic symptoms of vitamin B_{12} deficiency.

Vitamin B_{12} was isolated in 1948 and production of vitamin B_{12} from bacteria started soon after this, providing a reliable direct source of vitamin B_{12}. Vitamin B_{12} is produced by bacteria but not by plants or animals. Animals acquire this vitamin by a range of methods. Ruminants have bacteria in their four-compartment stomach that convert cobalt to vitamin B_{12}. Our closest relatives the chimpanzee and gorilla appear to consume sufficient termites to meet their needs[3] while termites themselves obtain vitamin B_{12} from intestinal bacteria.[4] Like other great apes, humans require an external source. Vitamin B_{12} is made commercially by bacteria and used to fortify foods or produce supplements both for humans and for many non-ruminant farmed animals such as chickens and pigs.[5]

By 1955, manufactured vitamin B_{12} from bacteria had been used successfully to treat spinal cord degeneration in a vegan teenager.[2] As vitamin B_{12} use from fortified foods and supplements has become more common among vegans, such severe problems have receded though, sadly, due to ignorance or misinformation, they have not entirely disappeared. The manufacture of vitamin B_{12} removed a critical potential weak point of vegan diets.

Other potential nutritional issues

Vegans generally have good intakes of most nutrients, including healthfully higher intakes of fiber, vitamin C and folate. Nutrients particularly associated with foods derived from animals within Western diets are protein, iron, zinc, calcium, vitamin D and long-chain omega-3 fats (EPA and DHA). Foods derived from animals can also be important sources of iodine and selenium if the local soil is low in these nutrients but farmed animals receive fortified feed or supplements.

Most nutritional authorities accept vegan diets as a valid choice. For example:

> It is the position of the Academy of Nutrition and Dietetics that appropriately planned vegetarian, including vegan, diets are healthful, nutritionally adequate, and may provide health benefits for the prevention and treatment of certain diseases. These diets are appropriate for all stages of the life cycle, including pregnancy, lactation, infancy, childhood, adolescence, older adulthood, and for athletes. Plant-based diets are more environmentally sustainable than diets rich in animal products because they use fewer natural resources and are associated with much less environmental damage. Vegetarians and vegans are at reduced risk of certain health conditions, including ischemic heart disease, type 2 diabetes, hypertension, certain types of cancer, and obesity … Vegans need reliable sources of vitamin B-12, such as fortified foods or supplements.[6]

The most critical recent position paper comes from the German Nutrition Society:

> On the basis of current scientific literature, the German Nutrition Society (DGE) has developed a position on the vegan diet. With a pure plant-based diet, it is difficult or impossible to attain an adequate supply of some nutrients. The most critical nutrient is vitamin B_{12}. Other potentially critical nutrients in a vegan diet include protein resp. indispensable amino acids, long-chain n-3 fatty acids, other vitamins (riboflavin, vitamin D) and minerals (calcium, iron, iodine, zinc and selenium). The DGE does not recommend a vegan diet for pregnant women, lactating women, infants, children or adolescents. Persons who nevertheless wish to follow a vegan diet should permanently take a vitamin B_{12} supplement, pay attention to an adequate intake of nutrients, especially critical nutrients, and possibly use fortified foods or dietary supplements. They should receive advice from a nutrition counsellor and their supply of critical nutrients should be regularly checked by a physician.[7]

These two statements differ not so much in the evidence on which they are based as in their stance as to what constitutes a vegan diet and their attitude to the need for dietary planning.

The statement that "it is difficult or impossible to attain an adequate supply of some nutrients" is true only if you define a vegan diet as a "pure plant-based diet," but a vegan diet is a diet free of foods derived from animals rather than a diet based only on plants. As vitamin B_{12} is produced by bacteria it is a valid part of a vegan diet and whether the bacterial vitamin B_{12} comes to humans via fortified foods or supplements or via other animals makes no difference nutritionally.

The first statement takes "appropriately planned" diets as its subject whereas the German statement focuses instead on potential problems in the absence of appropriate planning: If you don't plan a vegan diet then serious problems are indeed possible, but if you do plan then these problems can be avoided. So we need to consider what "appropriately planned" actually means and how the nutrients associated with foods derived from animals can be provided in vegan diets.

Vitamin B_{12} status in vegans in some studies is very poor indeed. The largest UK study of vegans found that 73% of vegans had blood B_{12} levels below 150 pmol/l compared with 24% of other vegetarians and 2% of omnivores.[8] Some other studies show somewhat higher prevalence of such low blood B_{12} levels in the general population: US adults (9%),[9] US adults over 65 (5%)[10] and UK adults over 65 (7.5%).[11] Three smaller studies of vegans who were using more fortified foods and supplements found the prevalence of low blood B_{12} levels (5%,[12] 7.5%,[13] 12%[14]) to be similar to that found in most studies of the general population. With consistent use of fortified foods and supplements, vegan B_{12} status can be excellent, but with limited use of such foods it can be very poor. Consensus guidelines for vegans recommend at least three micrograms per day spread over two or more meals or at least ten micrograms from a daily supplement or at least 2,000 micrograms from a weekly supplement.[15]

Vegan protein intakes are sometimes but not always lower than omnivore intakes and biological effectiveness of plant protein is about 10–20% lower than animal protein. However, vegan protein intakes are still generally within recommendations and can be readily boosted, both in quantity and quality, by increasing intake of legumes, including soya products.

Diets based almost exclusively on grains and legumes are normally due to poverty rather than choice and are associated with increased risks of iron deficiency anemia and zinc deficiency. However, the addition of foods containing vitamin C increases iron absorption from plants while the addition of foods such as carrots, onions and garlic *may* increase zinc absorption.[16] Bread-making, fermenting and sprouting all reduce phytates and increase absorption of zinc and iron. Western vegetarians do not show increased risk of iron deficiency anemia[17] and vegetarians and vegans show only a modest decrease in plasma zinc levels.[18]

In many Western countries, dairy products are the main source of dietary calcium. Vegan calcium intakes can be low, particularly if there is little use of fortified plant milks. A meta-analysis found vegans to have about 6% lower bone mineral density than omnivores, which is equivalent to about a 10% increase in fracture risk.[19] The worst bone mineral density results were found in vegans on high-raw diets. The EPIC-Oxford study found a 30% increased risk of fracture in vegans, but no excess risk in those vegans consuming at least 525 mg of calcium per day[20] which is easy to achieve by including some foods rich in calcium such as calcium-set tofu and fortified plant milks.

Sun exposure, with the sun well above the horizon, is the main source of vitamin D, but dietary intakes become important if such exposure is limited. Dietary intakes of vitamin D are lower for vegans than for omnivores but vitamin D supplementation is increasingly widely recommended for everyone[21] and vegan vitamin D supplements including cholecalciferol (vitamin D3) derived from lichen are widely available.

Iodine and selenium are not required by plants, so the amount in the plant depends on the amount in the soil. Some countries, including the UK, rely upon iodine in cattle feed to boost iodine supply, so those not receiving this indirect supplement need to find an alternative solution. Brazil nuts are rich in selenium and seaweeds are rich in iodine and supplements of both are readily available.

Blood levels of longer-chain omega-3s (EPA,DHA) are usually lower in vegans than in omnivores, though the second largest study to date found vegans to have higher EPA levels than omnivorous US soldiers but slightly lower DHA levels.[22] In this study vegan intakes of the plant omega-3 alpha-linolenic acid were unusually high (3.4 g per day) though individual intakes did not correlate with blood levels of EPA and DHA combined. The largest study[23] and several smaller studies found vegan levels to be notably lower. In the case of EPA, blood levels increase predictably in response to an increased ratio of alpha-linolenic acid to linoleic acid (plant omega-3 to omega-6)[24] but, in the absence of an external source, improved DHA levels may rely more on lower linoleic acid intakes or a direct intake of DHA. The conversion of plant fats to longer-chain fats has been found to be influenced by a mutation that has a different prevalence in traditionally vegetarian populations.[25] The difference in conversion does not seem large enough to be fundamental but indicates that some genetic adaptations have occurred.

It remains an open question whether at least some vegans would benefit from vegan EPA and DHA supplements, which are derived from micro-organisms. Pending further research, vegans may wish to consume at least 250 mg/day of EPA and DHA combined, which meets most current recommendations, as well as getting about 1% of calories from plant omega-3 and avoiding excessive omega-6 intake.

All these potential issues can be addressed by appropriate choices without relying on foods derived from animals.

A typical set of recommendations for vegan diets is as follows:

- Eat a wide variety of whole, lightly processed foods. Limit use of highly processed foods, particularly those containing partially hydrogenated (trans) fats or a lot of added salt or sugar.
- Include a variety of brightly colored fruit and vegetables – ideally 500 grams a day or more, including some green leafy vegetables and some carotene-rich foods such as carrots and sweet potatoes.
- Choose mostly whole grains (e.g. wholemeal bread) rather than refined grains (e.g. white or brown bread or white rice).
- Eat regular small amounts of nuts (20 to 30 g a day), preferably those rich in monounsaturates, such as almonds, cashews, hazels, and macadamias.
- Include two or more servings of legumes (peas, beans, lentils) a day, e.g. 250 ml soya milk or 100 g cooked legumes or tofu.
- Get at least 3 micrograms of vitamin B_{12} per day from fortified foods or supplements.
- Include a good source of plant omega-3s, e.g. a teaspoon of flaxseed oil OR two tablespoons of ground flaxseed OR two tablespoons of rapeseed oil per day. Limit oils high in omega-6 such as sunflower, corn, grapeseed, safflower, soy, sesame and peanut/groundnut.
- Consider including 250 mg per day of longer-chain omega-3s (EPA plus DHA).
- Get an adequate amount of
    - Iodine (iodized salt or an iodine supplement providing 75 to 150 micrograms per day)
    - Selenium (10 Brazil nuts a week or a supplement containing about 50 micrograms a day)
    - Vitamin D (sunlight when available, but a supplement could be useful if sun exposure is limited – the UK recommendation is 10 micrograms per day for everyone.)
    - Calcium (at least 500 mg per day from calcium-rich foods – such as fortified plant milks – or supplements, to give an overall intake above 700 mg)

Appropriate planning of vegan diets does require some attention and vitamin B_{12} is indeed the most critical consideration, but unless there are special circumstances it does not require advice from a nutrition counsellor and regular checks by a physician as suggested by the German Nutrition Society.

A longer, healthier life?

Much nutritional research now focuses on which diet choices support optimal health and longevity, rather than simply avoiding nutrient deficiencies.

One of the largest projects looking at the impact of diet risk factors on health is the Global Burden of Disease (GBD) Study.[1] This study looks at Disability Adjusted Lost Years (DALYs) attributable to risk factors that differ from their ideal values, e.g. eating less than 400 g vegetables per day. The estimated years lost by region depend on the prevalence of different diseases, the estimated impact of each risk factor on each disease, and the extent to which risk factors in that region differ from an "ideal" minimum risk level.

The GBD Study's ideal daily diet includes at least 300 g fruit, 400 g vegetables, 125 g whole grains, 16 g nuts and seeds, 450 g milk, 1,150 mg calcium, 250 mg of "seafood omega-3" (EPA plus DHA) and 12.5% of total energy from polyunsaturated fatty acids. It includes no more than 14 g of red meat, 7 g of processed meat, 30 g of sugar-sweetened beverages, 0.4% of energy from trans fatty acids and 3 g of sodium. The recommended plant foods account for about 1,000 kcal of energy intake, leaving plenty of room for additional food choices for the balance of the diet. Figure 1.1.1 shows the contribution of deviations from the ideal diet to global Disability Adjusted Lost Years (DALYs).

More fruit, less salt, more wholegrains, more vegetables, and more nuts and seeds are the top five recommended dietary changes globally, covering over

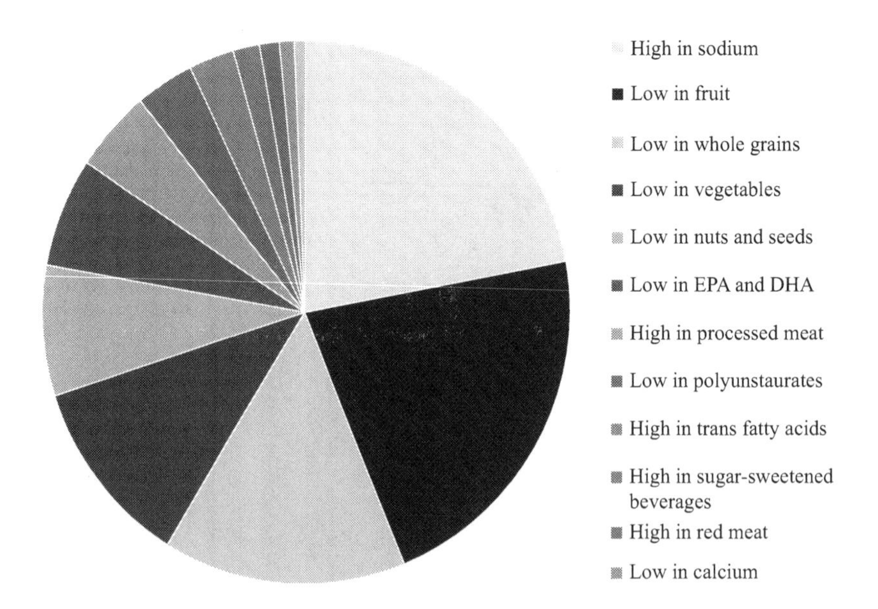

FIGURE 1.1.1 Relative impact of dietary risk factors on global DALYs

three-quarters of the total impacts from diet. These changes to improve healthy life expectancy globally are primarily about improving the quality of the non-animal part of the diet and are therefore relevant to all diet groups, from vegans to regular meat-eaters.

Globally, the largest expected benefits related to animal foods come from more long-chain omega-3 fats (EPA, DHA), which usually are obtained from fish, and from less processed meat but each of these impacts is smaller on a global scale than the five listed above. The projected net health impact of a global reduction in overall intake of animal-derived foods is small as positives and negatives cancel out.

In Western Europe (Figure 1.1.2), lack of seafood omega-3s becomes less important as a contributor to DALYs (largely because more are consumed in the average diet) and excessive processed meat becomes more important for the same reason, but the quality of the non-animal part of the diet continues to dominate the projected impact of diet choices.

From the standpoint of whether foods derived from animals are necessary for health we need to consider not just the overall impact of animal foods on health but whether the inclusion of *any* form of animal-derived food provides an irreplaceable benefit to health. Based on the Global Burden of Disease (GBD) Study, the main issue relates to EPA and DHA, which are usually derived from fish. However, EPA and DHA in fish come from single-celled

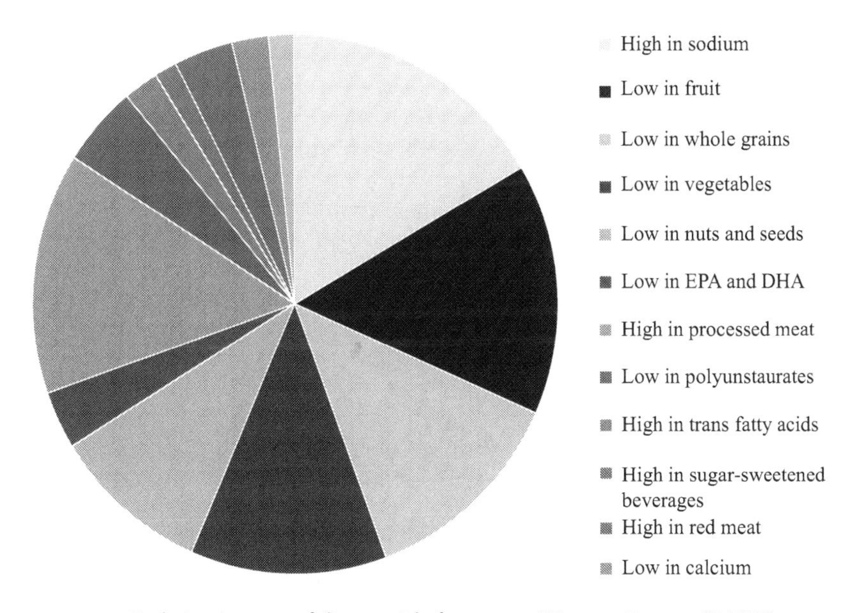

FIGURE 1.1.2 Relative impact of dietary risk factors on Western Europe DALYs

marine organisms and EPA and DHA can be derived directly from such sources bypassing the fish. Indeed, DHA in infant formula is currently derived from such single-celled organisms. Calcium can be provided without any use of animal milks as can other nutrients commonly derived from milk such as vitamin B_{12} and iodine.

While epidemiological models such as the GBD model provide useful insights, the most direct measure of vegan health comes from studies of actual vegans. The two largest studies to date in terms of the number of vegans are EPIC-Oxford (2,400 vegans)[26] and the Adventist Health Study 2 (5,500 vegans).[27] Both studies include varied diet groups, ranging from vegans to regular meat-eaters. Both study populations are also exceptionally healthy, making comparisons within these groups particularly relevant to exploring death rates. The EPIC-Oxford study participants had an age-standardized death rate 52% that of the general population of England and Wales[28] – equivalent to living about six years longer. A previous study of California Adventists estimated that they lived about seven years longer than the general population of California.[29]

Table 1.1.1 shows death rates relative to regular meat-eaters. A death rate of 0.9 is roughly equivalent to living one year longer than the regular meat-eaters. In these healthy populations, the impact of animal product intake is modest. Numbers in bold are statistically significant (less than 1 in 20 probability of differing from 1 by chance). Results are shown for earlier studies as well as for AHS-2 and for EPIC-Oxford in combination with the earlier Oxford Vegetarian study.

The key points from Table 1.1.1 are:

> All groups who restrict meat, including vegans, show the same or lower death rates compared with regular meat-eaters.
>
> The differences are modest: The vegan men and the pescetarian (fish but not meat) men in AHS-2 show an advantage of about 3 years in life expectancy while other groups show less advantage.

This provides very strong direct evidence that adult vegans can enjoy excellent longevity as would be expected from current nutritional knowledge.

TABLE 1.1.1 Comparative death rates in different diet groups compared with regular meat-eaters

	Vegan	*Other Vegetarian*	*Occasional/low meat-eaters*	*Pescetarian*
Pooled analysis[30]	1.00	**0.84**	**0.84**	**0.82**
EPIC-Oxford/Oxford Vegetarian Study[31]	1.00	0.93	0.93	0.91
AHS-2 All[32]	0.85	**0.91**	0.92	**0.81**
AHS-2 Men	**0.72**	0.86	0.93	**0.73**
AHS-2 Women	0.97	0.94	0.92	0.88

But what about ...

Children

Vegan children have been examined, though in smaller numbers than in the recent studies of death rates. The following summary is typical of the conclusions:

> The diets and growth of children reared on vegetarian diets are reviewed. Excessive bulk combined with low energy density can be a problem for children aged <=5 years and can lead to impaired growth. Diets that have a high content of phytate and other modifiers of mineral absorption are associated with an increased prevalence of rickets and iron-deficiency anemia. Vitamin B-12 deficiency is a real hazard in unsupplemented or unfortified vegan and vegetarian diets ... If known pitfalls are avoided, the growth and development of children reared on both vegan and vegetarian diets appears normal.[33]

So the conclusions on children are similar to those for adults.

Pregnant women

Vegan diets in pregnancy were recently reviewed,[34] concluding that:

> The evidence on vegan–vegetarian diets in pregnancy is heterogeneous and scant. The lack of randomised studies prevents us from distinguishing the effects of diet from confounding factors. Within these limits, vegan–vegetarian diets may be considered safe in pregnancy, provided that attention is paid to vitamin and trace element requirements.

Life vegans

There are a considerable number of life vegans, but so far there are far too few deaths to report any meaningful statistics.

Ex-vegans

The key gap in the scientific data on vegans is the absence of long-term randomized trials of vegan diets. The people who become vegan may differ in many ways from the general population, and no one is forced to remain vegan. This raises the possibility that those who stay vegan long enough to turn up in the results of scientific studies could simply be a minority who happened to find the diet suited them especially well. If many people were abandoning a vegan diet because they didn't feel healthy on it, then the results on long-term vegans could be misleading.

It is easy to find, through Google or through social contacts, individual ex-vegans who felt unhealthy on vegan diets and felt better after adding back some animal products. In some cases, their story fits with a known and avoidable issue such as vitamin B_{12}. In others, there often seems to be an element of trying too hard, e.g. excessive weight loss on high raw diets or gut or mineral absorption issues on diets centered on whole unprocessed seeds. The founder of The Vegan Society Donald Watson highlighted low salt as an issue with the early vegans as well as low vitamin B_{12}. The former issue has faded into the background while the latter is widely acknowledged, but in certain circumstances, particularly with raw food diets, very low salt intakes could be an issue. There are, however, sufficient cases that do not fall into neat categories that we cannot yet rule out the possibility that some individuals will not thrive on *any* vegan diet for some as yet unidentified reason.

It is estimated that there are twice as many ex-vegans in the USA (1.1%) relative to current vegans (0.5%), based on a survey of 11,399 people by Faunalytics in 2014, so there is room for selection to operate. As part of this survey, Faunalytics asked 1,115 former vegetarians, including 123 former vegans, their reasons for ceasing to follow that diet.[35]

The top reason for lapsing (vegans were not separated from other vegetarians) was not being satisfied with the food (293 people), followed by health (237), social issues (120), inconvenience (115), lack of motivation (56) and cost (56). Within the health category, 85 people mentioned nutrient concerns or deficiencies (of these 37 mentioned protein and 21 mentioned iron) and 56 mentioned feeling fatigued, lightheaded, weak or unhealthy. In total, 30 referred to not getting the benefits they expected and 22 mentioned pregnancy as a reason for stopping.

Those advocating vegan diets should be open-minded and curious about any reports of failure to thrive from vegans or ex-vegans. Before 1948 there was a huge "Factor X" in the form of vitamin B12 and there may be other, as yet unknown, lessons to be learnt. However, the scope for such a Factor X is limited, given the demonstrated health of large numbers of vegans and the fact that health does not dominate the reasons for ceasing to be vegan.

Conclusions

There are no essential nutrients that can be obtained only from foods derived from animals.

The Global Burden of Disease Study recommendations for improved health and longevity mainly relate to particular choices of *plant* foods and do not point to any unique benefit from foods derived from animals.

Animal-free (vegan) diets have been shown in multiple studies to support excellent longevity in adults.

Growth and development in vegan children is normal, provided known issues (too little vitamin B_{12} and too much bulk/fiber) are avoided.

Unexplained failure to thrive in vegans/ex-vegans is not common.

Foods derived from animals therefore cannot reasonably be claimed to be necessary for health though at the individual level genuine problems may arise and vegan advocates should be open to learning from these. Any ethical debate should therefore focus on other issues such as interactions between diet choices and poverty, human and animal suffering, and environmental sustainability.

Notes

1 Forouzanfar, Mohammad H., Lily Alexander, H. Ross Anderson, Victoria F. Bachman, Stan Biryukov, Michael Brauer, Richard Burnett et al. "Global, Regional, and National Comparative Risk Assessment of 79 Behavioural, Environmental and Occupational, and Metabolic Risks or Clusters of Risks in 188 Countries, 1990–2013: A Systematic Analysis for the Global Burden of Disease Study 2013." *The Lancet* 386, no. 10010 (2015): 2287–2323.
2 Frank Wokes, J. Badenoch and H. M. Sinclair, "Human Dietary Deficiency of Vitamin B12," *American Journal of Clinical Nutrition 3* (1955): 375–382.
3 Isra Deblauwe and Geert P. J. Janssens, "New Insights in Insect Prey Choice by Chimpanzees and Gorillas in Southeast Cameroon: The Role of Nutritional Value," *American Journal of Physical Anthropology 135*, no. i (2008): 42–55.
4 Edgar J. Wakayama, Jack W. Dillwith, Ralph W. Howard and Gary J. Blomquist, "Vitamin B12 Levels in Selected Insects," *Insect Biochemistry 14*, no. *2* (1984): 175–179.
5 Gérard Moine et al., "Ullmann's Encyclopedia of Industrial Chemistry," published online, October 15, 2011, DOI: 10.1002/14356007.o27_o09.
6 Vesanto Melina, Winston Craig and Susan Levin, "Position of the Academy of Nutrition and Dietetics: Vegetarian Diets," *Journal of the Academy of Nutrition and Dietetics 116*, no. *12* (2016): 1970–1980.
7 M. Richter et al., "For the German Nutrition Society (DGE) (2016) Vegan Diet. Position of the German Nutrition Society (DGE)." *Ernahrungs Umschau 63*, no. *04* (2016): 92–102.
8 Anne M. J. Gilsing et al., "Serum Concentrations of Vitamin B12 and Folate in British Male Omnivores, Vegetarians and Vegans: Results from a Cross-Sectional Analysis of the EPIC-Oxford Cohort Study," *European Journal of Clinical Nutrition 64*, no. *9* (2010): 933–939.
9 Katherine L. Tucker et al., "Plasma Vitamin B-12 Concentrations Relate to Intake Source in the Framingham Offspring Study," *The American Journal of Clinical Nutrition 71*, no. *2* (2000): 514–522.
10 John Lindenbaum et al., "Prevalence of Cobalamin Deficiency in the Framingham Elderly Population," *The American Journal of Clinical Nutrition 60*, no. *1* (1994): 2–11.
11 Robert Clarke et al., "Screening for Vitamin B-12 and Folate Deficiency in Older Persons," *The American Journal of Clinical Nutrition* 77, no. *5* (2003): 1241–1247.
12 Anna-Liisa Elorinne et al., "Food and Nutrient Intake and Nutritional Status of Finnish Vegans and Non-Vegetarians," *PloS one 11*, no. *2* (2016): e0148235.
13 R. Schüpbach, R. Wegmüller, C. Berguerand, M. Bui and I. Herter-Aeberli, "Micronutrient Status and Intake in Omnivores, Vegetarians and Vegans in Switzerland," *European Journal of Nutrition*, 56, no. 1 (2017): 283–293.
14 Ella H. Haddad et al., "Dietary Intake and Biochemical, Hematologic, and Immune Status of Vegans Compared with Nonvegetarians," *The American Journal of Clinical Nutrition 70*, no. *3* (1999): 586s-593s.
15 Stephen Walsh et al., "What Every Vegan Should Know About Vitamin B12," accessed October 1, 2017, www.veganhealth.org/articles/everyvegan.
16 Smita Gautam, Kalpana Platel and Krishnapura Srinivasan, "Influence of Combinations of Promoter and Inhibitor on the Bioaccessibility of Iron and Zinc from Food Grains," *International Journal of Food Sciences and Nutrition 62*, no. *8* (2011): 826–834.
17 Janet R. Hunt, "Bioavailability of Iron, Zinc, and Other Trace Minerals from Vegetarian Diets," *The American Journal of Clinical Nutrition 78*, no. *3* (2003): 633S–639S.
18 Meika Foster, Anna Chu, Peter Petocz and Samir Samman, "Effect of Vegetarian Diets on Zinc Status: A Systematic Review and Meta-analysis of Studies in Humans," *Journal of the Science of Food and Agriculture 93*, no. *10* (2013): 2362–2371.

19 Lan T. Ho-Pham, Nguyen D. Nguyen and Tuan V. Nguyen, "Effect of Vegetarian Diets on Bone Mineral Density: A Bayesian Meta-analysis," *The American Journal of Clinical Nutrition 90*, no. *4* (2009): 943–950.
20 P. Appleby, A. Roddam, N. Allen and T. Key, "Comparative Fracture Risk in Vegetarians and Nonvegetarians in EPIC-Oxford," *European Journal of Clinical Nutrition 61*, no. *12* (2007): 1400–1406.
21 Scientific Advisory Committee on Nutrition, "Vitamin D and Health," July 2016 www.gov.uk/government/uploads/system/uploads/attachment_data/file/537616/SACN_Vitamin_D_and_Health_report.pdf.
22 Barbara Sarter, Kristine S. Kelsey, Todd A. Schwartz and William S. Harris, "Blood Docosahexaenoic Acid and Eicosapentaenoic Acid in Vegans: Associations with Age and Gender and Effects of an Algal-derived Omega-3 Fatty Acid Supplement," *Clinical Nutrition 34*, no. *2* (2015): 212–218.
23 Magdalena S. Rosell et al., "Long-chain N–3 Polyunsaturated Fatty Acids in Plasma in British Meat-Eating, Vegetarian, and Vegan Men," *The American Journal of Clinical Nutrition 82*, no. *2* (2005): 327–334.
24 Stephen Walsh, *Plant Based Nutrition and Health* (St Leonards-on-Sea, East Sussex: The Vegan Society, 2003).
25 Kumar S. D. Kothapalli et al., "Positive Selection on a Regulatory Insertion-deletion Polymorphism in FADS2 Influences Apparent Endogenous Synthesis of Arachidonic Acid," *Molecular Biology and Evolution* (2016): msw049.
26 Gwyneth K. Davey et al., "EPIC–Oxford: Lifestyle Characteristics and Nutrient Intakes in a Cohort of 33 883 Meat-eaters and 31 546 Non-meat-eaters in the UK," *Public Health Nutrition 6*, no. *3* (2003): 259–268.
27 Michael J. Orlich et al., "Vegetarian Dietary Patterns and Mortality in Adventist Health Study 2," *JAMA Internal Medicine 173*, no. *13* (2013): 1230–1238.
28 Timothy J. Key et al., "Mortality in British Vegetarians: Results from the European Prospective Investigation into Cancer and Nutrition (EPIC-Oxford)," *The American Journal of Clinical Nutrition 89*, no. *5* (2009): 1613S–1619S.
29 Gary E. Fraser and David J. Shavlik, "Ten Years of Life: Is it a Matter of Choice?" *Archives of Internal Medicine 161*, no. *13* (2001): 1645–1652.
30 Timothy J. Key et al., "Mortality in Vegetarians and Nonvegetarians: Detailed Findings from a Collaborative Analysis of 5 Prospective Studies," *The American Journal of Clinical Nutrition 70*, no. *3* (1999): 516s–524s.
31 Paul N. Appleby et al., "Mortality in Vegetarians and Comparable Nonvegetarians in the United Kingdom," *The American Journal of Clinical Nutrition* (2015): ajcn119461.
32 Michael J. Orlich et al., "Vegetarian Dietary Patterns and Mortality in Adventist Health Study 2," *JAMA Internal Medicine 173*, no. *13* (2013): 1230–1238.
33 T. A. Sanders and Sheela Reddy, "Vegetarian Diets and Children," *The American Journal of Clinical Nutrition* 59, no. 5 (1994): 1176S–1181S.
34 Giorgina B. Piccoli et al., "Vegetarian Supplemented Low-protein Diets. A Safe Option for Pregnant CKD Patients: Report of 12 Pregnancies in 11 Patients," *Nephrology Dialysis Transplantation* (2010): gfq333.
35 *Faunalytics*, "Study of Current and Former Vegetarians and Vegans," July 2015, https://faunalytics.org/wp-content/uploads/2015/07/Faunalytics-Study-of-Current-and-Former-Vegetarians-and-Vegans-%E2%80%93-Qualitative-Findings1.pdf.

Bibliography

Appleby, P., A. Roddam, N. Allen and T. Key. "Comparative Fracture Risk in Vegetarians and Nonvegetarians in EPIC-Oxford." *European Journal of Clinical Nutrition* 61, 12(2007): 1400–1406.

Appleby, Paul N., Francesca L. Crowe, Kathryn E. Bradbury, Ruth C. Travis and Timothy J. Key. "Mortality in Vegetarians and Comparable Nonvegetarians in the United Kingdom." *The American Journal of Clinical Nutrition* (2015): ajcn119461.

Clarke, Robert, Helga Refsum, Jacqueline Birks, John Grimley Evans, Carole Johnston, Paul Sherliker, Per M. Uelandet *al.* "Screening for Vitamin B-12 and Folate Deficiency in Older Persons." *The American Journal of Clinical Nutrition* 77, 5(2003): 1241–1247.

Davey, Gwyneth K., Elizabeth A. Spencer, Paul N. Appleby, Naomi E. Allen, Katherine H. Knox and Timothy J. Key. "EPIC–Oxford: Lifestyle Characteristics and Nutrient Intakes in a Cohort of 33 883 Meat-eaters and 31 546 Non-meat-eaters in the UK." *Public Health Nutrition* 6, 3(2003): 259–268.

Deblauwe, Isra, and Geert P. J. Janssens. "New Insights in Insect Prey Choice by Chimpanzees and Gorillas in Southeast Cameroon: The role of Nutritional Value." *American Journal of Physical Anthropology* 135, 1(2008): 42–55.

Elorinne, Anna-Liisa, Georg Alfthan, Iris Erlund, Hanna Kivimäki, Annukka Paju, Irma Salminen, Ursula Turpeinen, Sari Voutilainen and Juha Laakso. "Food and Nutrient Intake and Nutritional Status of Finnish Vegans and Non-Vegetarians." *PloS one* 11, 2 (2016): e0148235.

Faunalytics. "Study of Current and Former Vegetarians and Vegans." July 2015. https://faunalytics.org/wp-content/uploads/2015/07/Faunalytics-Study-of-Current-and-Former-Vegetarians-and-Vegans-%E2%80%93-Qualitative-Findings1.pdf.

Forouzanfar, Mohammad H., Lily Alexander, H. Ross Anderson, Victoria F. Bachman, Stan Biryukov, Michael Brauer, Richard Burnettet *al.* "Global, Regional, and National Comparative Risk Assessment of 79 Behavioural, Environmental and Occupational, and Metabolic Risks or Clusters of Risks in 188 Countries, 1990–2013: A Systematic Analysis for the Global Burden of Disease Study 2013." *The Lancet* 386, 10010(2015): 2287–2323.

Foster, Meika, Anna Chu, Peter Petocz and Samir Samman. "Effect of Vegetarian Diets on Zinc Status: A Systematic Review and Meta-analysis of Studies in Humans." *Journal of the Science of Food and Agriculture* 93, 10(2013): 2362–2371.

Fraser, Gary E., and David J. Shavlik. "Ten Years of Life: Is it a Matter of Choice?" *Archives of Internal Medicine* 161, 13(2001): 1645–1652.

Gautam, Smita, Kalpana Platel and Krishnapura Srinivasan. "Influence of Combinations of Promoter and Inhibitor on the Bioaccessibility of Iron and Zinc from Food Grains." *International Journal of Food Sciences and Nutrition* 62, 8(2011): 826–834.

Gilsing, Anne M. J., Francesca L. Crowe, Zouë Lloyd-Wright, Thomas A. B. Sanders, Paul N. Appleby, Naomi E. Allen and Timothy J. Key. "Serum Concentrations of Vitamin B12 and Folate in British Male Omnivores, Vegetarians and Vegans: Results from a Cross-sectional Analysis of the EPIC-Oxford Cohort Study." *European Journal of Clinical Nutrition* 64, 9(2010): 933–939.

Haddad, Ella H., Lee S. Berk, James D. Kettering, Richard W. Hubbard and Warren R. Peters. "Dietary Intake and Biochemical, Hematologic, and Immune Status of Vegans Compared with Nonvegetarians." *The American Journal of Clinical Nutrition* 70, 3(1999): 586S–593S.

Ho-Pham, Lan T., Nguyen D. Nguyen and Tuan V. Nguyen. "Effect of Vegetarian Diets on Bone Mineral Density: A Bayesian meta-analysis." *The American Journal of Clinical Nutrition* 90, 4(2009): 943–950.

Hunt, Janet R. "Bioavailability of Iron, Zinc, and Other Trace Minerals from Vegetarian Diets." *The American Journal of Clinical Nutrition* 78, 3(2003): 633S–639S.

Key, Timothy J., Gary E. Fraser, Margaret Thorogood, Paul N. Appleby, Valerie Beral, Gillian Reeves, Michael L. Burret *al.* "Mortality in Vegetarians and Nonvegetarians: Detailed Findings from a Collaborative Analysis of 5 Prospective Studies." *The American Journal of Clinical Nutrition* 70, 3(1999): 516S–524S.

Key, Timothy J., Paul N. Appleby, Elizabeth A. Spencer, Ruth C. Travis, Andrew W. Roddam and Naomi E. Allen. "Mortality in British Vegetarians: Results from the European Prospective Investigation into Cancer and Nutrition (EPIC-Oxford)." *The American Journal of Clinical Nutrition* 89, 5(2009): 1613S–1619S.

Kothapalli, Kumar S. D., Kaixiong Ye, Maithili S. Gadgil, Susan E. Carlson, Kimberly O. O'Brien, Ji Yao Zhang, Hui Gyu Park*et al.* "Positive Selection on a Regulatory Insertion-deletion Polymorphism in FADS2 Influences Apparent Endogenous Synthesis of Arachidonic Acid." *Molecular Biology and Evolution* (2016): msw049.

Lindenbaum, John, Irwin H. Rosenberg, P. W. Wilson, Sally P. Stabler and Robert H. Allen. "Prevalence of Cobalamin Deficiency in the Framingham Elderly Population." *The American Journal of Clinical Nutrition* 60, 1(1994): 2–11.

Melina, Vesanto, Winston Craig and Susan Levin. "Position of the Academy of Nutrition and Dietetics: Vegetarian Diets." *Journal of the Academy of Nutrition and Dietetics* 116, 12 (2016): 1970–1980.

Moine, Gérard, Hans-Peter Hohmann, Roland Kurth, Joachim Paust, Wolfgang Hähnlein, Horst Pauling, Bernd–Jürgen Weimann and Bruno Kaesler. "Ullmann's Encyclopedia of Industrial Chemistry." Published online. October 15, 2011. DOI: doi:10.1002/14356007.o27_o09.

Orlich, Michael J., Pramil N. Singh, Joan Sabaté, Karen Jaceldo-Siegl, Jing Fan, Synnove Knutsen, W. LawrenceBeeson and Gary E. Fraser. "Vegetarian Dietary Patterns and Mortality in Adventist Health Study 2." *JAMA Internal Medicine* 173, 13(2013): 1230–1238.

Piccoli, Giorgina B., Rossella Attini, Elena Vasario, Pietro Gaglioti, Ettore Piccoli, Valentina Consiglio, Chiara Deagostini, Manuela Oberto and Tullia Todros. "Vegetarian Supplemented Low-protein Diets. A Safe Option for Pregnant CKD Patients: Report of 12 Pregnancies in 11 Patients." *Nephrology Dialysis Transplantation* (2010): gfq333.

Richter, M., H. Boeing, D. Grüne-wald-Funk, H. Heseker, A. Kroke, E. Leschik-Bonnet, H. Oberritter, D. Strohm and B. Watzl. "For the German Nutrition Society (DGE) (2016) Vegan Diet. Position of the German Nutrition Society (DGE)." *Ernahrungs Umschau* 63, 4(2016): 92–102.

Rosell, Magdalena S., Zouë Lloyd-Wright, Paul N. Appleby, Thomas A. B. Sanders, Naomi E. Allen and Timothy J. Key. "Long-chain N–3 Polyunsaturated Fatty Acids in Plasma in British Meat-eating, Vegetarian, and Vegan Men." *The American Journal of Clinical Nutrition* 82, 2(2005): 327–334.

Sanders, T. A. and Sheela Reddy. "Vegetarian Diets and Children." *The American Journal of Clinical Nutrition* 59, 5(1994): 1176S–1181S.

Sarter, Barbara, Kristine S. Kelsey, Todd A. Schwartz and William S. Harris. "Blood Docosahexaenoic Acid and Eicosapentaenoic Acid in Vegans: Associations with Age and Gender and Effects of an Algal-derived Omega-3 Fatty Acid Supplement." *Clinical Nutrition* 34, 2(2015): 212–218.

Schüpbach, R., R. Wegmüller, C. Berguerand, M. Bui and I. Herter-Aeberli. "Micronutrient Status and Intake in Omnivores, Vegetarians and Vegans in Switzerland." *European Journal of Nutrition* 56, 1(2017): 283–293.

Scientific Advisory Committee on Nutrition. "Vitamin D and Health." July2016. www.gov.uk/government/uploads/system/uploads/attachment_data/file/537616/SACN_Vitamin_D_and_Health_report.pdf.

Tucker, Katherine L., Sharron Rich, Irwin Rosenberg, Paul Jacques, Gerard Dallal, Peter W. F. Wilson and Jacob Selhub. "Plasma Vitamin B-12 Concentrations Relate to Intake Source in the Framingham Offspring Study." *The American Journal of Clinical Nutrition* 71, 2(2000): 514–522.

Wakayama, Edgar J., Jack W. Dillwith, Ralph W. Howard and Gary J. Blomquist. "Vitamin B12 Levels in Selected Insects." *Insect Biochemistry* 14, 2(1984): 175–179.

Walsh, Stephen. *Plant Based Nutrition and Health.* St Leonards-on-Sea, East Sussex: The Vegan Society, 2003.

Walsh, Stephen, P. Appleby, L. Baroni, G. Dallas-Chapman, B. Davis and W. Harris. "What Every Vegan Should Know About Vitamin B12." Accessed October 1, 2017. www.veganhealth.org/articles/everyvegan.

Wokes, Frank, J. Badenoch and H. M. Sinclair. "Human Dietary Deficiency of Vitamin B12." *American Journal of Clinical Nutrition*3(1955): 375–382.

1.2

AGAINST KILLING "HAPPY" ANIMALS

Andrew Fisher

Introduction

Although most of the billions of animals killed by humans for food annually are not reared and killed without suffering, let's assume for the purposes of argument that these animals could be happily raised and painlessly killed.[1] Would it then be morally acceptable to kill them for food? The purpose of this chapter is to present arguments against the permissibility of killing happy animals.

Preliminary assumptions

Whether it is morally permissible to kill animals for food depends at least partly on the nature of the animals in question. There is some reason to believe that the "Great Apes"[2] and some other mammals such as dolphins have a wide range of psychological capacities, including possibly self-awareness, a sense of continuity from the past and into the future, language, and planning for the future.[3] Being near-persons, killing these animals is morally worse, all else equal, than killing animals with less capacity for self-awareness.[4] In general, the greater the capacity for sentience and self-awareness an animal has, the greater the moral cost in killing it, all else equal. It is not clear to what extent farmed animals such as chickens, sheep, cows and pigs have self-awareness.[5] Given uncertainty, I will assume for the sake of argument that the happy animals to be killed are only minimally sentient, and are not self-aware. If it is wrong to kill minimally sentient happy animals, then *a fortiori* it is wrong to kill more sentient and self-aware animals in the same circumstances.

It is possible that killing happy animals harms not only the killed animals themselves, but indirectly (1) other animals in their social groups that witness the killings; (2) the humans that do the killing (or that have come to know the animals to be killed, even if they do not directly participate in the killing); and/or (3) other humans, through, for example, damage to the environment closely linked to the practice of killing happy animals. Since I want to focus on whether the killing of happy animals is *in itself* morally acceptable, I will assume that (1) other animals are not harmed at all; and (2) humans are not harmed at all. If it is wrong to kill happy animals on these assumptions, then *a fortiori* it is wrong to kill them if these assumptions are in fact false.

The prima facie case: killing animals harms them

Presumably, happy animals would be killed for food long before the end of their natural lifespan. According to one plausible account, the wrongness of killing is proportionate to the strength of "time-relative interests" in continuing to live that are thwarted by killing.[6] The strength of these interests depends, first, on the quantity and quality of future goods the animal is likely to experience but for being killed; and, second, the degree of psychological continuity between the animal now and the animal that will experience future goods.[7] These two factors, in turn, vary according to the nature of the animals in question. Although empirically uncertain, I have assumed for the sake of argument that the farmed animals in question are minimally sentient, so their interests in continuing to live are relatively weak. However, even animals with very limited psychological capacities have at least a basic interest in continuing to live an enjoyable life according to their natures. Killing them thwarts these interests and deprives them of future goods. All else equal, killing them harms them and so requires justification.

Meat-eating as an important component of a good human life?

One possible justification is that eating meat provides a comparable benefit to human meat-eaters. However, it is not at all clear that the benefits of meat-eating are morally comparable to the harm of depriving animals of future welfare. In modern societies with plentiful access to nutritious plant-based food, most meat-eating does not support basic human interests in survival, health or adequate food supply. Meat-eaters presumably like the taste of meat, and want variety in their diets. Eating meat may also be part of their identity or cultural practices, or serve some aesthetic interest.

Loren Lomasky has argued that dining well is an important part of a good human life, comparable to appreciating the fine arts, and that the quality of fine dining is diminished by the exclusion of meat, such that vegetarians are missing out on an appreciable good.[8] Therefore, the moral costs to animals must outweigh the (significant) enjoyments of meat eaters; and he suggests that they do not.

I am prepared to assume for the sake of argument that the benefits of meat-eating are significant in the way Lomasky suggests. Even so, I do not think it follows that meat-eating is justifiable. First, meat-eating is neither necessary nor sufficient for a good human life. Lomasky concedes as much. Even if meat-eating makes an already good life (much) better, the interests at stake for animals are fundamental in the sense that if animals do not continue to live, they cannot have even a minimally good life. Continuing to live is necessary (but not sufficient) for animal welfare; the same cannot be said for meat-eating with respect to human welfare.[9]

Second, even if eating well is an important human good, meat-eating need not be: Meat-eating is not necessary in order to achieve the aesthetic and cultural goods of dining which Lomasky describes. Vegetarian cuisine can constitute "fine dining": It can engage all the senses; bring people together around a table to converse and talk about the food; and provide variety and richness of color and texture.[10] Insofar as meat-eating is symbolic, or expressive of culture or identity, these goods can be expressed sufficiently eloquently without meat, since cultural practices, the appropriation of symbols, and markers of identity are flexible.[11] Insofar as forgoing the specific texture, smell and flavor of meat is regrettable to the meat-eater, plant-based meat-replacements or cultured meat made possible by new technologies could suffice.[12] Granted, forgoing meat limits the range of options for delicious and beautiful meals; but this is not sufficient to make eating well depend on eating meat. As long as there is a sufficient palette from which to create good meals (as is the case in modern societies), the loss of some colors from the palette is not regrettable.

Finally, if eating meat involved no moral costs (in the form of cutting animals' lives short), then the life-enriching pleasures of meat-eating would be uncontroversial. But there is good reason to think that killing happy animals carries a moral cost. Even if this were uncertain, but a serious possibility, it seems to me that, if a good and meaningful human life can be led without eating meat, the right thing to do is give the animals the benefit of the doubt and forgo meat.[13]

The replaceability argument

The most challenging argument I have so far come across in defence of eating happy animals is that the practice of killing these animals for food is permissible because but for the practice, they would not come into existence and so would not experience enjoyable lives at all. The harm of cutting an animal's life short is neutralized by the benefit conferred on the animal that is brought into existence (as a result) to replace the killed animal; so killing happy animals is permissible.[14]

The argument depends on a particular view within utilitarianism (the "Total View") according to which the morality of an act should be judged by its contribution to the total quantity of welfare, not by its effects on the welfare of any particular individual(s). The welfare of contingent beings (in this context, beings which would not have come into existence unless another was first killed) is taken into account as well as the welfare of already existing beings. Since killing animals causes contingent animals to exist and live enjoyable lives, killing happy animals does not reduce the total quantity of welfare in the world such that it is permissible to kill and replace sentient animals that do not count as self-aware persons.[15]

Note that the scope of the Replaceability Argument is very limited. Animals are replaceable in this way provided that killed animals are not self-aware persons; the animals lead pleasant lives and can be painlessly killed; surviving animals are not affected by the death of one of their number; and we could not rear these animals if we did not eat them, so killing them causes others to come into existence.[16]

Given these conditions, even if we accepted this argument, killing animals would be permissible in a very narrow set of circumstances that are currently very rare (if they exist at all). However, I think there are good reasons to reject the idea that animals are replaceable in this way. I will present utilitarian and theory-neutral responses to the replaceability argument.

Utilitarian responses to replaceability

Those committed to a utilitarian framework need not accept replaceability. For example, Tatjana Visak has developed a sophisticated utilitarian alternative that does not entail replaceability.[17] Her argument is based on the assumption that existing rather than never existing cannot benefit or harm (be better or worse for) an individual. This assumption is based on a plausible counterfactual account of harms and benefits: An event or state of affairs X benefits (harms) me if, due to X, I will be better off (worse off) than I otherwise would have been. However, since reasonable people disagree on this assumption, she treats it as a conditional: If this assumption holds, what follows?

The second strand of her argument is a defence of "person-affecting utilitarianism" as an alternative to "impersonal utilitarianism" (which includes the "Total View"). According to the impersonal view, the value of an outcome is determined by summing up all the welfare that the outcome entails. So outcome A is better (worse) than outcome B if and only if outcome A contains more (less) welfare than outcome B. This view considers the total quantity of welfare an outcome entails.

In contrast, the person-affecting view focuses on aggregate net benefit (the extent to which individuals are better or worse off in the outcome). In particular:

1. If outcome A is better (worse) than outcome B, then A is better (worse) than B for at least one individual [necessary condition].
2. If A is better (worse) than B for someone but worse (better) for no-one, then A is better (worse) than B [sufficient condition].

The assumption that existing rather than never existing cannot benefit or harm an individual, together with the person-affecting view, avoids replaceability: If we compare (A) the outcome in which an animal is not killed at a particular time but lives out its natural lifespan to (B) the outcome in which an animal is killed at a particular time and replaced by another animal, aggregate net benefit is greater in (A) than in (B). This is because killing an animal at a particular time harms him or her (by depriving him or her of an enjoyable future), but bringing a new animal into existence does not harm or benefit that animal (from the assumption). Since aggregate net benefit is greater in (A) than in (B), (A) is required, and (B) is forbidden (from the person-affecting view).

The upshot is that if one accepts Visak's assumption and favors a person-affecting view of utilitarianism, utilitarianism need not entail replaceability. However, reasonable people disagree on the relative merits of person-affecting versus impersonal views of utilitarianism.

For example, the impersonal view requires us to hold that if we could increase the number of beings with pleasant lives without making others worse off, we should. Many find this troubling, partly because it implies that couples should have children simply because the children are likely to lead enjoyable lives and no-one else is adversely affected. Person-affecting views (and the prior existence view in particular) deny there is value in increasing pleasure by creating additional beings, so plausibly imply that it is not wrong to fail to bring into existence a being whose life will be pleasant.

However, person-affecting utilitarianism has some implausible implications, too. For example, a couple that knowingly conceives a child that will have a miserable life does no wrong on this view. Existing as opposed to not existing cannot harm the child, so there is no direct reason against conceiving the child. On the impersonal view, the couple should not conceive the child, which is more plausible to many.

It seems that each of a range of utilitarian views on the ethics of creation of new sentient beings (including person-affecting and impersonal views) has some counter-intuitive implications, so it is understandable that there is reasonable disagreement on which is the correct view.[18]

Given that it seems difficult to resolve the replaceability issue from within the utilitarian perspective, I will now present two theory-neutral arguments against killing happy animals.

Theory-neutral arguments from the interests at stake at the time of killing

Jeff McMahan assesses the permissibility of "benign carnivorism" (killing happy animals) in terms of the *interests* at stake at the time of killing.[19] In particular, we compare the interests at stake when one happy animal will be killed to provide one meal each for 20 people: The animal has an interest in continuing to live an enjoyable life, which is a function of the amount of good its life would contain if it were not killed, as well as the degree to which it is psychologically connected to itself in the future. Twenty people each have an interest in gaining pleasure from eating meat. Importantly, this is the marginal pleasure they will get over and above eating plant-based foods, which is likely to be slight. It seems unlikely that the interests of 20 people in experiencing "a few minutes of slightly greater pleasure could outweigh all the good that an animal's life might contain over several years, even when that good is heavily discounted for the absence of significant psychological continuity within the animal's life."[20]

Defenders of benign carnivorism might object that the animals owe their existence to the practice, and we should take this fact into consideration. But the response is that there are no animal interests that favour benign carnivorism, since no animal has an interest in being caused to exist: Interests arise only once an individual exists.[21] Benign carnivorism evaluated with respect to the interests it affects, and evaluated at the point at which animals are about to be painlessly killed, is not justified because the animals' interest in continuing to live outweighs human interests in eating them.

A similar argument turns on the difference between vital and non-vital interests at stake when assessing killing happy animals.[22] Assuming sentient animals enjoy basic moral status (the status of being worthy of consideration for their own sakes in reasoning about the morality of actions which affect them), we must take their interests into account when our decisions impact on them. In particular, if animals count morally, we should not sacrifice their vital interests to promote the non-vital interests of ourselves. Raising and killing animals for food (even if raised happily and killed painlessly) overrides vital interests of animals to remain alive through their natural lifespan. Plausibly, at least for most people with ready access to nutritious plant-based alternatives, no similarly vital human interests (in life and health) are promoted by eating meat; therefore eating animals (even if humanely raised and painlessly killed) is wrong.

Conclusion

I have presented some reasons (both utilitarian and theory-neutral) to reject the Replaceability Argument and to think that killing happy animals is wrong. I have assumed for the sake of argument that the animals to be killed have minimal sentience and limited self-awareness. If it is possible that killing happy animals is wrong in the above circumstances and on this assumption, then it is more than possible that killing happy animals with greater capacities for self-awareness is wrong. Similarly, I have assumed for the sake of argument that no other animals and no humans are harmed in the killing process. If evidence shows these assumptions to be false, then the case against killing happy animals would be further strengthened.

In the context of empirical uncertainty, but the serious possibility that many happy animals have more than minimal psychological capacities, and maybe even some form of self-awareness, I suggest we should give animals the benefit of the doubt and desist from killing even happy animals for food.

Finally, it is important to note that, although the question of whether it is wrong to kill happy animals is interesting, we should not lose sight of the fact that the vast majority of farmed animals currently used by humans are raised on factory farms under cruel conditions. Practically and strategically (at least for farmed animal advocates) advocating against factory farming should be prioritised.

Notes

1 I will call animals reared without suffering and killed painlessly "happy animals", or just "animals" where the context makes it clear that I am referring to happy animals.
2 Chimpanzees, gorillas, orangutans, and bonobos.
3 Peter Singer, *Practical Ethics* (New York, NY: Cambridge University Press, 2011), 94–100.
4 Singer, *Practical Ethics*, 101.
5 For an overview of the science, see Marian Stamp Dawkins, *Why Animals Matter: Animal Consciousness, Animal Welfare and Human Well-being* (New York, NY: Oxford University Press, 2012).
6 Jeff McMahan, "Chapter 3: Killing," in *The Ethics of Killing: Problems at the Margins of Life* (New York, NY: Oxford University Press, 2002), 189–265.
7 McMahan, "Chapter 3: Killing."
8 Loren Lomasky, "Is it Wrong to Eat Animals?" *Social Philosophy and Policy* 30(1–2) (2013): 177–200.
9 I discuss the interests at stake for humans and animals later in this chapter.
10 Michael Gill, "On Eating Animals," *Social Philosophy and Policy* 30(1–2) (2013): 201–207.
11 Christopher Ciocchetti, "Veganism and Living Well," *Journal of Agricultural and Environmental Ethics* 25 (2012):405–417.
12 Ciocchetti, "Veganism and Living Well."
13 Gill, "On Eating Animals," 205.
14 Singer, *Practical Ethics*, 104–122.
15 Singer, *Practical Ethics*, 104–122.
16 Singer, *Practical Ethics*, 120.
17 Tatjana Visak, "Do Utilitarians Need to Accept the Replaceability Argument?" in *The Ethics of Killing Animals*, eds. Tatjana Visak and Robert Garner (Oxford, England: Oxford University Press, 2015).
18 Shelley Kagan, "Singer on Killing Animals," in *The Ethics of Killing Animals*, eds. Tatjana Visak and Robert Garner (Oxford, England: Oxford University Press, 2015). Kagan prefers the impersonal approach and in particular the total view, since he thinks it is the least unacceptable of the various alternatives. He concedes that replaceability (of sentient beings) does seem to follow from the total view.
19 Jeff McMahan, "Eating Animals the Nice Way," *Daedalus* (Winter 2008), 66–76.
20 McMahan, "Eating Animals the Nice Way," 70.
21 McMahan, "Eating Animals the Nice Way."
22 Mark Rowlands, "Chapter 4: Morally Wrong and Terminally Stupid," in *Animal Rights: All That Matters* (London, England: Hodder & Stoughton, 2013).

Bibliography

Ciocchetti, Christopher. "Veganism and Living Well." *Journal of Agricultural and Environmental Ethics* 25(2012): 405–417.

Dawkins, Marian Stamp. *Why Animals Matter: Animal Consciousness, Animal Welfare and Human Well-being*. New York, NY: Oxford University Press, 2012.

Gill, Michael. "On Eating Animals." *Social Philosophy and Policy* 30(1–2) (2013): 201–207.

Kagan, Shelley. "Singer on Killing Animals." In *The Ethics of Killing Animals*, edited by Tatjana Visak and Robert Garner. Oxford, England: Oxford University Press, 2015.

Lomasky, Loren. "Is it Wrong to Eat Animals?" *Social Philosophy and Policy* 30(1–2) (2013): 177–200.

McMahan, Jeff. "Eating Animals the Nice Way." *Daedalus* (Winter 2008): 66–76.

McMahan, Jeff. "Chapter 3: Killing." In *The Ethics of Killing: Problems at the Margins of Life*, 189–265. New York, NY: Oxford University Press, 2002.

Rowlands, Mark. "Chapter 4: Morally Wrong and Terminally Stupid." In *Animal Rights: All That Matters*, 35–50. London, England: Hodder & Stoughton, 2013.
Singer, Peter. "Taking Life: Animals." In *Practical Ethics*, 94–122. New York, NY: Cambridge University Press, 2011.
Visak, Tatjana. "Do Utilitarians Need to Accept the Replaceability Argument?" In *The Ethics of Killing Animals*, 117–135, edited by Tatjana Visak and Robert Garner. Oxford, England: Oxford University Press, 2015.

1.3

FOOD ETHICS AND JUSTICE TOWARD ANIMALS

Corine Pelluchon

Introduction

The way I feed myself affects the possibility for other human and non-human beings to live and flourish, because the foods I consume have an impact on them. Ethics refers to the limits I set upon my right to use whatever is good for my own preservation for the sake of future and current persons and animals. Ethics is at stake when I am eating. This is especially clear when it comes to meat or other animal products.

The existence of animals and their own needs calls into question and raises issues of justice, since we share the planet with them, yet we usually live as if all other sentient beings were the mere objects of our ends. Making eating the paradigm of a philosophy that takes materiality and corporeality seriously implies replacing the concept of the individualistic if not atomistic subject that still grounds political liberalism with another conception of an embodied, relational self whose life is deeply connected to the lives of other non-human beings. This provides the philosophical foundation for a political theory in which the duties of the State are no longer reduced to security and the reduction of unfair inequalities between human beings. The consideration of all the dimensions that enable animals to flourish, and not only to survive, enters politics.

In the first part, I shall consider how our relationships with animals raise issues of justice and how they affect the way we understand and frame the political community. Are we entitled to speak of zoopolitics and what exactly does it mean? How can principles of justice be elaborated that take into account the interests of animals instead of organizing society solely upon speciesist prejudices?

In the second part, I shall focus on the moral traits that could encourage individuals to change their food habits. The challenge is to fill the gap between theory and practice and to overcome the following paradox. On the one hand, animal ethics is intended to provide an alternative to the human-centered or anthropocentric traditional ethics. It is a creative academic field and there are numerous public debates revolving around animal issues, including the ethological discoveries that have blurred the frontiers between us and them. Almost no one believes nowadays that animals are mere machines. On the other hand, the plight of animals worldwide has never been more serious than it is today. People continue to eat meat and wear fur and leather. Norms and knowledge have been unable to fill the gap between theory and practice. This is why a virtue-based approach to ethics can help us resolve this paradox.

The animal question as a political question

The animal question as an inquiry into the arbitrary foundations of our justice

The animal question is a political question for several major reasons. The first is that our relationships with animals, the way we treat them, are a mirror of ourselves. Animal factories do not only raise moral issues linked to human cruelty, but they also raise issues of justice. For instance, the fact of putting sentient beings in cages reflects the absolute power we confer to ourselves. More generally, we do not recognize that the ethological norms of animals set limits to our right to exploit them. We still believe that we have the right to do to them whatever we please and that human beings are the absolute source of legitimacy.

Fighting against violence toward animals is a political priority because they are suffering, but also because such violence sheds light on the unjust foundations of our justice. Not only is society organized as if animals were our slaves, but the way we treat them reveals who we are. This is especially clear when we eat their flesh. They were born to live according to the ethological needs of their species. And yet, they are never respected from the beginning till the end of their miserable and short existence. What we do when eating flesh is to incorporate this suffering.

Moreover, the transformation of farmed animals into factory "biomachines" means that breeding, which is a relation with a living being, is equated to industry. Pursuit of profit which explains the reduction of manufacturing costs is not compatible with animal welfare and respect for the conditions of workers in industrialized farms or in slaughterhouses who, in addition to a physically demanding labor, have to cope with the suffering of animals.

Violence toward animals in industrialized farms reflects the injustice of capitalism whose victims are both humans and non-humans. Eating meat makes us accomplices of such a system. On the contrary, adopting a vegan diet is not only a way of saying, like Plutarch, that eating meat is morally problematic because it implies killing a being whose life is as important for him as ours is for us.[1] It has also a political meaning as it advocates another way of living with other beings, opposes the meat industry and its allies and imposes a transition toward another way of consuming and producing.

To be sure, more and more people can no longer tolerate the violence toward humans and animals and the principle of domination that characterizes speciesism and capitalism. Unfortunately, their voices are not heard in political debates and they remain on the margins of politics. This is a democratic problem.

The focus of our interest is less on the legal status on animals than on the foundations of our ethics and politics, which the animal question helps us to critically examine. Traditional humanism is based upon a subject defined as an autonomous rational entity, that is to say upon an elitist and anthropocentrist standard that can explain the discrimination against all the beings that are not like it. The question we ask is whether another way of thinking that would abandon this speciecist perspective to create a new humanism is possible and whether we can envisage a State and a concept of justice that takes into account the interests of both non-human animals and human beings.[2] Such inquiry is critical and implies that we ask how animals can enter politics.

The case for a political theory including animals' interests

In order to provide a political theory that would allow us to determine the rules of a fair coexistence between human beings and animals, the meanings of *zoopolis* and of sentience must be clarified. Justice for animals requires the development of a political theory of animal rights that would dictate our concrete obligations toward them. However, we must acknowledge that even if we are not the sole source of legitimacy, animals have rights because we accept to consider them as sentient beings. Thus a political theory that accounts for animal rights requires a new philosophy of the subject. For all these reasons, I will mention the merit of Donaldson's and Kymlicka's political theory and also its difficulties.[3]

Traditionally, animals are seen as moral patients who are granted negative rights. But once we understand sentience as agency that refers to the ability to express one's individual preferences, and not only the ethological norms that supposedly set limits upon our right to use and abuse animals, we must accept that animals are moral and individual subjects. Animals are also political subjects, because they coexist with us. And as our interests and theirs are not the same, this community is political: It is a *zoopolis*.[4] However, the difficulty is to figure out *how* animals can enter politics. Are they moral and political subjects in the same way as our fellow citizens?

The originality of Donaldson and Kymlicka in their 2011 book is to give content to the concrete obligations we have towards animals. These obligations, which enable them to speak of positive rights for animals, are drawn from our relations with them. Free living animals have their own societies and are not willing to live with us. They do not have the same needs nor the same rights as domesticated animals. We have to feed our dog, but there is no duty to feed the lion in the jungle or to save the free living animals from predators. Lastly, we do not have the same obligations towards liminal animals, such as mice or foxes, which colonize our gardens without being invited, even if we cannot deny their right to exist.

However, I do not agree with Donaldson and Kymlicka when they apply human categories to animals. Animals are political subjects even if they are not deliberative agents. However, they cannot be considered as citizens because they do not imagine themselves as members of a political community. Moreover, we cannot deny the asymmetry between us and them. Even if taking into account the agency of animals and their capacity to have interests and to communicate them sets limits upon our rights, they need our help because we are those who ascribe rights to them.

My own approach of the animal question as a political question takes the asymmetry between human beings and animals seriously.[5] Justice means that the foundations of these rights are not speciesist and both our interests and those of animals are considered. Moreover, the expression of animal suffering and needs, understood as basic universal needs and as individual needs, are the point of departure for negotiation. Human beings negotiate and the laws that emerge do not express the nature of the different beings, but they determine the rules of coexistence between animals and humans. To promote greater justice toward animals requires considering their needs and interests when elaborating our policies.

Taking into account these interests should be one of the goals of the State. Our politics are always a zoopolitics, since our activities have an impact on animals. We are always connected to other persons, be they past, present or future and we are also always connected to other animals. The subject is no longer atomistic as in the philosophies that ground political liberalism. A new philosophy based on an embodied, relational subject, provides a new foundation for politics and integrates animals, future generations and ecology into politics. This extends the realm of politics beyond the conciliation of the freedom of living individuals and the reduction of unfair inequalities. A major innovation of this political theory is that it respects the subjectivity of animals seen as vulnerable selves and as dependent moral and political agents. Political theory is here intertwined with ontology.[6]

However, to make such a *zoopolis* a reality, we need to supplement representative democracy, which aims at representing the interests of current human individuals, so that it represents animals' interests as well. A possible approach is to supplement the representative system by a committee of associations engaged in animal protection that examine the political propositions and ensure that policies are not contrary to animals' interests. Such a committee would have the power to reject inappropriate proposals and suggest changes to these proposals.

Because we live in a speciesist and pluralistic society it is impossible to immediately eliminate animal exploitation. The process may take time and will be gradual. In the short term, we have to determine and implement concrete measures for reducing animal suffering, that are acceptable to the larger public. The point is to promote a society in which it becomes easier to live well with less suffering for animals and to promote a view of justice that makes the killing of animals more and more exceptional.

The long-term goal is to educate individuals and convince them to place nonhuman beings within the sphere of their moral consideration to the point that the killing of animals becomes problematic. When consideration of the interests of animals becomes a moral standard for most of us, it will change the foundations of right and the way natural right is understood – just as it happened with slavery. This would allow us to build a *zoopolis* that would not be a democracy with and for animals, but a democracy where human rights do not exclude justice towards animals.

This social and political evolution assumes that animals are considered as vulnerable, individual selves and it implies that respect for them is a condition of respect for ourselves and a way of being. Conversely, animal exploitation reflects a way of being that is characterized by arrogance and domination. This suggests the relevance of a virtue-based approach in animal ethics and emphasizes the link between all forms of domination, whether they are exerted on animals or on human beings.

From theory to practice: The case for a virtue-based approach in animal ethics

Responsibility and vulnerability as the ability to be hurt by animal suffering

Taking animals' interests seriously and feeling responsible for them to the point that we decide to frame a political community in which their interests do count is not only a matter of rules and institutions. It also requires a deep change in the way we represent and feel ourselves. This change goes beyond the overcoming the anthropocentric foundations of our ethics and politics. It goes hand in hand with a new philosophy of the subject, whose relational dimension implies that responsibility defines our identity. "I am" insofar as I answer the call of the other and there is a call, coming from animals seen as individuals, asking us who we are. For Levinas, the other was the other human being. Here an animal can also be the other and my responsibility extends to it.

Responsibility changes our person from within. It also refreshes the meaning of freedom which no longer refers to the ability to make choices and to change them. Moreover, responsibility cannot be equated with obligation nor duty, since it does not derive from a choice. It is the answer to the call of the other. Lastly, it is linked to our vulnerability.[7] Only a vulnerable person can feel responsible for the other.[8] Vulnerability expresses the frailty of any sentient being, animal or human, who experiences damage in a subjective way, but it also confers the ability to care for another being. In its rationality, responsibility goes beyond empathy, since we can be responsible for persons who are not in front of us or who do not yet exist. In this sense, responsibility is a trait specific to humans. Animals may feel empathy, but nobody has ever seen an animal mourning the disappearance of a species. However, we human beings need to feel some emotions (such as fear or shame) so that our responsibility causes us to act.

Vulnerability is the ability to be wounded because one suffers in one's flesh, but also because one is responsible for the other. Pity is the source of morality; it is a pre-reflexive identification with any being experiencing pain and suffering. It differs from empathy, which maintains the distinction between me and the other and pertains to our representation. Pity, as Rousseau says, is the innate repugnance we feel when we are in the presence of a sentient being who is suffering.[9] Pity is not responsibility nor justice, but is it possible to speak of responsibility and justice when there is no pity?

What happens with animals today, especially in the meat industry, resembles war. It is not only a war against animals, but also a war within ourselves since we choke the voice of pity. "War is waged over the matter of pity," Derrida said.[10] We call for morality and justice, but the worldwide plight of animals demonstrates that our morality and justice are mere words.

People continue to eat meat even though they know how cows, pigs or chicken are bred and slaughtered. The cognitive dissonance[11] that characterizes individuals who continue doing something they know to be wrong affects the construction of our identity. To reduce this dissonance – as well as to actively avoid situations and information likely to increase it – individuals strive for consistency and make use of strategies such as denial or rationalization which for example lead them to see meat simply as food and not as the flesh of an animal. People who eat meat use strategies to stay blind to the carnage of animals. On the contrary, those who accept the reality – even if it largely escapes our perceptive capacity – experience a shock and it becomes impossible for them to feed on animals.

Vulnerability as the exposure to the suffering of the others, including animals, is the first step. Without such an exposure and the feeling of shame for what humans inflict on animals, it is impossible to change our lifestyles. Thus, taking animals suffering seriously is not merely a matter of argumentation and calculation. It requires a deep change in the way we think, feel and live. This interior revolution in turn implies that we shed the entirety of our social attributes in order to fully grasp what we share with other sentient beings and accept to be hurt by their suffering.

Respect for animals as the result of a way of being

A particular selfhood and a certain kind of relation between self and others are required to stop considering that animals exist only for the sake of human beings or that we can exploit them without restraint. As Plumwood said, a more promising approach to ethical questions is to stop focusing on rights, obligations and duties, and to pay more attention to some moral concepts such as respect, care, concern, compassion, gratitude, wonder, responsibility.[12] This shift from norms to moral agents and their concrete motivations does not mean that virtues or moral traits are reduced to feelings and emotions.

Virtue is concerned with feelings and emotions and involves dispositions to particular emotional reactions, which may be enjoyable or painful. It also supposes a perceptive capacity with regard to animals, such as the capacity to experience wonder in relation to them. Virtue, as seen in *The Nicomachean Ethics*, is a complex unity of dispositions (*hexis*) that explains why the person flourishes when acting and feeling for the right reasons.[13] It implies deliberation and prudence and supposes a unity of virtues. Thus, to possess a virtue is to be a person with a complex mindset and the most significant aspect of this mindset is the wholehearted acceptance of a particular range of considerations as reasons for action.

When applied to our relationships with nature and animals, virtue ethics implies self-transformation. Gratitude, respect for the other living beings, compassion, humility express the acceptance of our finitude and that we feel that we are not an empire within an empire, as Spinoza said.[14] More precisely, our awareness of the interactions between beings and species changes the way we understand ourselves. As seen in Næss's last book, *Life's Philosophy*,[15] emotions are the consequences of such a knowledge that is not only intellectual. Likewise, Spinoza's *amor intellectualis dei* means that knowledge has a psychological, existential and affective effect. This is particularly clear with *hilaritas*, which Næss refers to. *Hilaritas* is a global joy that is quite different from *titillatio*. It derives from the intellectual love for God or from the knowledge of the interactions between individual things and beings and their contribution to the beauty of the whole or the biotic community.

What is interesting for us in these references to Spinoza and Næss is not that they may support holism or eco-centrism, nor that they may be perceived as advocating a mystical union between individuals and nature or God. The point is instead that respect for others, including animals, is a way of being. It does not derive from duty or norms, but it is a way of looking at them and being with them. The lesson of virtue ethics is not perfectionism, that is to say the effort to be perfect and to be proud of this achievement. It rather consists in asserting that moral traits constitute ways of being with others that derive from our own relationship with ourselves.

This is what I call an ethics of consideration.[16] *Cum sideris* means to look at the stars and to pay attention to the things and beings you are looking at. Consideration is not reduced to contemplation, but it is a way of being and being together, even in daily life. It starts with oneself and the acceptance of one's finitude, as seen in the book of B. de Clairvaux, *On consideration*. Such a way of being is also expressed in the way we eat, work, interact with the others, whether they are human or non-human. Consideration requires humility and, as it suggests a certain attitude toward the others and encompasses both moral, psychological and even aesthetical emotions and some representations, it provides a way to move from theory to practice, from thought to action.

Somebody who is not exposed to the suffering of others and uses strategies to avoid being harmed by their suffering and by injustice cannot experience consideration. Likewise, consideration is the opposite of domination, which implies arrogance, fosters the exploitation of the most vulnerable and makes one blind to the otherness of the other.

Concluding considerations

Eating reflects who we are. We are not what we eat, but what we eat and how we eat, the fact that we worry about where our food comes from and what was needed to produce it, reflects what we allow ourselves to do and what we prohibit ourselves from doing. Some believe that their right to be is absolute. In a phenomenology of nourishment, in which our existence considered in its materiality is the life of a relational subject, this right is limited by the existence of others and by our worry not to compromise their lives. Eating expresses our consideration.

Notes

1 Plutarch, "On the Eating of Flesh," *Moralia*, vol. I., translated by F. C. Babbitt (Whitefish: Kessinger Publishing, 2005).
2 C. Pelluchon, *Les Nourritures: Philosophie du corps politique* (Paris, France: Le Seuil, 2015). Nourishment. A Philosophy of the Political Body, translated by J. E. H. Smith (London: Bloomsbury, 2019).
3 S. Donaldson and W. Kymlicka, *Zoopolis: A Political Theory of Animal Rights* (Oxford, England: Oxford University Press, 2011).
4 Donaldson and Kymlicka, *Zoopolis*, 65–69.
5 C. Pelluchon, *Manifeste animaliste. Politiser la cause animale* (Paris, France: Alma, 2017).
6 Pelluchon, *Les Nourritures*, 240–267.
7 E. Levinas, *Totality and Infinity. An Essay on Exteriority*, translated by A. Lingis (Pittsburgh, PA: Duquesne University Press, 1969), 82–84.
8 E. Levinas, *Of God Who Comes to Mind*, translated by B. Bergo (Palo Alto, CA: Stanford University Press, 1992).
9 J.-J. Rousseau, *Discourse on the Origin of Inequality*, translated by D. Cress (Indianapolis, IN: Hackett Classics, 1992).
10 J. Derrida, *The Animal that Therefore I am,* translated by D. Wills (New York, NY: Fordham University Press, 2008), 108.
11 L. Festinger, *A Theory of Cognitive Dissonance* (Palo Alto, CA: Stanford University Press, 2009); M. Guibert, *Voir son steak comme un animal mort* (Montréal, Canada: Lux, 2015).
12 V. Plumwood, "Nature, Self, and Gender: Feminism, Environmental Philosophy, and the Critique of Rationalism," *Ecological Feminism* 6/ 1 (Spring 1991): 3–27.
13 Aristotle, *The Nicomachean Ethics*, translated by C. D. C. Reeve. (Indianapolis, IN: Hackett Classics, 2014).
14 B. Spinoza, *Ethics, Book III*, translated by E. Curley (London, England: Penguin Classics, 2005).
15 A. Næss, *Life's Philosophy: Reason and Feeling in a Deeper World* (Athens, GA: The University of Georgia Press, 2008), 171–175.
16 C. Pelluchon, *Éthique de la considération* (Paris, France: Le Seuil, 2018).

Bibliography

Aristotle. *The Nicomachean Ethics*. Translated by C. D. C. Reeve. Indianapolis, IN: Hackett Classics, 2014.

Clairvaux, Bernard de. *On Consideration*. Translated by G. Lewis. Oxford, England: Oxford University Press, 1906.

Derrida, Jacques. *The Animal That Therefore I am*. Translated by D. Wills. New York, NY: Fordham University Press, 2008.

Donaldson, Sue and Kymlicka, Will. *Zoopolis: A Political Theory of Animal Rights*. Oxford, England: Oxford University Press, 2011.

Festinger, Leon. *A Theory of Cognitive Dissonance*. Palo Alto, CA: Stanford University Press, 2009.

Guibert, Martin. *Voir son steak somme un animal mort*. Montréal, Canada: Lux, 2015.

Levinas, Emmanuel. *Totality and Infinity. An Essay on Exteriority*. Translated by A. Lingis. Pittsburg, PA: Duquesne University Press, 1969.

Levinas, Emmanuel. *Of God Who Comes to Mind*. Translated by B. Bergo. Palo Alto, CA: Stanford University Press, 1998.

Næss, Arne. *Life's Philosophy: Reason and Feeling in a Deeper World*. Athens, GA: The University of Georgia Press, 2008.

Pelluchon, Corine. *Les Nourritures: Philosophie du Corps Politique*. Paris, France: Le Seuil, 2015. *Nourishment. A Philosophy of the Political Body*. Translated by J. E. H. Smith. London, England: Bloomsbury, 2019.

Pelluchon, Corine. *Manifeste animaliste Politiser la cause animale*. Paris, France: Alma, 2017.

Pelluchon, Corine. *Éthique de la considération*. Paris, France: Le Seuil, 2018.

Plumwood, Val. "Nature, Self, and Gender: Feminism, Environmental Philosophy, and the Critique of Rationalism." *Ecological Feminism*, 6/1 (Spring 1991): 3–27.

Plutarch. "On the Eating of Flesh." *Moralia*, vol. I. Translated by F. C. Babbitt. Whitefish: Kessinger Publishing, 2005.

Rousseau, Jean-Jacques. *Discourse on the Origin of Inequality*. Translated by D. Cress. Indianapolis, IN: Hackett Classics, 1992.

Spinoza, Benedictde. *Ethics*. Translated by E. Curley. London, England: Penguin Classics, 2005.

1.4

ANIMALS AS HONORARY HUMANS

Bob Fischer

Introduction

Many philosophers who defend veganism are not defending *strict* veganism. Although they think that you shouldn't support animal agriculture, they also think there's nothing wrong with eating roadkill, or meat from a dumpster.[1] Granted, they claim, there are a whole host of reasons not to support animal agriculture, but if an animal's body is available due to an accident, or if consumption *wouldn't* support animal agriculture, then eating is morally permissible. Let's call the diet that these arguments require *philosophical veganism*. Many vegans wouldn't feel comfortable eating as philosophical veganism says they may. Indeed, they think it would be *wrong* for them to consume found animal products, whatever their source.

Some are happy to explain away this intuition. Neil Levy, for example, argues that it comes from "sacralizing" a particular diet, which involves, among other things, seeing that diet as intrinsically valuable.[2] Instead, he thinks, we should see veganism as a means to the end of reducing animal suffering, and while we might still maintain a strict dietary regime, we should do so largely for pragmatic reasons: e.g., it's easier to follow strict rules rather than ones that admit exceptions.

Levy may be right. Before we cave, however, it's worth exploring whether anything can be said in defense of the intuition behind strict veganism. Is it possible to justify the strict dietary standards to which many vegans hold themselves?

My goal here is to sketch an affirmative answer, though also to highlight some of its costs. In short, the thought is that we can justify strict veganism if we modify the relational approach to animals that Cora Diamond recommends.[3] On Diamond's view, a certain way of relating to animals—namely, as fellow creatures—grounds various moral obligations to them. However, she leaves some wondering why we should relate to them as she suggests, and she grants that her relational approach doesn't deliver strict veganism: "It does normally, or very often, go with the idea of a fellow creature, that we do eat them."[4] In what follows, I put forward a different way of relating to animals—as honorary humans—and I show how we might justify opting into that relationship. I conclude by considering the costs of this account relative to Levy's proposal.

Diamond

Let's start with Diamond's project. Her goal is to criticize the arguments that Peter Singer and Tom Regan made famous. These arguments begin with the observation that "marginal cases"—e.g., infants, those with severe cognitive disabilities, the comatose, etc.—deserve moral consideration. Nevertheless, they lack some of the more sophisticated cognitive capabilities that the rest of us enjoy. So, what explains their moral importance must be some simpler trait, such as *being sentient*, or *being the subject of a life*. The next step is to observe that animals also have that trait, and so it can only be mere prejudice—i.e., speciesism—that leads us to deny them moral consideration.

Diamond rejects the idea that any non-moral trait can, in and of itself, explain why an entity matters morally. Instead, she thinks that moral importance depends on a prior disposition to see an entity as falling into a category that is *already* moral. So, for example, to see an entity as human is to see it as an entity about which it would be bizarre to ask, "Why are you relieving its pain?" If we see an entity as human, we don't see her as a mere entity. Instead, we see her as *her*, as *someone*, and as a someone whose suffering provides a *pro tanto* reason to offer aid (even if that reason can, in some cases, be overridden easily).[5] If Diamond is right, then a remarkable conclusion follows: Singer/Regan-style arguments are either self-undermining or irrelevant. They're self-undermining if they lead us to abandon the thick categories that underwrite our moral responses, such as *human* and *animal* and *thing*. If we make that move, then we're left with nothing but the search for consistency, which we can just as well achieve by denying moral consideration to marginal cases. They're irrelevant if they work primarily because they involve relaying the many awful things we do to animals, and not because of the philosophical framework on which each relies. In that case, the arguments are implicitly invoking our way of seeing animals—namely, as beings to which we shouldn't be cruel—and not because they offer the right account of when a thing deserves moral consideration.

Diamond's argument for this view is subtle, and I won't discuss the details here. The short version, however, is this. We don't eat our dead, and our sense that we shouldn't is not easily explained in terms of either maximizing utility or respecting rights. But rather than give up the taboo, we should look for a different foundation. And that foundation, she thinks, is that prohibitions of this sort are part and parcel of how we relate to one another as human—i.e., as beings who aren't food.

I'm largely sympathetic to Diamond's approach, but many haven't been. Some contest the claim that it's difficult to explain why we don't eat our dead; others complain that Diamond's approach licenses any morality you might like, no matter how racist, sexist, or xenophobic. (After all, if morality boils down to how we choose to relate to things, then can't we configure the moral universe however we want?) I don't think that the explanatory challenge is so easily dismissed, nor that the charge of wild relativism is quite fair. However, I won't argue for either point here.[6] Instead, I want to grant that we should abandon Diamond's metanormative theory, but then try to reach a position like it via another route.

The strict vegan's principle

Strict vegans need a principle like this:

> You ought to treat an animal however you ought to treat a human being with comparable cognitive capacities.

Essentially, this principle says that you ought to treat animals as honorary humans, and it rules out a number of things that philosophical veganism permits. So, for instance, if it's wrong to eat dead human beings, it's also wrong to eat dead animals.[7] At the same time, this principle doesn't commit us to saying that we ought to give animals the right to vote, since we don't give similar rights to humans with similar cognitive abilities. Still, there are cases where it would be bad for animals to be treated as honorary humans: We think we ought to institutionalize human beings who have cognitive capacities that are comparable to those of sparrows, but it would be bad for sparrows to be institutionalized. So, we need to add a caveat to the principle:

> You ought to treat an animal however you ought to treat a human being with comparable cognitive capacities *unless it would be bad for that animal to be so treated.*

Let's call this *the Strict Vegan's Principle* (SVP). We could refine it further, but this is good enough for present purposes. How might you defend it?

The case proceeds in two steps. First, we need to view the SVP as part of a way of relating to animals. Second, we need an argument for relating to animals in the particular way that the SVP recommends.

Norms and relationships

Let's begin with a detour through sexual ethics. It seems plain that some relationships involve the presumption that you'll respect certain norms. If you're in a monogamous romantic relationship, you act wrongly if you have an affair. This isn't because it's inherently wrong to have sex with someone other than the person to whom you are romantically attached, or just because such behavior will upset (or otherwise harm) your partner. Granted, the consequences for your partner matter. But even if your infidelity went unnoticed, you would still be acting wrongly: part of what it is to have a monogamous relationship is to submit yourself to a certain moral norm. In having an affair, you fail to meet it.

Suppose we grant this. Still, it doesn't follow that we should commit ourselves to monogamous romantic relationships. There are two issues here: The first concerns what's wrong *given a particular relational structure* (e.g., partner in a monogamous romantic relationship); the second concerns whether we have reason to select *that* relational structure from the many that are available. And if we decide that we don't have particularly good reasons to select it, we aren't bound by its norms. If, for example, we choose to have open relationships, it won't be wrong to have affairs.

I propose that we think of *strict veganism* as a particular relationship that you might have toward animals, just as monogamy is a particular relationship that you might have with another person. Moreover, I propose that the SVP—or something much like it—is the central norm that structures strict veganism. In virtue of opting into a monogamous relationship, you are morally obligated not to become romantically or sexually involved with anyone other than your partner. Likewise, in virtue of opting into strict veganism, you are morally obligated to treat animals as the SVP requires.[8]

Strict veganism is one way to construe the vegan's rejection of speciesism: She affirms a broader moral community, but not a new ethic that guides it. Instead, she attempts to have the same ethic with more beings. In other words, the vegan proposes a revision to our default view about the scope of the moral community, but not to the way of relating that structures it. And just as new principles answer to our prior moral commitments when it comes to intra-human matters, the same is true concerning how humans should relate to animals. It might be attractive to say that we should maximize utility, respect rights, or avoid causing extensive, unnecessary harm. But if, paradoxically, that principle suggests treating either animals in a way forbidden by the relevant way of relating, there's a presumption in favor of the way of relating, not the principle. To opt into strict veganism is not just to endorse the SVP; it's also to give it a certain priority, just as we give priority to the norms that structure how we relate to other human beings. Granted, priority isn't a trump card: we might be led to abandon the SVP, and so strict veganism, based on weighty moral considerations. But they would need to be roughly as weighty as those that would lead to change how we honor our dead.

Why strict veganism?

Of course, advocates for strict veganism need to explain why we should be so related to animals. There are really two questions here. On the one hand, why should we revise the current relationship we have with animals? On the other, why should we have the particular relationship with animals that is strict veganism?

The answer to the former question is familiar. Where our most basic interests are concerned—e.g., in bodily integrity and continued existence—there don't seem to be any morally relevant differences between human and nonhuman animals. Hence, it would be arbitrary to limit moral concern to our kind.

The answer to the latter question is more complex. The issue here is why we ought to take up a certain type of relationship with animals, one that comes along with various absolute prohibitions that go well beyond *Don't cause unnecessary harm*. These prohibitions concern, among other things, what can be done to corpses and how lives can (and can't) be traded off against one another. There are two moves to make here.

First, the advocate for strict veganism can make a burden-shifting move. Again, when it comes to basic interests, there don't seem to be any morally relevant differences between human and nonhuman animals. But when we realized that there aren't any morally relevant differences between the interests of *racial* others, we didn't consider whether we should create a new set of norms for interacting with them—we just extended our insider practices to former outsiders. So, if the species boundary is morally irrelevant, why shouldn't we do the same thing here? That is, when we realize that species membership is morally irrelevant, the presumption shouldn't be that we need to create a new set of norms for interacting with those of other species. Instead, the default should be extending our insider practices to former outsiders.[9]

Of course, the differences between racial others feel quite minor when compared to the differences between members of various species. So any presumption in favor of extending our insider practices will be quickly challenged. This brings us to the second move, which involves explaining why the default is worth preserving.

Here's one such explanation. Strict veganism is a risk averse strategy for managing our tendency to make moral compromises: It errs on the side of condemning otherwise permissible actions instead of permitting genuinely impermissible ones. We should prefer it to philosophical veganism because our cognitive architecture looks to be stacked against animals, because eating animals appears to be associated with less accurate and morally worse beliefs about animals, and because we're better at following simple, inflexible rules.

There is a large literature on how humans cognize animals, and I can only hint at its contents here. However, it does seem clear that our cognitive architecture predisposes us to think about, and act toward, animals in inconsistent ways. Hal Herzog points out that our responses to animals are often affect-driven, and that our affective responses to animals are largely based on factors like cuteness and repulsiveness, not the capacity to suffer.[10] Moreover, as Daniel Wegner and Kurt Gray argue, we tend to think that animals matter insofar as they have experiences like pleasure and pain, and we're inclined to attribute the capacity to have such experiences insofar as we judge them to be similar to us.[11] Unfortunately for many animals, we don't seem to think we're that similar. However, as Bastian et al. show, when we *do* think about the similarities between animals and humans, we're more likely to regard them as morally important.[12]

Additionally, there seem to be connections between eating animals, or even viewing them as food, and our willingness to treat them poorly. Loughnan et al. and Bratanova et al. provide evidence that categorizing animals as food is correlated with the degree to which we take them to be sensitive to pain, and is similarly associated with their perceived moral status.[13] Bilewicz et al. make the case that omnivores are more likely to regard human beings as unique than are those who vegetarians, finding that members of the former group are less willing to attribute various psychological characteristics to animals than the latter.[14] They hypothesize that this is one way that omnivores disengage morally from the animals they eat.

Finally, we might agree with Levy that this way of relating to animals is more sustainable. As Rozin et al. and Rothgerberger argue, we're better at following simple, strict rules than we are at following more complex ones that admit exceptions.[15] The simplicity here isn't syntactic: It concerns how easy it is for us to apply the rule. That's why, as Rothgerberger shows, those who follow a *No meat* rule lapse less often than do those who follow an *Only humane meat* rule—the former is much easier to assess than the latter.[16] And the advocate for strict veganism might point out that *Don't do to an animal what you wouldn't do to a person* does very well on ease of applicability. We've all spent our lives cultivating habits of thought and feeling concerning how we should engage with other people, and we are generally pretty good making the right judgments (even if behavior sometimes lags behind).

When we put these points together—the first about our cognitive architecture, the second about how eating is related to the way we think about animals, and the third about sustainability—the advocate for strict veganism can say that relating to animals as honorary humans protects animals from aspects of ourselves that foster moral compromises, and is therefore preferable to any way of relating to animals that isn't sensitive to these dimensions of our psychology. Does this quasi-pragmatic justification undermine the moral force of the SVP? Absolutely not. You might have a quasi-pragmatic justification for opting into a monogamous relationship—e.g., you don't trust yourself not to become destructively jealous in an open relationship—but this has no bearing on whether, once in a monogamous relationship, you're morally bound by its norms. So if the above is compelling, we have good reason to say that you ought to treat animals as honorary humans. And if we should treat animals as honorary persons, then we can say what strict vegans want to say: Namely, that it's *pro tanto* wrong to eat *any* animal products, whatever their origin.

A caveat and a worry

I conclude with two thoughts. The first is a caveat. We should note that at least some of the norms that govern our interactions with one another, and hence with animals, are quite contingent. Famously, Herodotus described how the Greeks and Callatians were horrified by one another's funeral practices: the former burned their dead; the latter ate them. And while we might not accept the version of cultural relativism that Herodotus takes this story to support, it does seem implausible that one way of showing honor is morally superior to the other.[17] As a compromise, we might conclude that there are many ways to show respect for deceased persons (as well as their loved ones still alive), and so while there may be moral reasons to have some standard practices in each locale, there may not be moral reasons to have any particular set.

In the present context, what this means is that strict vegans have a *pro tanto* reason to show respect for animals by not eating their bodies *because they don't eat their dead*. But in a culture that celebrated life by consuming what remains after death, respect for the deceased wouldn't provide any reason to abstain. So strict veganism will get you the SVP, but since the content of the SVP depends on context, the results of the SVP will vary.[18]

Second, a worry. In general, we're strongly inclined to err on the side of caution when attributing sentience to humans, and it takes very strong reasons to justify harming a conscious human being.[19] But now consider some numbers from the Dutch entomologist Arnold van Vliet:

> In 2007, over 7 million cars [in the Netherlands] traveled about 200 billion kilometers. If we assume for simplicity that every month the average is the same for all cars, then 16.7 billion kilometers are traveled a month. In just the license plates, 3.3 billion bugs are killed per month. The front of the car is at least forty times as large as the surface of the plate. This means that cars hit around 133 billion insects every month. In half a year, that is 800 billion insects.[20]

Two hundred billion kilometers is about 124 billion miles. By contrast, the US Department of Energy estimated that Americans drove roughly *three trillion* miles in 2013,[21] and one can only guess at the worldwide statistic. Plainly, cars kill a staggering number of insects across the globe. If we ought to err on the side of caution when attributing sentience, as well as demand strong reasons to harm sentient beings, then driving a car begins to look highly suspect.[22] And, of course, driving is just the tip of the iceberg. Think of the stunning number of insects killed by insecticides for the sake of agriculture, as well as the insects exterminated in homes and schools and offices. Much of human life is predicated on the permissibility of these practices, so if we embrace strict veganism, we would be committed to condemning many aspects of modern life.

Perhaps we can live with this conclusion, or perhaps there's some way of dodging it. Some will be willing to embrace the conclusion that our obligations are much more demanding than we thought previously. Others might argue that the odds of insect sentience are, while non-zero, quite low. What's more, they might argue that if insects are conscious at all, their mental lives are sufficiently simple that they have relatively few interests. And, crucially, people might think that insects don't have an interest in continued existence, since it seems highly unlikely that they have much in the way of beliefs about the future. It isn't obvious that these moves are sufficient to block all the counterintuitive implications—the sheer number of insects may swamp any discounting that these kinds of considerations justify—but they still seem to be worth exploring.

Note, though, that Levy's pragmatic approach doesn't face this issue at all. For Levy, the goal is to reduce animal suffering, and if he thinks that the odds of insect sentience are low enough, he can happily endorse strict rules about larger land animals without endorsing strict rules about bugs. He can explain away—rather than try to validate—the conviction that we have moral reasons to abstain from eating roadkill or found animal products, and so needn't go down the path taken here.

This isn't an argument for Levy's pragmatism. It is, however, a reminder that strict veganism generates puzzles that we don't necessarily need to solve.[23] The argument for strict veganism begins with the intuitions of many ordinary vegans, so it's critical to determine how much weight those intuitions can bear. Are they, for instance, secure enough to justify accepting the conclusion that we have much more extensive obligations to insects than we might have thought? I'm not sure.

Notes

1 See, e.g., Jordan Curnutt, "A New Argument for Vegetarianism," *Journal of Social Philosophy 28*, no. *3* (1997): 153–172; Peter Singer's remarks in Richard Dawkins, "Peter Singer—The Genius of Darwin: The Uncut Interviews," The Richard Dawkins Foundation for Reason and Science, accessed January 5, 2014, http://richarddawkins.net/videos/3951-peter-singer-the-genius-of-darwin-the-uncut-interviews; Stuart Rachels, "Vegetarianism," *The Oxford Handbook of Animal Ethics*, ed. R. G. Frey and Tom L. Beauchamp (New York, NY: Oxford University Press, 2011), 877–905; and the now-common qualification in many pro-vegan essays, such as Tristram McPherson, "A Moorean Defense of the Omnivore?" *The Moral Complexities of Eating Meat*, ed. Ben Bramble and Bob Fischer (New York, NY: Oxford University Press, 2015), 118–134, that the argument is only supposed to show that eating animal products is *usually* wrong.

2 Neil Levy, "Vegetarianism: Toward Ideological Impurity," *The Moral Complexities of Eating Meat*, ed. Ben Bramble and Bob Fischer (New York, NY: Oxford University Press, 2015), 172–184.

3 Cora Diamond, "Eating Meat and Eating People," *Philosophy 53*, no. *206* (1978), 465–479.

4 Diamond, "Eating Meat and Eating People," 475.

5 As she puts it: "We cannot point and say, 'This thing (whatever concepts it may fall under) is at any rate capable of suffering, so we ought not to make it suffer'. [...] That 'this' is a being which I ought not to make suffer, or whose suffering I should try to prevent, constitutes a special relationship to it" (1978, 470).

6 For fuller presentations of these criticisms, see Peter Singer, "Utilitarianism and Vegetarianism," *Philosophy and Public Affairs 9* (1980), 325–337; and Daniel Dombrowski, "Is the Argument from Marginal Cases Obtuse?," *Journal of Applied Philosophy 23*, no. *2* (2006), 223–232. For defenses of Diamond against both criticisms, see Andrew Gleeson, "Eating Meat and Reading Diamond," *Philosophical Papers 37* (2008), 157–175.

7 Likewise, it provides tools to resist a number of anti-vegan arguments. Consider, for instance, Peter Singer's Replaceability Argument, which trades on a version of utilitarianism according to which the happiness lost due to an animal's painless death can be replaced by the creation of a new happy animal (i.e., total welfare isn't affected by the first animal's death as long as we bring another one into existence; see his *Practical Ethics* (New York, NY: Cambridge University Press, 1976)). Given this version of utilitarianism, total welfare would be increased insofar as meat-eating benefits us; so, we ought to eat happy animals. However, if we assume the SVP, no such argument would be compelling: We would never buy an argument for the conclusion that we should slaughter orphaned infants because the ones with which we'll place them will be just like them.

8 To be clear: There *might* be decisive moral reasons to have either monogamous romantic relationships or none at all. If so, then it's indeed wrong to have affairs, but not because

having them conflicts with the norms inherent to non-monogamous relationships. Instead, it's because we ought not to have such relationships at all, and hence aren't permitted to act in ways that are compatible with those relationships. This is roughly akin to saying that you have decisive moral reasons not to become a thief, and so even if your taking my wallet is perfectly compatible with your relating to me as thief-to-victim, you still shouldn't rob me. (This is *not* to suggest that the reasons for opting out of a life of crime are comparable to those for opting out of a polyamorous life.) But even if there are decisive moral reasons to have either monogamous romantic relationships or none at all—which seems unlikely—it isn't how they play out with respect to animals. Again: my project is motivated by the fact that the standard arguments for veganism deliver *philosophical* veganism, not strict veganism.

9 Granted, we can realize that our norms *shouldn't* be extended to former outsiders: e.g., perhaps they were norms that helped preserve the insider / outsider distinction. But we often make such discoveries by listening to those we once excluded, and animals can't speak for themselves in the same way that racial others can. So, the default here should be a simple extension. Thanks to Andy Lamey for this point.

10 Hal Herzog, *Some We Love, Some We Hate, Some We Eat: Why It's So Hard to Think Straight about Animals* (New York, NY: Harper Perennial, 2011).

11 Daniel Wegner and Kurt Gray, *The Mind Club: Who Thinks, What Feels, and Why It Matters* (New York, NY: Viking, 2016).

12 B. Bastian, S. Loughnan, N. Haslam, and H. Radke, "Don't Mind Meat? The Denial of Mind to Animals Used for Human Consumption," *Personality and Social Psychology Bulletin* vol. *38* (2012), 247–256.

13 S. Loughnan, N. Haslam, and B. Bastian, "The Role of Meat Consumption in the Denial of Moral Status and Mind to Meat Animals," *Appetite 55* (2010), 156–159; B. Bratanova, S. Loughnan, and B. Bastian, "The Effect of Categorization as Food on the Perceived Moral Standing of Animals," *Appetite 57* (2011), 193–196.

14 M. Bilewicz, R. Imhoff, and M. Drogosz, "The Humanity of What We Eat: Conceptions of Human Uniqueness Among Vegetarians and Omnivores," *European Journal of Social Psychology 41* (2011), 201–209.

15 Hank Rothgerber, "Can You Have Your Meat and Eat It Too? Conscientious Omnivores, Vegetarians, and Adherence to Diet," *Appetite 84* (2015), 196–203; P. Rozin, M. Markwith, and C. Stoess, "Moralization and Becoming a Vegetarian: The Transformation of Preferences into Values and the Recruitment of Disgust," *Psychological Science 8* (1997), 67–73.

16 This study fits with a recent study by the Humane Research Council on the lapse rate of vegans vs. vegetarians—the latter seem not to fare as well as the former. Perhaps *Just plants* is simpler than *No meat*—again, not because of the syntax, but because the former is more consistent with the motivations for the position than is the latter. Here, the simplicity isn't in the rule itself, but in the interplay between the rule and the justification for adhering to it. For details see, Humane Research Council, "Study of Former and Current Vegetarians and Vegans: Initial Findings," December 2014, https://faunalytics.org/wp-content/uploads/2015/06/Faunalytics_Current-Former-Vegetarians_Full-Report.pdf.

17 Admittedly, not everyone agrees; see Mathew Lu, "Explaining the Wrongness of Cannibalism," *American Catholic Philosophical Quarterly 87* (2013), 433–458.

18 We might hope that there are independent reasons not to imitate the Callatians, at least when it comes to animals. But to look for independent reasons is to give up the project that I'm pursuing here.

19 Abortion is not a counterexample to either claim: At some stages, it's clear that the fetus isn't conscious; at others, the debate is over whether the woman's reasons are strong enough to justify killing the fetus. So the generalizations in the main text hold true, but people disagree about other relevant issues.

20 Stephen Messenger, "Trillions of Insects Killed by Cars Every Year, Says Study," Treehugger. July 10, 2011. www.treehugger.com/cars/trillions-of-insects-killed-by-cars-every-year-says-study.html.

21 Federal Highway Administration, "Annual Vehicle Miles Traveled in the U.S.," last modified August 2016, www.afdc.energy.gov/data/10315.
22 For an argument for the view that insects are conscious, see Andrew Barron and Colin Klein, "What Insects Can Tell Us about the Origins of Consciousness," *Proceedings of the National Academy of Science of the United States of America 113* (2016), 4900–4908.
23 Another such problem, which is at least as bad as the problem of insects: Free living animal suffering.

Bibliography

Barron, Andrew, and Colin Klein. "What Insects Can Tell Us About the Origins of Consciousness." *Proceedings of the National Academy of Science of the United States of America* 113 (2016): 4900–4908.

Bastian, B., S. Loughnan, N. Haslam, and H. Radke. "Don't Mind Meat? The Denial of Mind to Animals Used for Human Consumption." *Personality and Social Psychology Bulletin* 38(2012): 247–256.

Bilewicz, M., R. Imhoff, and M. Drogosz. "The Humanity of What We Eat: Conceptions of Human Uniqueness Among Vegetarians and Omnivores." *European Journal of Social Psychology* 41(2011): 201–209.

Bratanova, B., S. Loughnan, and B. Bastian. "The Effect of Categorization as Food on the Perceived Moral Standing of Animals." *Appetite* 57(2011): 193–196.

Crisp, Roger. "Utilitarianism and Vegetarianism." *International Journal of Applied Philosophy* 4 (1988): 41–49.

Curnutt, Jordan. "A New Argument for Vegetarianism." *Journal of Social Philosophy*, 28 (1997): 153–172.

Dawkins, Richard. (n.d.) "Peter Singer—The Genius of Darwin: The Uncut Interviews." The Richard Dawkins Foundation for Reason and Science. Accessed January 5, 2014. http://richarddawkins.net/videos/3951-peter-singer-the-genius-of-da rwin-the-uncut-interviews.

Diamond, Cora. "Eating Meat and Eating People." *Philosophy* 53, 206(1978): 465–479.

Dombrowski, Daniel. "Is the Argument from Marginal Cases Obtuse?" *Journal of Applied Philosophy* 23, 2(2006), 223–232.

Federal Highway Administration. "Annual Vehicle Miles Traveled in the U.S." Last modified August2016. www.afdc.energy.gov/data/10315.

Gleeson, Andrew. "Eating Meat and Reading Diamond." *Philosophical Papers* 37(2008): 157–175.

Herzog, Hal. *Some We Love, Some We Hate, Some We Eat: Why It's So Hard to Think Straight about Animals*. New York, NY: Harper Perennial, 2011.

Humane Research Council. "Study of Former and Current Vegetarians and Vegans: Initial Findings." December2014. https://faunalytics.org/wp-content/uploads/2015/06/Faunalytics_Current-Former-Vegetarians_Full-Report.pdf.

Levy, Neil. "Vegetarianism: Toward Ideological Impurity." In *The Moral Complexities of Eating Meat*, edited by Ben Bramble and Bob Fischer, 172–184. New York, NY: Oxford University Press, 2015.

Loughnan, S., N. Haslam, and B. Bastian. "The Role of Meat Consumption in the Denial of Moral Status and Mind to Meat Animals." *Appetite* 55(2010): 156–159.

Lu, Mathew. "Explaining the Wrongness of Cannibalism." *American Catholic Philosophical Quarterly* 87(2013): 433–458.

McPherson, Tristram. "A Moorean Defense of the Omnivore?" In *The Moral Complexities Eating Meat*, ed. Ben Bramble and Bob Fischer, 118–134. New York, NY: Oxford University Press, 2015.

Messenger, Stephen. "Trillions of Insects Killed by Cars Every Year, Says Study." Treehugger. July 10, 2011. www.treehugger.com/cars/trillions-of-insects-killed-by-cars-every-year-says-study.html.
Rachels, Stuart. "Vegetarianism." In *The Oxford Handbook of Animal Ethics*, edited by R. G. Frey and Tom L. Beauchamp, 877–905. New York, NY: Oxford University Press, 2011.
Rothgerber, Hank. "Can You Have Your Meat and Eat It Too? Conscientious Omnivores, Vegetarians, and Adherence to Diet." *Appetite* 84(2015): 196–203.
Rozin, P., M. Markwith, and C. Stoess. "Moralization and Becoming a Vegetarian: The Transformation of Preferences into Values and the Recruitment of Disgust." *Psychological Science* 8(1997): 67–73.
Singer, Peter. *Practical Ethics*. New York: Cambridge University Press, 1976.
Singer, Peter. "Utilitarianism and Vegetarianism." *Philosophy and Public Affairs* 9(1980): 325–337.
Wegner, Daniel and Kurt Gray, *The Mind Club: Who Thinks, What Feels, and Why It Matters*. New York, NY: Viking, 2016.

1.5

NONHUMAN ANIMALS' DESIRES AND THEIR MORAL RELEVANCE

Robert Patrick Stone Lazo

It is common, which is not to say that it is universal, to treat, in research and in ethics, the psychic lives of nonhuman animals with little charity. Chiefly, this is methodological: In the interest of parsimony, scientists want to give the simplest picture they can to explain any given phenomenon. Mental episodes seem far from parsimonious, requiring as they do so many other ontological commitments that we oftentimes do not have a clear idea of how to either verify or falsify. We have clearer ideas about how to verify or falsify a stimulus-response or instinct-based understanding of nonhuman animal behavior, so we prefer these. Now this is just methodological behaviorism, distinct from philosophical behaviorism, which claims that all mental episodes are constituted only by their associated behavior. But there has been a tendency to confuse these two in ethical discourse over nonhuman animals, mistaking methodological behaviorism for the much stronger philosophical thesis when evaluating whether or not nonhuman animals have psychic lives.

The importance of this confusion for ethical discourse, particularly in the utilitarian tradition, cannot be overstated. There is a general tendency to assume that nonhuman animals do not have those psychic capacities we deem morally relevant. On the opposite side of this debate, authors have perhaps given too much to nonhuman animals to compensate for the behaviorist camp. Little of this has actually been argued. More often, authors assume they have the right answer, or that the question has been settled, ignoring the complexities of the question (i.e. the sheer number of species encompassed by the word "animal"). It should be obvious that whether or not nonhuman animals have temporally extended desires is as live a question as it has ever been. It is not for me to settle that question here; that is empirical work that I do not have the knowledge or the expertise to do.[1] No, I mean here to discuss whether or not what the behaviorists say about some theoretical nonhuman animal, stipulating ahead of time a number of their characteristics, actually entails what they, or some of them, say it does. In particular, I mean to examine whether or not the behaviorist can deny future-oriented desires to their theoretical nonhuman animal, and with them the at least indirect desire to continue living.

This chapter is broken up into four parts. First, it will present one of these behaviorist's argument, or, as we shall see, the lack thereof. Second, I will do my utmost to construct the best argument I can for his position, relying heavily on the works of Wilfrid Sellars and Robert Brandom. Third, I will offer a description of a desire. Fourth, and finally, I will offer my argument for why awareness of the future as future in a desire is not as morally relevant as some suppose.

But before beginning properly, I would like to briefly reiterate that I will not be evaluating the empirical data about nonhuman animals in this chapter. An overarching thesis to this chapter is that whether or not nonhuman animals actually have future-oriented desires cannot be answered by armchair reflection.[2] My remarks here are only meant to show how nonhuman animals might, without language, and without an awareness of themselves existing into the future, still have desires that are future-oriented in a morally significant way. Whether or not they do have such desires is for others to determine, though I expect the answer will be in the affirmative in many more cases than the behaviorist expects.

The behaviorist's argument

In "Death, Pain, and Animal Life," Christopher Belshaw argues that, because nonhuman animals do not have categorical desires, they do not have a desire for continued existence.[3] To be a little more particular, Belshaw argues that nonhuman animals do not have a desire for continued life because they do not have temporally extended lives; that is, they exist, really and truly, "in the moment."[4] Killing a nonhuman animal cannot, then, frustrate any desires for the future, as they have none to be frustrated. Death is not inherently evil for a nonhuman animal; any evil in death must be secondary, as from pain that accompanies it. Killing a nonhuman animal might still be morally wrong – because the killing causes pain, or would disrupt an ecosystem, or would cause suffering for humans – but it is never morally wrong because it ends the life of the nonhuman animal.

Of course, Belshaw's thesis here is only a small part of a larger project, but there is no reason for us to consider that project here. What matters is this central claim, that if an individual does not have a conception of themselves existing through time, then death cannot be morally bad for that individual.[5] The curious thing is, I don't believe Belshaw ever argues for this.[6] He accepts it at face value. He might seem to argue for it, as when he takes on the objections that death can be bad for individuals who currently desire death, and that it can be bad even for those who live in the moment, but really he just reiterates the claim, slightly modified, to respond to the first case, and quite literally reiterates the claim to respond to the second.[7] He does not go into why he accepts it, what reasons he has for it. Perhaps he takes it to be self-explanatory, or, more likely, there are arguments in the background that he takes to have already been played out.

The following section will attempt to construct a strong version of what I take to be the central argument from this background, given Belshaw's focus on an awareness of the future. I will not be focusing on Bernard Williams, who introduced the term "categorical desire," but rather on Wilfrid Sellars, whose essay, "Empiricism and the Philosophy of Mind," offers an excellent example of the general behaviorist argument on which I take Belshaw to be relying. Sellars was an ardent proponent of behaviorism, and, while the school of thought of course came before him, he offered many of the arguments that have since become central to the behaviorist tradition in justifying its assumptions and methodologies.

Language, concepts, and nonhuman animals

Before we can discuss Sellars in relation to Belshaw, we must, of course, understand what Sellars has to say. I will not be able to do justice to his larger project, and I will only be able to sketch out his arguments here, without considering objections. For the sake of the chapter, I must accept him at face value.[8]

There are two arguments in particular that I believe are important in the present context, the first because it is needed to make sense of the second. These are: (1) Sellars's argument against what he calls "the Myth of the Given"; and (2) his argument that language and concepts arrive simultaneously on the scene. We begin (1) with a definition of the myth. The myth supposes that there is a set of sentences that are both noninferentially known and that can be used as premises in arguments.[9] In other words, there is some set of sentences that we can both know without relying on any argument whatsoever (i.e. that are self-authenticating), and that act as a foundation for all of our other knowledge, which is built up off of them by way of argumentation. The classic examples of these sorts of self-authenticating, but still useful, sentences are analytic statements, which are true by virtue of their meaning (e.g. "2 + 2 = 4"), and observation reports, which are true because an individual cannot be wrong about how some x to them at time t (e.g. "This looks red to me now").[10] Members of this set (or these sets) can be construed as nonconceptual; this has particularly been the case with observation reports. Sellars means to deny the very possibility of any such set of sentences.

The argument here is more straightforward than one might assume, at least once one sees it.[11] First, a technical point: To say that any set of sentences is nonconceptual is to confuse what that sentence is about, which might be nonconceptual, with the sentence about it, which inherently involves conceptualization. Second, and this is the heart of the critique, Sellars subscribes to an inferentialist theory of meaning, a sort of meaning-as-use theory, where the meaning of a statement is, at least in part, determined by the inferences that sentence is used to make. As Brandom summarizes it,

> for *any* judgment, claim, or belief to be contentful in the way required for it to be *cognitively, conceptually*, or *epistemically* significant, for it to be a potential bit of *knowledge* or *evidence*, to be a *sapient* state or status, it must be able to play a

> distinctive *role in reasoning*: It must be able to serve as a *reason for* further judgments, claims, or beliefs, hence as a *premise* from which they can be *inferred*. That the role in reasoning, in particular, what those judgments, claims, or beliefs can serve as reasons or evidence *for*, is an essential, and not just an accidental, component of their having the semantic content that they do.[12]

There can never be a set of sentences that could be called noninferential, because part of the meaning of all sentences, even a set that theoretically only ever appears as premises and never as conclusions – and Sellars denies that such a set exists[13] – comes from how they are used to make inferential moves. Thus, there cannot in principle be a set of sentences that can be properly said to stand on their own. The use of any such set is not a language game one could play without playing any others.

Here we move into (2), Sellars's argument that language and concepts only arrive together.

> [T]o be the expression of knowledge, a report must not only *have* authority, this authority must *in some sense* be recognized by the person whose report it is … For if the authority of the report "This is green" lies in the fact that the existence of green items appropriately related to the perceiver can be inferred from the occurrence of such reports, it follows that only a person who is able to draw this inference, and therefore who has not only the concept *green*, but also the concept of uttering "This is green" … could be in a position to token "This is green" in recognition of its authority.[14]

To put this a little differently, in order for a token of "This is green" to have authority for the individual tokening it, and not only for those who witness it – as one might know there is something green in front of a computer that has been programmed to say "This is green" when presented with green objects – the individual must have the second-order recognition that performances of "This is green" are reliable indicators that there is a green thing present. Even more fundamentally, to know that a tokening of "This is green" is a reliable indicator, one must also know more generally what it is to follow a rule of the sort "Sentence token x is a reliable indicator of y." The following of a rule presupposes the implicit practice of following a rule.[15]

To have any concept at all is to suppose that one has a battery of other concepts; this is another implication of Sellars's inferentialist theory of meaning, since to know what a sentence means is to know how it is used, which presupposes conceptions about using something, which in turn rely on conceptions of particular examples of using something. To suppose that one has access to the rules of inference without having access to the rest of language is to buy into yet another version of the myth of the given. It is to assume that there is some set of practices, namely, practices involved in inference, that we have access to prior to any other.[16] But, as Sellars's argument against the myth shows, one can never have access to any part of language without having access to all of it. If part of the meaning of a concept is how it is used in inferences, one cannot have a concept without already having access to the so-called "space of reasons," the logical space of giving and accepting reasons, and, again, access to this space presupposes one has access to "giving," "accepting," and "reasons." Language cannot be acquired piecemeal, and since concepts rely on language, since part, or perhaps all, of the meaning of our concepts comes from how they are used in language, concepts and language must come fully formed together.[17]

To understand the importance of the above for Belshaw, let us just briefly consider a typical argument for nonhuman animals having future-oriented desires. The first step is to admit that they have desires at all, which Belshaw is willing to do.[18] Understandably, too, as it must surely be ridiculous to say that the fox caught in a trap, struggling to free herself, does not desire to free herself, or that a dog begging for a bite of food does not desire that food. From here, the move might seem an innocuous one: If the fox, or the dog, or the nonhuman animal generally is trying to do something, then they must have beliefs about the state of the world and how they interact with the world. The success or failure of their actions then provides further evidence for or against their beliefs. Given that a number of these beliefs must be about causal relations in the world (e.g. "If I do this, then this will happen"), which inherently involve the future as future, nonhuman animals capable of having beliefs at all must also be aware of the future as future. Nonhuman animals can have access to the space of reasons, and so to concepts, without any recourse to language.

The problem with this sort of argument, however, is that it is yet another version of Sellars' myth. Desire here construed is expected to play the role, first, of noninferential knowledge and, second, of a premise from which one may draw conclusions. Or, to be a little more precise, the satisfaction of the desire is supposed to be both noninferentially known and, simultaneously, a premise from which one can draw conclusions.[19] This is straightforwardly a case of the myth, an effort to move from a supposedly nonconceptual mental episode to a conceptual one with something analogous to language, but that somehow does not rely on the same kind of framework that our language does. In other words, it is to suppose nonhuman animals have a sort of language that is not a language. How one is to make sense of this idea is totally unclear, except that one might have a private language that does not depend on social interactions. I assume here that the behaviorist accepts arguments against private language, and so ignore the possibility.[20]

From here, it is supposed to follow that nonhuman animals who do not have language cannot have desires that extend into the future, since, without language, they are limited to desires that do not rely on having concepts (e.g. desire for food, desire to escape present pain, etc.). This is not to say that they don't have desires, or that we do not have certain obligations to nonhuman animals, such as to not cause them pain. It is only to say that they do not have future-oriented desires that would make death an evil for them in a way relevantly similar to the evil of death for humans, and that this has moral implications for how we ought to treat nonhuman animals.

I think something very much like this is what lies behind Belshaw's arguments against nonhuman animals having future-oriented desires, though he does not explicitly articulate it in the article.[21] I will accept the two arguments from Sellars, but I nevertheless believe there is a serious confusion at the base of this last move, which I will explain in the final section. For the moment, let us look more closely at a desire.

Temporal extension and the desires of nonhuman animals

The behaviorist is right to draw a distinction between desires like the desire to avoid present pain and like the desire to put off some present pleasure for the sake of greater pleasure later. I mean to argue here, however, that they are wrong to think that only the latter sort of desires involve the future in a way that morally matters. In fact, it seems likely – in the actual world, at least – that to fulfill any given desire at all entails temporal extension. Please allow a brief, pre-emptive defense against certain straw men arguments against my view before we move on to the argument proper.

I do not mean to say that the content of any desire entails temporal extension. Only the content of a desire with conceptual content can involve temporal extension. For that matter, there are serious reasons to think that discussions of the "content" of a desire, like the desire to escape present pain, run the risk of confusing themselves. There is a temptation to say that one "knows the content" of such a desire the way one knows the content of a conceptually articulated desire. To discuss the content of a desire is to run the risk of confusing our sentences about that content with the content itself, the same move Wittgenstein ridicules when he says, "It can't be said of me at all (except perhaps as a joke) that I *know* I'm in pain. What is it supposed to mean – except perhaps that I *am* in pain?"[22]

I also do not mean that a desire entails temporal extension for the one having the desire. This should be obvious enough, not only because it runs precisely counter to what I have already accepted, that one needs language in order to have concepts, such as the concept of the future as future, but also because there are plenty of instances where we don't accept an entailment like this. My desire for more money does not mean I am aware, in having that desire, that it involves, since there can be only so much money in the economy at any given time, others having less. Similarly, nothing about an individual's desire for a steak means that individual must be aware that this desire involves the death of an animal.

All I mean is that, for the outside observer with language, and so concepts, the fulfillment of any given desire in the actual world entails that the individual with that desire exists through a stretch of time. Obvious enough, since, for the fox to escape a trap, or for the cow to walk across the meadow to eat some grass on the other side, the individual cannot be said to fulfill that desire in the very moment of having it. It is only by moving through time that the desire is fulfilled.

The moral significance of theoretical nonhuman animals' desires

It might be true that our recognizing entailment of the sort described above relies on our having access to the space of reasons, but since this is now about our awareness of time, and not about nonhuman animals', it no longer matters. We have changed from an internal consideration of desires to an external evaluation of them. From this, I argue, it follows that nonhuman animals have temporally extended desires in a way that matters for the ethics of killing them. Death can be an evil for nonhuman animals, since it can frustrate their desires for the future that we understand as temporally extended, whether they are aware of that extension or not. In general, we accept this principle of the external evaluation of desires. To illustrate, let me offer a couple of examples.

First, suppose that I want to go for a walk. It's early autumn, the weather is lovely, and I would like to clear my head and enjoy the outdoors for a bit. Now, there is a great deal we know about the mechanics of walking, but there seems to be a great deal more we don't know, or we are at least missing some crucial piece of the puzzle. Hence we have yet to create a bipedal robot capable of walking over inconsistent terrain. So I can have a desire to go for a walk without understanding the physical mechanics of doing so.[23]

Now let us suppose that an alien – we'll call them Fred – has traveled across the galaxy to observe earth life, and one of their observations is that the key to how we walk is x. Fred has a sense of humor, and, as both an experiment and a joke with themselves, they decide to temporarily interfere with x just as I decide to go for that walk. Surely we must say that Fred has frustrated my desire even if I have no knowledge or awareness of x.

Perhaps, though, this example is a bit too science-fictional, or perhaps it relies on the fact that I have concepts. Allow a second, then. Consider the ravens kept in the Tower of London. They have all had their wings clipped. Just because the ravens do not have any awareness of the sort afforded by language that their wings have been clipped, thus preventing them from flying, that does not mean that their desire to fly has not been thwarted. It does not mean that it was not morally wrong to have clipped their wings.

The reason why temporal extension was supposed to be different was because it relies on concepts to make sense of it. This was the distinction between thwarting a nonhuman animal's desire to avoid present pain and thwarting a human desire to forgo some present pleasure for greater pleasure in the future. But we have now shown how one might externally evaluate desires to include a temporal element without recourse to the Myth of the Given, by relying on the conceptual power of the observer rather than on the one having the desire.[24] Belshaw is wrong in saying that desires had by those without language cannot be temporally extended.

There are a few final points I would like to make. The argument I have presented is not as strong as I would like it to be. First, it relies on its opponent admitting that nonhuman animals do have desires. While I take this to be obvious, I do not want to deny that there might be good reasons to suppose otherwise within the frameworks of Sellars and others. Second, it only applies to that class of nonhuman animals that do have desires, which it could turn out ignores a great many species, like starfish, honey bees, etc. Third, human interests could still override the interests of nonhuman animals not to die, particularly because this desire must still be incidental, and not "in the heads," so to speak, of nonhuman animals. It is this one that most bothers me. Fourth, I have failed to specify what nonhuman animal desires might particularly be morally relevant. There is no way around this: Only experimental work can give an answer.

There are still some advantages, though. We have, after all, disallowed behaviorists from claiming that the desires of nonhuman animals are only grounded in the present. This view also has a so far unmentioned advantage. It explains why we have moral commitments to nonhuman animals while they do not have moral commitments to us, or at least do not have those commitments in the way that we do. Language has given us access to the moral sphere thanks to our entry into the space of reasons, but, since nonhuman animals do not have this same access, they do not have moral commitments to us or to one another.[25]

One last advantage: We have taken the question of whether or not nonhuman animals have temporally extended desires out of the hands of the theorist and put it into the hands of the scientist. Whether or not nonhuman animals have desires is an empirical question. This adds a serious element of uncertainty to my conclusions, and so also worry, as I have just said, but that is just the way it is. We have also, I hope, cleared up a bit of the confusion between methodological and philosophical behaviorism, showing how, while the latter might mean nonhuman animals do not have desires at all, the former is committed to no such thing. Just the opposite, it must commit itself to answering whether or not they do.

Notes

1 I am not sure we even know how to state the question; we are, after all, still arguing whether or not mental states even exist.

2 I should also say that these questions should properly only be asked of individual species, and not of the fantastically broad, and so unhelpful in scientific research, category "nonhuman animals." But since I am here worried only about very broad possibilities rather than actualities, I do not think my use of the category is problematic.

3 Christopher Belshaw, "Death, Pain, and Animal Life," in *The Ethics of Killing Animals*, ed. Tatjana Visak and Robert Garner (Oxford, England: Oxford University Press, 2015), 37.

4 Of course, he then goes on to point out that he is not committed to all nonhuman animals living in the moment: "perhaps elephants, chimpanzees, some birds and cetaceans have a grasp of the future relevantly similar to ours, such that death is bad for them." But there seems also to be a class of nonhuman animals for whom there is no reason to suppose they have future-oriented desires, and in between these two categories there must be a group under debate (e.g. dogs, cats, salmon, etc.). Belshaw, "Death, Pain, and Animal Life," 38–39.

5 Belshaw, "Death, Pain, and Animal Life," 37.

6 He doesn't, for that matter, offer much of any reason why we should think most nonhuman animals do not have future-oriented desires in a morally relevant way.

7 Belshaw, "Death, Pain, and Animal Life," 37–38.

8 I do not accept Sellars's arguments for reasons related to externalist critiques of internalist theories of mental content and epistemic justification, but it is not my purpose to attack

these arguments. I mean to show why, even if these arguments are correct, the behaviorist is still wrong about nonhuman animals' desires.

9 Wilfrid Sellars, *Empiricism and the Philosophy of Mind* (Cambridge, MA: Harvard University Press, 1997), 15–17. Sellars's argument here is explicitly directed at various forms of classical empiricism, particularly sense–datum theories, but the structure of the myth can be generalized to cover any foundationalist system.

10 Or these are the interpretations offered by those who subscribe to the myth.

11 Again, Sellars is particularly arguing against traditional empiricist conceptions of the given, so I choose here to focus on generalizations from his arguments rather than on the arguments he presents. The same will be true when considering (2).

12 Robert Brandom, *Perspectives on Pragmatism* (Cambridge, MA: Harvard University Press, 2011), 87.

13 Sellars, *Empiricism and the Philosophy of Mind*, 78.

14 Sellars, *Empiricism and the Philosophy of Mind*, 74–75.

15 Brandom, *Perspectives on Pragmatism*, 13.

16 Sellars, *Empiricism and the Philosophy of Mind*, 65.

17 One might be tempted to say that this then requires a vicious infinite regress, the "Hegelian serpent of knowledge with its tail in its mouth," but all that is required is that one has a habit of tokening "This is green" in the appropriate circumstances; one is not required to say that those past tokenings expressed knowledge. Sellars, *Empiricism and the Philosophy of Mind*, 77–78.

18 Belshaw, "Death, Pain, and Animal Life," 39.

19 Brandom, *Perspectives on Pragmatism*, 73.

20 Though I would point out that the denial of private language is not the same as the denial of private episodes. These episodes could be nonlinguistic, like Wittgenstein's "not a Something, but not a Nothing, either!" Ludwig Wittgenstein, *Philosophical Investigations*, 4th ed., trans. G. E. M. Anscombe and Joachim Schulte (Oxford, England: Blackwell Publishing, 2007), 109[e].

I ought also to say that I think there are good reasons to think that nonhuman animals gain knowledge about the world in this general way that do not rely on language and so escape falling into the same trap as instances of the Myth of the Given..

21 Whether or not Belshaw actually subscribes to something like this I do not know, but it is the best argument I know that is used to support his position.

22 Wittgenstein, *Philosophical Investigations*, 96[e].

23 Here the behaviorist might be tempted to turn to "know how" instead of "know that" to explain how we can want to go for a walk, whereas there can be no "know how" for temporal extension. But it is precisely the trait of "know how" that one lacks awareness of how one is doing something (it took someone watching baseball players catch fly balls to understand how they were doing it), which is what the behaviorist has thus far been relying on to argue that a nonhuman animal's desire whose fulfillment is temporally extended is not morally relevant to the death of that animal.

24 Notice that this parallels how a computer's tokening of "This is green" is a reliable indicator that a green thing is present for us, even though it is not a reliable indicator for the computer, since the computer does not have the requisite concepts.

25 Since I disagree with Sellars, I do not believe this formulation is the right one, but I do believe something very much like it explains the asymmetry of moral obligations between nonhuman animals and humans.

Acknowledgements

Thanks to Andrew Stewart, for his comments on this chapter and our rambling conversations on its contents: His help and patience are greatly appreciated. Thanks, as always, to Andrew and Clair Linzey, for their continuing kindness and support.

Bibliography

Belshaw, Christopher. "Death, Pain, and Animal Life." In *The Ethics of Killing Animals*, edited by Tatjana Visak and Robert Garner, 32–50. Oxford, England: Oxford University Press, 2015.

Brandom, Robert. *Perspectives on Pragmatism*. Cambridge, MA: Harvard University Press, 2011.

Sellars, Wilfrid. *Empiricism and the Philosophy of Mind*. Cambridge, MA: Harvard University Press, 1997.

Wittgenstein, Ludwig. *Philosophical Investigations*, 4th ed. Translated by G. E. M. Anscombe, P. M. S. Hacker, and Joachim Schulte. Oxford, England: Blackwell Publishing, 2007.

1.6

WHY VEGETARIANISM WASN'T ON THE MENU IN EARLY GREECE

Simon Pulleyn

According to the Hebrew Bible, the earth before the flood was fully populated with humans and creatures of all kinds.[1] But animals were not to eat each other, and humans were not to eat animals: God explicitly states that his creatures are to nourish themselves from the plant kingdom, not the animal.[2] After the flood, God gave a new and wider dispensation that specifically permitted the eating of animals.[3]

Whilst this part of Genesis was written by the source known as P ("Priestly") towards the end of the sixth century BC, it is plainly meant to reflect the earliest condition of mankind. To this extent, it is an example of a Golden Age story: It evokes that happy era before everything went wrong, before Eve ate the apple, when the wolf lay down with the lamb and the earth brought forth its produce spontaneously.

But the story tells us something else. The explicit attribution of a vegetarian diet to early humans means that one cannot simply conclude that eating meat was a natural function of all pre-modern societies, to whose minds the alternative had not occurred.

A century or so after this passage in Genesis was written, the pre-Socratic Greek philosopher Empedocles (*ca.* 495–435 BC) told a fantastical tale of a period before Zeus and the Olympians when the older gods were propitiated with incense and honey and with paintings of living creatures rather than with the slaughter of bulls.[4] He is our earliest Greek source explicitly to attribute abstinence from sacrifice to primitive mankind.

Two centuries or more before this, Hesiod (*fl.* 700–675 BC) tells a similar Golden Age story. In his *Works and Days*, he speaks of a "golden race" of humans. Crops grew of their own accord in those days.[5] But nothing is said about the existence or exploitation of animals at that time.

But in his *Theogony*, composed before *Works and Days*, Hesiod had told a story designed to explain why it was that humans got the best cuts of meat after a sacrifice whereas the gods were left with fat and bones. The explanation was a trick played by Prometheus on Zeus at Mecone at a very early stage in history.[6] In this story, animal sacrifice was treated as a given. What fell to be explained was not *why* humans sacrificed but why they sacrificed *in the way that they did.*

Several writers have traced in detail the hostility of various Greek philosophical schools to animals and then followed the arc through to neo-Platonists and the beginnings of ethical vegetarianism in the modern sense.[7] But the views of Plato or Aristotle or the Stoics concerning the mental apparatus of animals can scarcely be taken to explain why sacrificing and eating them started in the first place. We may assume that the average Greek when cutting the throat of a sacrificial victim and later eating its butchered flesh did not give much thought to philosophical ideas about mind and reason. If he thought anything, it would have been, "This is what we do. This is what is handed down." My aim here is to attempt to sketch in outline some sort of explanation for this state of affairs.

The most important thing to grasp is the nature of the ideas that Greeks had about where things in general came from. They had, of course, no Bible: On account of this, it is often said that they had no revealed religion or scriptures. In the strict sense, this is correct. But they did have Homer and Hesiod, who told detailed stories about some aspects of creation and hinted at others.

The purpose of Hesiod's *Theogony* is to describe where the gods came from. What we do not find there is a divine creator who made heaven and earth and all that therein was. Hesiod describes the first reality as Χάος, which is best thought of not as a turbulent mass but as a yawning chasm. This chasm simply γένετο ("came into being").[8] No agency is described. In the same way, Earth and Tartarus (a sort of nether abyss) and Eros ("love"), Darkness and Night just "came into being". Nobody made them. Darkness and Light coupled sexually to give birth to Ether and Day.[9] By parthenogenesis, Earth bore Heaven, Sea and Mountains. Having brought forth Heaven in this way, Earth copulated sexually with it to give birth to Ocean, the Titans and the Hundred-Handers.[10] Among the Titans is Cronus, whose son is Zeus.

But of the origin of human beings Hesiod says nothing explicit in the *Theogony*. He mentions them in the same breath as giants, but with no indication of their ancestry.[11] If he knew of the much later story that humans were born from the blood of the giants,[12] he does not tell it. Similarly, when he tells in the *Theogony* of nymphs called Meliai ("ash trees"), he does not say that humans were born from them.[13] This is striking because, in his later poem *Works and Days*, he does mention a race of humans (metaphorically identified with bronze) as being made from ash trees.[14] But the story is told very much in passing – the origin of humans is not uppermost in the context of a tale describing their gradual moral decline.

In the same story, we are told that the golden race was "made" by the gods as a group during the reign of Cronus.[15] The race of silver was also "made" by the gods but terminated by Zeus in his anger that they did not give him due worship.[16] The races of bronze, of heroes and of iron are all said to have been "made" by Zeus.[17]

This idea of Zeus directly making human beings is peculiar and does not surface in Homer. He is called πατὴρ ἀνδρῶν τε θεῶν τε ("father of gods and men");[18]

but that is a figure of speech. Zeus was plainly not father of all the gods – Poseidon, Demeter, Hera and Hestia were his siblings, for example. Nor was he father of all men, even in the sense of some Adam through whom all others came into being. The Greeks had no equivalent of Adam.

If anything emerges from this curious parade of myths, it is that the early Greeks had no clear account of where they came from. They existed. The gods existed. The gods did not live in a world of spirit opposed to the human world of matter. The gods were like humans in most respects save that they were ageless and deathless,[19] had *ichor* in their veins instead of blood,[20] and ate ambrosia instead of meat and drank nectar instead of wine.[21]

What then of the animals? It will not be a surprise to learn that, even as Greek mythology is not greatly concerned with human origins, so it has even less to say about the origin of animals. There was no creator who ordained their cosmic purpose. But animals are ubiquitous in the early Greek hexameter poems of Homer and Hesiod: They appear as predators, prey, guardians, transport and, of course, food for human beings.

If there is one animal that dominates the *Iliad*, it is the horse.[22] The poem tells of the battles of great warriors, Greek and Trojan. These battles could not be fought without chariots, which are pulled by horses. Homeric heroes use their chariots as transport: Instead of fighting from them, they use them to get to a place, then park them, get out and fight on the ground.[23] The Homeric poems must have been shaped at a time when fighting from a chariot, so clearly depicted in numerous Assyrian reliefs,[24] was no longer current in Greece; the true use having been forgotten, the chariot is reduced to a run-about.

The poem dwells on the attributes of horses. Just after the famous Catalogue of Ships in *Iliad* II, there is a brief summary of who was best among the men and horses on the Greek side:

> By far the best horses were those of Pheretiades. Eumelus drove them and they were swift-footed, fast as birds. Their coats were the same colour and they were of the same age and they were matched in height to a hair. Apollo of the silver bow had raised them in Pereia – both mares, bringers of the rout of Ares.[25]

It is striking that the horses come before the men. That by itself is an indicator of the regard in which they were held in the poetic tradition. It is also striking that their history is given. It is characteristic of the epic style to dwell on the ancestry of a warrior at the moment of this death[26] or that of an inanimate object at a moment of significance;[27] the same is true of the horses here. And horses are significant enough that a god spends his time looking after them. This is doubtless a reflection of the enduring link between horses, wealth and power in the world in which the poems took shape.[28]

The prominence of the horse in Homer is doubtless a reflection of the fact that it was the taming of this creature that allowed the Indo-Europeans to sweep across so much of Eurasia around 2000 BC carrying all before them.[29] Just as Hannibal's elephants terrorized a world that did not know of them, Indo-European horses must have been a decisive advantage against infantry.

Brahmanic texts from India describe a hierarchy of value in the victims of Vedic sacrificial rituals on a descending scale starting with humans, followed by horses, cattle, sheep and goats.[30] The prominence of the horse in this list is striking. Vedic horse sacrifice (*aśvamedha*) was of high antiquity and prestige and finds some parallels in the Roman rites of the *October equus*. [31]

In the *Iliad*, by contrast, there is no depiction of the sacrifice or eating of a horse. When Achilles slaughters Lycaon into the river Scamander, he refers to the Trojans' customs of throwing live horses into the same river.[32] Whilst this is doubtless a sacrificial act, it would appear to be thought of as non-Greek at this stage.[33] Later texts attest horse sacrifice, but it is rare.[34] At any event, in these sacrifices involving precipitation into water, the horses are never eaten.

In Alan Bennett's play *A Question of Attribution*, Arthur Chubb is looking at Titian's famous painting *The Allegory of Prudence*. He asks, "Was he fond of animals?" Without pausing to think, Sir Anthony Blunt, the learned historian of art replies, "Titian? I've no idea. Shouldn't think so for a moment. People weren't." Chubb replies: "Rembrandt was. Rembrandt liked dogs." Blunt's reply is as withering as it is swift: "Rembrandt's dogs, Titian's age. I can see you've been down at the Purley Public Library again."[35]

We might ask in the same way whether Homer liked dogs. The death of Odysseus's dog Argus is described in a way that seems pathetic in itself rather than being designed merely to say something about Odysseus.[36]

Odysseus has been away for 20 years and has now returned in disguise. He comes with the swineherd Eumaeus to the latter's hut. There Odysseus sees his faithful dog Argus, reared by him and once a fine hunter. Now he is filled with fleas and lies in the dung. When he hears the sound of humans approaching, he lifts his head and pricks up his ears. Unlike Eumaeus, he recognizes Odysseus and wags his tail. Odysseus is moved to tears, which he surreptitiously wipes away. He asks who this fine dog can be, now brought so low. Eumaeus replies that the dog belonged to Odysseus. Now his master is dead and the maidservants do not look after Argus. Eumaeus and Odysseus go into the hut. What happens next is described in two lines of almost unbearably pathetic brevity:

> But the doom of black death overtook Argus as soon as he saw his master in the twentieth year.[37]

That the passage is developed at such length is of itself significant. In the epic sensibility, more is more: The longer an episode, the greater its importance. The way in which Argus has gone from being an honoured animal of his master to lying in the dung is the kind of reversal that is central to later Greek tragedy. We might expect a happy reunion – Argus bounds up to his master and is patted and restored to honour. But this cannot happen without Odysseus giving himself away.

The strength of Odysseus's feeling is shown by his tears – the same tears he weeps when Phemius sings of the doom of his companions at Troy. And the dog simply dies. It is as though he has been hanging on for all this time. Upon seeing his master, he lets slip his precarious hold on life.

This passage, so full of feeling for an animal, is unusual in Greek literature. When dogs are among the first victims of the plague in *Iliad* I, the fact is reported without sentimentality and almost in passing: Dogs and donkeys died first,[38] then humans.[39] The function of dogs in the *Odyssey* is clear enough: They are there to guard[40] and to hunt.[41] It is not clear why Greek dogs have been taken to Troy – most likely they are there to keep guard. But they are there and they are dying, just like the horses that have also been brought by boat by the Greek contingent.

Dogs, like horses, are only rarely mentioned as sacrificial animals. The idea of their being eaten is never mentioned in Homer. In later Greek, the ritual killing of dogs is very rare and limited to apotropaic and purificatory rituals.[42] This might be because they are trusted companions. But it might equally be because of some taboo. Achilles twice refers to dogs as bywords for shamelessness[43] – doubtless because they do in public that which embarrasses humans. Helen uses the word of herself,[44] in a passage where she suggests that she willingly followed Paris to Troy.[45]

This is not the place to rehearse old debates concerning the origins of sacrifice or what it meant for Greeks at different periods. What matters for our purpose is that sacrifice was closely linked with the eating of meat. Sacrificial ritual was obligatory when certain kinds of animals were needed as food for a meal, large[46] or small.[47] But a sacrifice might also be offered on the occasion of coming safely to land after a sea voyage.[48] Animals were regularly sacrificed in fulfilment of a vow in return for some perceived act of divine assistance.[49] An oath would usually be sealed by a sacrifice, although that might involve the animal being thrown into the sea after slaughter, to be eaten only by fishes.[50]

Sacrifice was pervasive. It was the chief source of such meat as was eaten, and Greeks took it for granted as part of the settled order of things. Inscriptional evidence from later periods gives an idea of the enormous cost of sacrificial oxen. A single ox of the cheaper sort could cost what one person in fourth-century Athens would expect to spend on food in five months.

Homer is more interested in prestige than economics. He routinely depicts the slaughter of large numbers of oxen,[51] or oxen mixed with large numbers of other animals.[52] In real life, most archaic Greeks will have eaten cheese, milk, nuts and fish. Meat was available, but not on as lavish a scale as depicted in Homer. There is a sense in the epics of a disappeared world in which men of heroic stature ate heroically.[53]

There are, nevertheless, indications in the formulaic language of the poems that the stereotypical human source of nutrition was not meat. Mortals (βροτοί) are referred to in the *Iliad* as those who ἀρούρης καρπὸν ἔδουσιν ("eat the fruit of the ploughland").[54] In the *Odyssey*, men[55] are referred to three times as ἀλφησταί.[56] If, as seems likely, this means "eaters of grain", then the conclusion appears to be that the consumption of grain is so characteristic of humans that it can form a standard epithet. Elsewhere, barley[57] (or barley mixed with flour), is described as the "marrow" of mankind. Presumably, the idea is that it somehow contributes to the making of bone-marrow. When Odysseus arrives at the land of the Cyclopes, he puts together a scouting party to discover οἵ τινες ἀνέρες εἶεν ἐπὶ χθονὶ σῖτον ἔδοντες ("what sort of corn-eating men these were upon the earth").[58]

Perhaps this is a recollection of some early idea of mankind as living solely from crops, not unlike what is found in Genesis. Whilst this affords us a glimpse of a potentially different earlier state of affairs, the fact remains that sacrifice is fundamental to the world of archaic Greece.

We cannot say with certainty whether or not early Greeks ever ate meat that had not been sacrificed. Eating the flesh of animals killed in the hunt is described with no mention of ritual.[59] Unless this is the result of some compression of narrative, we might assume that meat could be eaten without first being sacrificed. The same appears also to be true of eating fish and birds.[60] The correct formulation would appear to be that, if an animal was of a species that was usually sacrificed, Greeks did not usually eat its flesh without first sacrificing it.[61] Other animals might be eaten unceremoniously.

That tells us something quite important: The eating of living creatures is not simply the concomitant of having a sacrificial system. It was doubtless true that this is how most people ate most of the flesh that came their way. But it does not explain the eating of meat in the first place. To ask why some animals were sacrificed before being eaten and some not is to pose an unanswerable question. What the existence of meat-eating without sacrifice strongly suggests is that the meat-eating came first and its wrapping up as part of a religious system came second.

So one might conclude that Greeks killed and ate animals not because it was divinely mandated – there was no revealed religion; not because it was a by-product of the sacrificial system – some meat was not sacrificed; not because they did not have plentiful supplies of other food – vegetables, nuts and cheese were in plentiful supply.

Polyphemus eats Odysseus's men because he is stronger than them and he can do so. The expression used (ὡπλίσσατο δόρπον) is that of an ordinary man who "prepares" an ordinary "meal".[62] The sense one is left with is that there was a ladder of creation and if A is stronger than B, then A is apt to kill and perhaps eat B. This is true in exceptional cases such as Polyphemus; it is institutionally true in the case of humans eating animals, whether sacrificed or not.

A considerable part of the modern discourse concerning eating animals has turned on the question of whether they have souls. If a creature has a soul, then it might be immoral to kill it save in circumstances of self-defence. This underlies the denial of many Christians that animals have souls. If they do, hamburgers are forbidden. What was the position in early Greece?

Humans in Homer are said to have a thing called ψυχή. Students are taught to translate this as "soul". The problem with this is that "soul" has for modern observers a host of meanings that might not be present in the earliest stages of the

Greek language. In Homer, both humans *and* animals have ψυχή and both lose it at death.[63] So whatever it was, the ψυχή was common to all living things.

In large part (arguably even exclusively), they preserve an archaic view of the world in which the ψυχή was no more and no less than the vaporous breath which is lost or taken away at the moment of death,[64] passes the barrier of the teeth[65] and flits around feebly in Hades.[66] This is in keeping with the sense of the etymologically cognate verb ψύχω which has to do with breath and cooling.[67]

So whilst both humans and animals have ψυχή, and whilst both lose it at death, this was no more than a physical fact. There was no theological reason to suppose a kinship on the basis of shared ψυχή and more than on that of shared blood. There was thus no reason why anyone should think to challenge the sacrifice and eating of animals on this basis. However, the use of the word ψυχή was to change over time.[68]

Xenophanes tells of Pythagoras (*ca.* 530 BC) taking pity on a dog that was being beaten and commanding that the beating should cease because in this dog was the ψυχή of what had in a previous life been a friend of his and he recognized its voice.[69] Whilst this seems bizarre – possibly even a joke – the underlying point is arresting: One cannot say that an animal is inferior by reason of its possessing an "animal soul" because all "soul" is materially the same wherever it is found. Plainly Pythagoras saw the ψυχή differently from the epic tradition since its commonality to all living things has become a reason for abstinence from flesh.

Earlier but less definite signs of this shift in meaning can be seen in the Spartan poet Tyrtaeus (*ca.* 650–600 BC), who uses the verb φιλοψυχέω ("to love one's ψύχη")[70] in such a way as to suggest that he sees the ψυχή as more like "life" in the full modern sense rather than mere breath. Other more developed indications may be found in the pre-Socratic philosophers of the sixth century BC. [71] In other words, outside Homer, and by the middle of the seventh century BC, ψύχη takes on a meaning less like "breath" and more like "life" in the sense of something to be enjoyed and reflected upon.

This being so, one can see how a defence of animals from sacrifice and eating could be mounted by Pythagoras. One might say that his reasoning would only apply in a system that believes in reincarnation. Whilst that is how it arose, the perception that the stuff from which souls are made is likely to be the same wherever it is found is not limited to world-views that accept metempsychosis.

To conclude, the early Greeks were not unsophisticated. They knew that animals had breath and mental apparatus. To some extent, this made them kin with humans. But their sense of a soul had not developed to a point where it caused the eating of animals to seem problematic. And for all that Orphics, Pythagoreans, and neo-Platonists put the contrary case from time to time, the fact is that early notions of sacrifice being normative were so deeply imbedded by Homer's own day that nothing was likely to change it and in persisted until the end of antiquity and beyond. As late as the 1970s, it was reported that the sacrifice of a sheep or goat was a regular part of Christian Sunday worship in Soviet Armenia.[72]

Notes

1 Genesis 1: 24–6. All references to the Genesis are from A. Tal, ed., *Biblia Hebraica Quinta – Band 1: Genesis* (Stuttgart, Germany: Deutsche Bibelgesellschaft, 2016).
2 Genesis 1: 29–30.
3 Genesis 9: 3.
4 Empedocles 31 B 128, cited in H. Diels and W. Kranz, *Die Fragmente der Vorsokratiker* (Berlin, Germany: Weidmann, 1964).
5 Hesiod *Opera et Dies* 118–9. All references to Hesiod are from F. Solmsen, R. Merkelbach and M. L. West, *Hesiodi Theogonia Opera et Dies Scutum Fragmenta Selecta.* Third edition (Oxford, England: Clarendon Press, 1990).
6 Hesiod *Theogonia* 535–69.
7 See for example, R. Sorabji, *Animal Minds and Human Morals* (London, England: Duckworth, 1993), *passim.*
8 Hesiod *Theogonia* 116.
9 Hesiod *Theogonia* 125.
10 Hesiod *Theogonia* 133–53.
11 Hesiod *Theogonia* 50.
12 Ovid *Metamorphoses* 1. 154–62 cited from W. S. Anderson, *Ovidius: Metamorphoses* (Stuttgart and Leipzig, Germany: Teubner, 1998); M. L. West, *Hesiod: Theogony, Edited with Prolegomena and Commentary* (Oxford, England: Clarendon Press, 1966), 173.
13 Hesiod *Theogonia* 187; West, *Hesiod: Theogony*, 221.
14 Hesiod *Opera et Dies* 143–5; M. L. West, *Hesiod: Works and Days, Edited with Prolegomena and Commentary* (Oxford, England: Clarendon Press, 1978), 187.
15 Hesiod *Opera et Dies* 109–111.
16 Hesiod *Opera et Dies* 137–9.
17 Hesiod *Opera et Dies* 143–4, 157–8, 173 d–e.
18 Homer *Iliad* 1. 544. All references to Homer are from T. W. Allen, *Homeri Opera.* 5 vols (Oxford, England: Clarendon Press, 1912–15).
19 Homer, *Odyssey* 5, 218; cf. *Iliad* 1, 290.
20 Homer, *Iliad* 5, 340.
21 Homer, *Odyssey* 5, 199 vs. 196–7; cf. *Iliad* 5, 341.
22 J. Dumont, *Les Animaux dans L'antiquité Grecque* (Paris, France: Harmattan, 2001), 52–60.
23 Homer, *Iliad* 3, 261–5; O. Murray, *Early Greece.* Second edition (London, England: Fontana, 1993), 36.
24 P. Collins, *Assyrian Palace Sculptures* (London, England: British Museum, 2008), 39.
25 Homer, *Iliad* 2, 763–7, translation mine.
26 Homer, *Iliad* 6, 21–8.
27 Homer, *Iliad* 2, 100–109.
28 A. R. Burn, *The World of Hesiod* (London, England: Kegan Paul, 1936), 160–1.
29 J. P. Mallory and D. Q. Adams, *The Oxford Introduction to Proto-Indo-European and the Proto-Indo-European World* (Oxford, England: Oxford University Press, 2006), 154.
30 J. Puhvel, "Victimal Hierarchies in Indo-European Animal Sacrifice," *AJP 99* (1978): 354.
31 C. B. Pascal, "October Horse," *HSCP 85* (1981): 261–91.
32 Homer, *Iliad* 21, 132.
33 N. J. Richardson, *The Iliad: A Commentary, Volume VI: Books 21–24* (Cambridge, England: Cambridge University Press, 1993), 65.
34 R. Parker, *On Greek Religion* (Ithaca, NY: Cornell University Press, 2011), 138 n. 59.
35 A. Bennett, *Plays*, Volume 2 (London, England: Faber and Faber, 1998), 325. The play itself having been performed originally in 1989.
36 Homer, *Odyssey* 17, 290–327.
37 Homer, *Odyssey* 17, 326–7, translation mine.
38 Homer, *Iliad* 1, 50.

39 Homer, *Iliad* 1, 51.
40 Homer, *Odyssey* 14, 21, 29; 17. 200; cf. the fantastical robotic guard-dogs of Alcinous at *Odyssey* 7, 91–4.
41 Homer, *Odyssey* 19, 429.
42 Parker, *On Greek Religion*, 159 nn. 133, 134.
43 Homer, *Iliad* 1, 159, 225.
44 Homer, *Iliad* 3, 180.
45 Homer, *Iliad* 3, 174; cf. Herodotus *Historiae* 1. 4. 2, cited from N. G. Wilson, *Herodoti Historiae* (Oxford, England: Clarendon Press, 2015).
46 Homer, *Odyssey* 8, 59–61.
47 Homer, *Odyssey* 14, 73–7, 414–34.
48 Homer, *Odyssey* 3, 5, 159.
49 Homer, *Iliad* 6, 305–310.
50 Homer, *Iliad* 19, 266–8; cf. G. S. Kirk, *The Iliad: A Commentary, Volume I: Books 1–4* (Cambridge, England: Cambridge University Press, 1985), 310–11.
51 Homer, *Odyssey* 3, 8.
52 Homer, *Odyssey* 8, 59–60; 17. 180–1; 20. 250–1
53 J. Griffin, *Homer on Life and Death* (Oxford, England: Clarendon Press, 1980), 19.
54 Homer, *Iliad* 6, 142; 21, 465.
55 The word used in always ἀνήρ which denotes man as opposed to woman. But it is unlikely that anything turns on this point.
56 Homer, *Odyssey* 1, 349; 6, 8; 13, 261.
57 Homer, *Odyssey* 2, 290; 20, 108.
58 Homer, *Odyssey* 9, 89.
59 Homer, *Odyssey* 10, 180–4.
60 Homer, *Odyssey* 12, 330–2; cf. 4, 368–9.
61 Parker, *On Greek Religion*, 131.
62 Homer, *Odyssey* 9, 291; cf. *Iliad* 11, 86.
63 Homer, *Iliad* 16, 453 (man); *Odyssey* 14, 426 (pig).
64 Homer, *Iliad* 24, 754.
65 Homer, *Iliad* 9, 408–9.
66 Homer, *Iliad* 23, 100–101.
67 Homer, *Iliad* 20, 439–40; *Odyssey* 24, 348; M. Clarke, *Flesh and Spirit in the Songs of Homer* (Oxford, England: Oxford University Press, 1999), 144–7.
68 There might be signs of it doing so with the poems themselves: D. L. Cairns, "Myths and Metaphors of Mind and Body," *Hermathena* 175 (2003): 41–75.
69 Xenophanes 21 B 7 D-K cited from Diels and Kranz, *Die Fragmente der Vorsokratiker*.
70 Tyrtaeus *fr*. 10. 17–18 cited from M. L. West, *Iambi et Elegi Graeci ante Alexandrum Cantati*, second edition (Oxford, England: Clarendon Press, 1989–92); Clarke 1999: 298.
71 Clarke 1999: 288–9.
72 W. Burkert, *Homo Necans*, trans. Peter Bing (Berkeley, CA: California University Press, 1983), 8–9.

Bibliography

Allen, T. W. *Homeri Opera*. 5 vols. Oxford, England: Clarendon Press, 1912–1915.

Alt, A., O. Eisfeldt, P. Kahle and R. Kittel, eds. *Biblia Hebraica Stuttgartensia*. Second edition. Stuttgart, Germany: Deutsche Bibelgesellschaft, 1984.

Anderson, W. S. *Ovidius: Metamorphoses*. Stuttgart and Leipzig, Germany: Teubner, 1998.

Bennett, A. *Plays*, Volume 2. London, England: Faber and Faber, 1998.

Burkert, W. *Homo Necans*. Translated by Peter Bing. Berkeley, CA: California University Press, 1983.

Burn, A. R. *The World of Hesiod*. London, England: Kegan Paul, 1936.

Cairns, D. L. "Myths and Metaphors of Mind and Body." *Hermathena* 175(2003): 41–75.
Clarke, M. *Flesh and Spirit in the Songs of Homer*. Oxford, England: Oxford University Press, 1999.
Collins, P. *Assyrian Palace Sculptures*. London, England: British Museum, 2008.
Diels, H. and W. Kranz. *Die Fragmente der Vorsokratiker*. Berlin, Germany: Weidmann, 1964.
Dumont, J. *Les Animaux dans L'antiquité Grecque*. Paris, France: Harmattan, 2001.
Griffin, J. *Homer on Life and Death*. Oxford, England: Clarendon Press, 1980.
Kirk, G. S. *The Iliad: A Commentary, Volume I: Books 1–4*. Cambridge, England: Cambridge University Press, 1985.
Mallory, J. P. and D. Q. Adams. *The Oxford Introduction to Proto-Indo-European and the Proto-Indo-European World*. Oxford, England: Oxford University Press, 2006.
Murray, O. *Early Greece*. Second edition. London, England: Fontana, 1993.
Parker, R. *On Greek Religion*. Ithaca, NY: Cornell University Press, 2011.
Pascal, C. B. "October Horse." *HSCP* 85(1981): 261–291.
Puhvel, J. "Victimal Hierarchies in Indo-European Animal Sacrifice." *AJP* 99(1978): 354–362.
Richardson, N. J. *The Iliad: A Commentary, Volume VI: Books 21–24*. Cambridge, England: Cambridge University Press, 1993.
Solmsen, F., R. Merkelbach and M. L. West. *Hesiodi Theogonia Opera et Dies Scutum Fragmenta Selecta*. Third edition. Oxford, England: Clarendon Press, 1990.
Sorabji, R. *Animal Minds and Human Morals*. London, England: Duckworth, 1993.
Tal, A., ed. *Biblia Hebraica Quinta – Band 1: Genesis*. Stuttgart, Germany: Deutsche Bibelgesellschaft, 2016.
West, M. L. *Hesiod: Theogony, Edited with Prolegomena and Commentary*. Oxford, England: Clarendon Press, 1966.
West, M. L. *Hesiod: Works and Days, Edited with Prolegomena and Commentary*. Oxford, England: Clarendon Press, 1978.
West, M. L. *Iambi et Elegi Graeci ante Alexandrum Cantati*, Second edition. Oxford, England: Clarendon Press, 1989–1992.
Wilson, N. G. *Herodoti Historiae*. Oxford, England: Clarendon Press, 2015.

1.7

THE ETHICS OF EATING IN "EVANGELICAL" DISCOURSE

1600–1876

Philip Sampson

Introduction

This chapter addresses a seeming contrast. A number of authors have identified a "(minority) Christian tradition" which has had a disproportionate influence on the growth of animal advocacy in the UK. This tradition includes the heirs to the magisterial Reformations: From the "strong Protestants or Puritans" to the eighteenth century Evangelical Revival and the Calvinistic Baptists of the nineteenth century. I have elsewhere referred to this tradition as "evangelical," a term I shall use in this chapter.[1] By contrast, there is no equivalent literature linking this tradition with the modern vegetarian movement.[2] In fact, quite the opposite; most accounts regard Christianity in general, and its evangelical form in particular, as hostile to vegetarianism.[3] Now, on the face of it, those advocating for animals might be expected to disapprove the inevitable cruelty associated with eating them. So, how can we account for this disparity? This chapter sets out to address this question from within the evangelicals' own discourse, focusing upon the disparity identified above, rather than their place in the history of theology.

An obvious response to this perceived discrepancy is that it is an anachronistic misunderstanding. The evangelical discourses associated with animal advocacy date from the early sixteenth century to the later nineteenth. The term "vegetarian" dates from the late 1830s and took some time to come into popular use; "vegan" was coined in 1944. Plainly, evangelicals before the later nineteenth century could not have contributed to the debate about either vegetarianism or veganism, for the simple reason that they were mainly dead. This response, however, misses the heart of the matter. Evangelicals before the mid-nineteenth century may not have heard of vegetarianism, but terms such as "vegetable diet" and "Pythagoreanism" were in common use. Moreover, evangelical authors did refer to diet in a number of related ways; for example, their discussion of gluttony. So the contrast remains.

I will start by outlining the influence of evangelical discourse on the tradition of animal advocacy, before moving to the question of diet.

Evangelical discourse and the tradition of animal advocacy

The evangelicals believed that the world was made to praise the glory of God. This is clearly set out in the *Westminster Confession* of 1647: "It pleased God the Father, Son, and Holy Ghost *for the manifestation of the glory of His eternal power, wisdom, and goodness*, in the beginning, to create, or make of nothing, the world, and all things therein ..." Moreover, God "does uphold, direct, dispose, and govern all creatures, actions, and things ... *to the praise of the glory of His wisdom, power, justice, goodness, and mercy*."[4]

Everything was made by and for God; everything belongs to God; and the purpose of the cosmos is that everything shall glorify God.[5] All evangelicals before the twentieth century took this for granted, and it is the pivot to their discourse.

C. H. Spurgeon puts flesh on the Westminster bones:

> The "earth" ... is to be made vocal everywhere with praise. Ye dragons, and all deeps ... Terrible beasts or fishes, whether they roam the earth or swim the seas, are bidden to the feast of praise ... they are commanded ... to yield their tribute to the creating Jehovah. *They pay no service to man; let them the more heartily confess their allegiance to the Lord.* [6]

As creation was designed to praise its creator, this is its chief end and highest good. The works of God, says Matthew Henry, "are all made in wisdom, for they are all made to answer the end they were designed to serve, the good of the universe ... to the glory of the universal Monarch."[7] This praise was designed without violence or predation; in Eden, says William Hinde, "creatures were first of all at peace amongst themselves, all very good in themselves ..."[8]

Moreover, creation enjoys fulfilling its purpose. In 1721, John Ray called this praise the animals' "entertainment," and Jonathan Edwards refers to a spider's "pleasure and recreation" in swinging on its web.[9] Creation is, as it were, a great choir of praise to God, each creature delighting to contribute its own distinctive note and tone to the harmony. And humans are called to conduct it; we may, as John Owen exhorts, call in all our "fellow creatures to the work" of praise.[10] But cruelty makes creation groan and disrupts its glorifying of God (Rom. 8.22). A creature created to glorify God has been reduced to suffering flesh, and people who do this are, says Solomon, wicked (Prov. 12.10). It is easy to understand that this discursive structure would lead to a strong antipathy to animal cruelty.

But the real world is not peaceable. It is full of violence and suffering among animals, not least caused by humans. So why shouldn't human animals also be cruel? Wolves eat lambs; why shouldn't we?

The evangelicals' response was that the present nature of the world is not normal, so it cannot be normative. Moreover, the cause of its abnormality is human sin. It is therefore egregious wickedness to exacerbate suffering that we have caused in the first place. So, even in a world marked by suffering, the evangelicals concluded that cruelty towards animals is an evil. There are two steps in this discursive structure.

Adam's transgression

First, the world we see around us is not the peaceable kingdom that God created. Suffering is a distortion of the norm. This is a consequence of the doctrine of the Fall of humankind, and we find a narrative account in Genesis Chapter 3, where its impact on diet is noted: "Thorns also and thistles shall it bring forth to thee; and thou shalt eat the herb of the field; In the sweat of thy face shalt thou eat bread ..."[11] The central feature of the Fall narrative is that humans declared themselves autonomous, no longer bound to worship their creator and serve the creation, but existing to satisfy their own desires. The harmony of creation was destroyed, and animal suffering began. Thus John Wesley argues that "death came to those creatures also that 'did not sin after the similitude of Adam's transgression.' And not death alone came upon them, but all of its train of preparatory evils; pain, and ten thousand sufferings".[12]

Compounding our fault

Second, if animal suffering is our fault, we should not make it worse by being cruel ourselves. As John Dod and Robert Cleaver lamented in 1612, "have our sinnes in Adam brought such calamities upon ... [animals], and shall we add unto them by cruelty in our owne persons?".[13]

The case against animal cruelty in evangelical discourse pivots upon the chief end of creation being to praise its creator. The fact of suffering in the world, for which humans are culpable, actually strengthened this case, rather than providing a reason to indulge cruelty. Such a theology of kindness naturally leads to veganism. So why were the evangelicals not even vegetarian? To understand this, we must now turn to their discourse of diet.

The evangelical discourse of diet

First, what, in the peaceable kingdom of Eden, did creatures eat? The Genesis creation narrative is clear. Humans and animals alike were vegans: "Then God said, 'Behold, I have given you every herb bearing seed, which is upon the face of all the earth, and every tree, in the which is the fruit of a tree yielding seed; to you it shall be for meat'" (see Gen. 1.29–30).

Animals were created to praise their creator as *living* creatures, not as dead or dying ones on a plate. As George Walker put it in 1641: "in the state of innocency man had no power over living creatures to kill, and eat them; neither did one beast devoure another and feed on his flesh; but the food of man was only herbes and fruits of trees; and the food of beasts and birds was the greene herb and grasse of the field."[14]

Even though such food was considered inferior peasant fare in the sixteenth century, John Calvin regarded it as a Michelin-starred diet:

> It is not to be doubted that this [diet of herbs and seeds] was abundantly sufficient for their highest gratification. …[God] promises a liberal abundance, which should leave nothing wanting to a sweet and pleasant life.[15]

This was God's diet for a world that praises him as it was created to do. But things changed with the Fall.

Diet for a fallen planet

We have seen that an immediate consequence of the Fall was the beginning of animal suffering. A second consequence was to compromise the harmony that had previously provided their vegan diet; from now on, "thorns and thistles," and "sweat of the brow" would accompany our daily bread (Gen. 3.18–19). Many evangelical authors traced the origin of predation to this period of increasing disorder. Some animals progressively replaced their vegan diet with the flesh of weaker species. Thus Calvin asks:

> Whence comes the cruelty of brutes, which prompts the stronger to seize and rend and devour with dreadful violence the weaker animals? There would certainly have been no discord among the creatures of God, if they had remained in their first and original condition. When they exercise cruelty towards each other, and the weak need to be protected against the strong, it is an evidence of the disorder which has sprung from the sinfulness of man.[16]

Calvin takes for granted that the human calling in this new disordered state of creation is to protect animals from the "dreadful violence" of rending and devouring, not to join in. But most evangelicals believed that the Fall did not, of itself, lead to a carnivorous diet for humans. Before Noah's flood, says Adam Clarke, "animal food was not in use" by humans. Matthew Henry tells us that this was the majority view among evangelicals in the eighteenth century.[17]

Eating flesh for the support of their lives

This changed with a catastrophe narrative known as Noah's Flood; all living things were destroyed except those on Noah's ark. This did not signify a quasi-ontological shift as the Fall had been, but a contingent growth in violence, culminating in a new covenant between God and *all living creatures*. It included a change in lawful diet: "Every moving thing that liveth shall be food for you; even as the green herb, have I given you all things. But flesh with the life thereof, which is the blood thereof, shall ye not eat" (Gen. 9.3–4). This explicitly echoes the covenant of Genesis 1.29, cited above, which specified a plant-based diet for humans. But whereas Genesis 1 pronounces the vegan diet to be "very good," that judgement is conspicuously absent about the carnivorous diet of Genesis 9. There is no indication that a *living* creature can praise God in its dying or death. Indeed, evangelical discourse framed animal death in exactly the opposite way.

The Noahic covenant implicitly accepts that predator animals will eat prey, and it extends human diet to include animal flesh. But this extension of diet was very negatively framed. Far from being "good," it was the direct consequence of violence and wickedness. Many evangelicals thought of it as an evil necessary for survival in what we would now call a stone-age culture. The Fall had affected the harmony of nature, and the flood had destroyed all the crops. Thus Matthew Henry says: "They should be allowed to eat flesh for the support of their lives." John Trapp explains:

> God of his goodness grants here to mankind, after the flood, the use of flesh and wine, that the new and much weakened world might have new and more strengthening nourishment. For it is not to be doubted but that, by the deluge, a great decay was wrought … in the earth with its fruits …[18]

To summarize:

1. The chief end and highest good of creation is to glorify God. The originary diet of both humans and animals was vegan.
2. The Fall signified a quasi-ontological shift in creation occasioned by human sin, but lawful human diet remained vegan.
3. However, the harmony of creation was compromised and violence grew, culminating in the Noah narrative.
4. The Noahic covenant responded to a growth in violence and food scarcity by permitting some flesh eating for the sake of survival.
5. The new carnivorous diet, unlike the vegan diet, was conspicuously *not* pronounced "good."

There is no foundation within this discourse to oppose, as modern vegetarians do, eating animal flesh *in principle*. Consequently, evangelicals before the twentieth century did not advocate for "vegetarianism" as they did against cruelty.

Evangelical disquiet about flesh eating

However, the fact that vegetarianism cannot be sustained by evangelical discourse does not imply that diet appears therein as neutral or indifferent. On the contrary, evangelical discourse sustains a profound disquiet about meat consumption, with practical consequences. We can identify several discursive foci.

A testimony against sin

First, eating flesh reminds us of our sin which, for the evangelicals, was not a comfortable experience. Ralph Venning observed that:

> The way in which we now use the creatures bears witness against sin. When we eat flesh we [bear witness against sin], for there was no such grant in the first blessing; since sin our appetite has been more carnivorous.[19]

When an animal is killed, says Andrew Bonar:

> awful violence [is] done to one so pure, so tender, and so lovely. We shrink back from the terrible harshness of the act, whether it be plunging the knife into the neck of the innocent lamb, or wringing off the head of the tender dove. ... A testimony against sin ascends up into the ears of the Lord of Sabaoth [Hosts].[20]

Sucking blood like a wolf

Second, whilst lawful, flesh eating corrupts the appetite. "Before the fall," says Matthew Henry, "humans had no more desire to eat flesh than a sheep has to suck blood like a wolf."[21] The Fall, observes Luther, led us to prefer meat over the "the delightful fruits of the earth."[22] Andrew Willet scathingly warns that a lust for flesh caused humans to eat so gluttonously that meat "came out their Nostrils."[23] In fact, Luther goes further, claiming that the perversion of flesh eating corrupts the human body as well as the appetite: Vegetables, says Luther, "would *not* have resulted [in] that leprous obesity [we see today], but physical beauty and health ..."[24]

The corruption of the appetite by flesh eating informs the evangelical discussion of the restrictions on Israel of eating animal fat or roasted meat (Lev. 3.16–17, 22–25; 8.16&31; 1Sam. 2.15). As we have seen, Calvin did not take an ascetic approach to diet. In the case of a vegetable diet, he celebrated the "highest gratification" of the appetite, and "a liberal abundance" of food, "which should leave nothing wanting to a sweet and pleasant life." His view of a carnivorous appetite is different. Animal fat, he says, is forbidden to faithful Israel because it is "the part which might have been most attractive to the greedy" and its ban is "a restraint upon their gluttony."[25] John Gill concurs: It is forbidden because "men used to count it delicious ..."[26] This is not mere asceticism; the evangelicals roundly rejected the view that spiritual merit is obtained through dietary restrictions. Rather, their comments are informed by the corruption of our appetites which makes us prefer meat over "the delightful fruits of the earth." We should, cautions George Walker, "take heed, and beware that wee do not in any case abuse any of God's creatures to sin and vanity, to feed our owne vaine appetite, to satisfy our sinfull desires and pleasures, and to serve our corrupt fleshly lusts."[27]

Horrid cruelty

Flesh eating, then, reminds us of our sin, and corrupts our appetite. The third aspect of disquiet towards meat is that it involves killing a living creature made to sing God's praise. This extended even to "species of Toades and Spiders"; to destroy them, says John Bulwer, is to take away "one note of his [God's] harmony."[28] Moreover, this killing inevitably involves cruelty, and animal cruelty is wicked. Thus, though it is lawful to "destroy serpents, hurtful beasts and noisome creatures" for our own safety or survival: "Yet to doe it with cruelty … and *without sense of our owne sins and remorse for them,* is a kind of scorne and contempt of the workmanship of God our creator …".[29] Cruelty must be minimized consistent with survival, a principle they found in both the Noahic covenant and the Mosaic law.

In the Noahic covenant, blood must be drained from an animal before he/she is butchered. Even as permission is given to eat flesh, a barrier is erected against the cruel "excesses of cannibal ferocity in eating flesh of living animals."[30] The ban on blood, says Albert Barnes

> is to prevent the horrid cruelty of mutilating or cooking an animal while yet alive and capable of suffering pain. The draining of the blood from the body is an obvious occasion of death, and therefore the prohibition to eat the flesh with the blood of life is a needful restraint from savage cruelty.[31]

This prohibition of cruelty was systematized in the Mosaic covenant. As Mary Douglas has argued, the Levitical law tightly regulated the meat supply within a framework of Temple sacrifice, ensuring high standards of animal care and slaughter, and outlawing animal mutilation altogether.[32]

In summary then, flesh eating is permitted for survival in times of scarcity but is deeply problematic. It reminds us of our sin; it corrupts our appetite; and it inevitably involves cruelty and oppression which must be kept to a minimum.

An experimental religion

So far, I have outlined the abstract relationships within evangelical discourse as they bear upon animal advocacy and diet. But evangelicals emphasized that Christianity must transcend what Anthony Burgess called the "brain-knowledge" of "general arguments and abstracted reasons" to affect the heart and life. Evangelical Christianity is, he says, "experimental" or, in modern parlance, *existential.* The Christian who has mere "brain-knowledge," is no more satisfied spiritually than "a painted fire content you in a cold winter." Moreover, they that "have such an experimental working, as that it hath influence upon their lives and conversations, it makes some alteration and change there."[33] Or, as the Epistle of James more pithily says, "faith without works is dead" (Jas. 2.20). Those who merely speak the *language* of truth without living it in everyday life attracted censure; in the case of clergy, this censure could be harsh. For example, William Cowper regarded hunting for sport as cruel, and calls the "cassock'd huntsman" "a designing knave, [a] mere church juggler, hypocrite, and slave."[34]

This wider discursive formation, of the *abstract* with the *existential,* generated both evangelical subjects and the everyday life they sought to live. Flesh eating is framed by sin, corruption and cruelty; and this frame is no abstract "brain-knowledge", but as much a reality as the meat between the teeth (Num. 11.33). If evangelical discourse is not to be mere talk, if it is a life to be lived, it might almost put us off our hamburger. In fact, there is evidence that it did just that, despite the substantial difficulties associated with avoiding meat before the twentieth century.

Details of the diet of pre-twentieth century evangelicals are scarce, and their dietary non-conformity is often complex. But we know that John Wesley intermittently stopped eating animal flesh from the mid-1730s, "for reasons that had everything to do with his attitude towards animals".[35] Spurgeon stopped doing so in later life; and the Booth family, leaders of the Salvation Army, were largely vegetarian. In 1691, the elderly Richard Baxter summarized his evangelical disquiet towards meat, saying that,

> Though God allows us to take away the lives of our fellow creatures and to eat their flesh, *to show what sin hath brought on the world, and what we deserve ourselves* ... yet all my daies it hath gone, as against my nature, with some regret; which hath made me the more contented that God hath made me long renounce it through the necessity of nature, in my decrepite age.[36]

The evangelical discourse of diet cannot sustain a vegetarian ideology, but the two are related by a parallel thematic structure: The avoidance of cruelty; a deep suspicion of carnivorous desire; a visceral dislike of flesh eating; and a belief that we should live with integrity, not in word only.

But if the evangelical discourse of diet is unable to sustain a vegetarian ideology, even less can it sustain an argument for meat consumption in the developed world of late modernity, especially given the hideous cruelties of the industrial meat industry and its impact upon the poor. In such a context, evangelical discourse favors a de facto, if not ideological, veganism.

An evangelical spirituality of diet

I have discussed both abstract and experimental features of the evangelical discursive formation which generated a distinctive subjectivity, and the world of which it speaks. However, evangelical discourse supplements the parallels with vegetarianism. At its center is praise to the creator God, and this transforms the parallels with vegetarianism as they move from the domain of human ethics to that of divine imperative. I will briefly outline the resultant spirituality of diet before concluding.

Grief and mercy

Animal cruelty is an evil, for it subverts the praise of God. Indeed, for many evangelicals this rendered it demonic.[37] Moreover, the evangelicals asserted the reality of judgement upon such devilry. John Calvin advised his congregation that God will "condemn us for unkind folk if we pity not the brute beast."[38] For Calvin, to be condemned by God was the worst possible thing that could happen to a person.

Evangelicals, then, were certainly sensitive to God's censure on wickedness, but their distinctive spirituality of diet emphasized not judgement but *grief*; not condemnation but *kindness*. This spirituality is embedded in the discourse of living animals created to praise God. Eating such an animal *in order to survive in a fallen world* should remind us of our sin, and with every mouthful we wish to God that we did not have to kill a creature made to praise its creator. The lamb is slaughtered, the poor creature dies, says Spurgeon, "not for any guilt that it has ever had, but to show me that I am guilty, and that I deserved to die like this."[39] Thomas Draxe, preaching in 1612, emphasizes that our sin should cause us to *grieve*: "If the poor dumbe creature, (bird or beast) bee in any paine and miserie, let us … be sorry for it, and be greeved for our own sinnes …"[40]

Carol Adams has recently noted this aspect of Christian spirituality:

> The Christian practice of care acknowledges the importance of *grieving*. We aren't going to live fully in this world and love truly without experiencing grief. To fear the grief from knowledge of animals' lives (and deaths) is to deprive oneself of one of the true gifts of being in relationship with the world, caring enough to feel for another, and then allowing that caring to change your life.[41]

The second distinctive aspect to the evangelical spirituality of diet was mercy. Christians should be merciful because God is merciful. Hannah More describes a cruel coachman who was much moved by the "compassion the great God of heaven and earth had for poor beasts" as he read his Bible. He reformed his life, saying "Doth God care for horses, and shall man be cruel to them?"[42] Moreover, animals are not ours. A good man will be merciful to "his beast," says Matthew Henry, "because it is *God's* creature …"[43]

For Spurgeon, as for many other "experimental" evangelicals, our words are tested by our hearts, and our behavior towards animals reveals the state of our heart. So let Spurgeon have the last lines: "The man who truly loves his Maker becomes tender towards all the creatures his Lord has made. In gentleness and kindness our great Redeemer is our model." On the other hand the "man of dead heart towards God has a heart of stone towards the Lord's creatures, and cares for them only so far as he can make them minister to his own wealth or pleasure … no person really penitent for sin can be cruel [to animals]."[44]

Notes

1 Keith Thomas, *Man and the Natural World* (London, England: Penguin, 1984), 180, 154. For discussion, see Philip Sampson, "Lord of Creation or Animal Among Animals?" in Andrew Linzey and Clair Linzey, eds., *The Routledge Handbook of Religion and Animal Ethics* (Routledge, 2018). This usage is quite different from the modern movement called "evangelicalism."

2 By "vegetarian," I mean an ideological assertion that it is, in principle, unethical to kill an animal for the purpose of eating it. A "vegan" ideology extends this to exploitation of animals more generally.

3 For example, Peter Singer, *Animal Liberation* (Wellingborough, England: Thorns, 1976), 226. For discussion, see Philip Sampson, *Six Modern Myths* (Downer Grove, IL: IVP, 2001), ch. 3.
4 Westminster Assembly (1643–1652), *The Humble Advice of the Assembly of Divines* (London, England: Evan Tyler, 1647), ch. 4; my emphasis. This Confession was the basis for most subsequent statements of the evangelical faith.
5 For intertextual canonical sources, see Col. ch. 1; Eph. ch. 1.
6 C. H. Spurgeon, *The Treasury of David* (Easton, PA: Niche, 2011), vol. 7, Ps. 148.7 (my emphasis).
7 Matthew Henry, *Commentary on the Whole Bible* (London, England: Marshall, Morgan and Scott, 1959), Ps. 104.19–30.
8 William Hinde, *A Faithful Remonstrance of the Holy Life and Happy Death of John Bruen of Bruen-Stapleford, in the County of Chester, Esquire* (London, England: R. B. for Philemon Stephens, and Christopher Meredith, 1641), 31.
9 John Ray, *Three Physico-Theological Discourses* (London, England: William and John Innys, 1721), 36. Jonathan Edwards, "To Judge Paul Dudley, October 31, 1723," in *The Works of Jonathan Edwards*, ed. George S. Claghorn (New Haven, CT: Yale University Press, 1998), vol. 16, 45.
10 John Owen, "The Strength of Faith: Sermon II" in *Works* (New York, NY: Robert Carter, 1851), vol. 9, 52–3. See Ps. 96.
11 Gen. 3.18–19. I have used the 1611 Authorized Version of the Bible.
12 John Wesley, "The General Deliverance," in *Sermons on Several Occasions* (Grand Rapids, MI: Christian Classics Ethereal Library, 1872), vol. II, Sermon 60, para 2.ii.5.
13 John Dod and Robert Cleaver, *A Plaine and Familiar Exposition of the Eleventh and Twelfth Chapter of the Proverbs* (London, England: Thomas Man, 1612), 142.
14 George Walker, *God Made Visible in His Works* (London, England: G. M. for John Bartlet, 1641), 232–3.
15 John Calvin, *A Commentary on Genesis*, ed. John King (Edinburgh, Scotland: Calvin Translation Society, 1847), 100.
16 John Calvin, *Commentary on the Book of the Prophet Isaiah*, trans. William Pringle (Edinburgh, Scotland: Calvin Translation Society, 1850), vol. 1, ch. 11.6.
17 Adam Clarke, *Commentary on the Bible* (New York, NY: J. Emory and B. Waugh, 1831), vol. 1, Gen. 9.4; Henry, *Commentary on the Whole Bible*, Gen. 9.4.
18 Henry, *Commentary on the Whole Bible*, Gen. 9, preface and v. 3. John Trapp, *Commentary on the Old and New Testaments* (London, England: Dickinson, 1867), vol. 1, 40.
19 Ralph Venning, *Sin, the Plague of Plagues* (London, England: John Hancock and T. Parkhurst, 1669), 75–6.
20 Andrew Bonar, *A Commentary on Leviticus* (Edinburgh, Scotland: Banner of Truth, 1966), 27. Bonar compares this to the torturing to death of Jesus of Nazareth.
21 Henry, *Commentary on the Whole Bible*, Genesis ch. 9.1–7
22 Martin Luther, *Works*, ed. Jaroslav Pelikan (Saint Louis, MO: Concordia, 1958), vol. 1, 72.
23 Andrew Willet, *Hexapla in Exodus*, (London, England: John Haviland, 1608), 619, Q. LVXIII. Willet is here quoting the Dominican Oleaster with approval who, in turn, intertexts Num. 11.20.
24 Luther, *Works*, my emphasis.
25 John Calvin, *Harmony of the Law*, trans. Charles William Bingham (Edinburgh, Scotland: Calvin Translation Society, 1852), vol. 2, Lev. 3.16f.
26 John Gill, *Exposition of the Entire Bible* (London, England: William Woodward, 1811), Lev. 7.25 quoting Maimonides with approval.
27 Walker, *God Made Visible in His Works*, 160–1.
28 J. Bulwer, *Anthropometamorphosis: Man Transform'd, or the Artificial Changeling* (London, England: William Hunt, 1653), B6. N.b. Bulwer was a supporter of Archbishop Laud, but shared this view of creation with his Puritan opponents.
29 W. Walker, *God Made Visible in His Works*, my emphasis.

30 Robert Jamieson, Andrew Faussett and David Brown, *Commentary Critical and Explanatory on the Whole Bible* (Grand Rapids, MI: Zondervan, 1999), Gen. 9.4.
31 Albert Barnes, *Notes on the Whole Bible* (Grand Rapids MI: Baker, 1996), Gen. 9.4.
32 Mary Douglas, *Leviticus as Literature* (Oxford, England: Oxford University Press, 1999).
33 Anthony Burgess, *Spiritual refining* (London, England: A. Miller for Thomas Underhill, 1652), 1, 6, 128, 15.
34 William Cowper, "The Progress of Error," in *Table Talk and Other Poems* (London, England: John Sharp, 1825), 29–50 at 33.
35 Roy Hattersley, *John Wesley* (London, England: Abacus, 2002), 92. But this is complex; see J. Wesley, "Letter to Dr. Gibson, Bishop of London, London, June 11, 1747," in *The Letters of John Wesley*, ed. John Telford (London, England: Epworth Press, 1931), vol 2.
36 R. Baxter, *The Poor Husbandman's Advocate to Rich Racking Landlords*, ed. F. J. Powicke (Manchester, England: Manchester Univ Press,1926), ch4.3, 39, my emphasis.
37 E.g., Thomas Hodges, *The Creatures Goodness as They Came out of God's Hand* … (London, England: Tho. Parkhurst, 1675), 40.
38 John Calvin, *Sermons on Deuteronomy,* trans. Arthur Golding (London, England: Henry Middleton for John Harison, 1583), 770.
39 C. H. Spurgeon, "The Blood," in *The Complete Works* (Harrington, DE: Delmarva, 2013), vol. 5, Sermon 228, I.2.
40 Thomas Draxe, *The Earnest of our Inheritance* (London, England: F K for George Norten, 1613), 26–7.
41 Carol Adams, "Witnessing Against Violence," Sarx, accessed February 27, 2016, http://sarx.org.uk/human-dominion/witnessing-against-violence/.
42 Hannah More, "History of Tom White, the Postboy," in *The Complete Works of Hannah More* (New York, NY: Harper, 1840), vol. 1, 224–231at 226.
43 Henry, *Commentary on the Whole Bible*, Prov. 12.10; my emphasis.
44 C. H. Spurgeon, "A Word for Brutes against Brutes," in *The Sword & Trowel* vol. 3, June 1873 (Pasadena, CA: Pilgrim, 1975), 333.

Bibliography

Adams, Carol. "Witnessing Against Violence." Sarx. Accessed February 27, 2016. http://sarx.org.uk/human-dominion/witnessing-against-violence/.

Barnes, Albert. *Notes on the Whole Bible*. Grand Rapids MI: Baker, 1996.

Baxter, Richard. *The Poor Husbandman's Advocate to Rich Racking Landlords*. Edited by F. J. Powicke. Manchester, England: Manchester University Press, 1926.

Bonar, Andrew. *A Commentary on Leviticus*. Edinburgh, Scotland: Banner of Truth, 1966.

Bulwer, John. *Anthropometamorphosis: Man Transform'd, or the Artificial Changeling*. London, England: William Hunt, 1653.

Burgess, Anthony. *Spiritual Refining*. London, England: A. Miller for Thomas Underhill, 1652.

Calvin, John. *Sermons on Deuteronomy*. Translated by Arthur Golding. London, England: Henry Middleton for John Harison, 1583.

Calvin, John. *A Commentary on Genesis*. Edited by John King. Edinburgh, Scotland: Calvin Translation Society, 1847.

Calvin, John. *Commentary on the Book of the Prophet Isaiah*, vol.1. Translated by William Pringle. Edinburgh, Scotland: Calvin Translation Society, 1850.

Calvin, John. *Harmony of the Law*. Translated by Charles William Bingham. Edinburgh, Scotland: Calvin Translation Society, 1852.

Clarke, Adam. *Commentary on the Bible*. New York, NY: J. Emory and B. Waugh, 1831.

Cowper, William. "The Progress of Error." In *Table Talk and Other Poems*, 29–50. London, England: John Sharp, 1825.

Dod, John and Robert Cleaver. *A Plaine and Familiar Exposition of the Eleventh and Twelfth Chapter of the Proverbs*. London, England: Thomas Man, 1612.

Douglas, Mary. *Leviticus as Literature*. Oxford, England: Oxford University Press, 1999.

Draxe, Thomas. *The Earnest of our Inheritance*. London, England: F. K. for George Norten, 1613.

Edwards, Jonathan. "To Judge Paul Dudley, October 31, 1723." In *The Works of Jonathan Edwards*. Edited by George S. Claghorn, vol.16. New Haven, CT: Yale University Press, 1998.

Gill, John. *Exposition of the Entire Bible*. London, England: William Woodward, 1811.

Hattersley, Roy. *John Wesley*. London England: Abacus, 2002.

Henry, Matthew. *Commentary on the Whole Bible*. London, England: Marshall, Morgan and Scott, 1959.

Hinde, William. *A Faithful Remonstrance of the Holy Life and Happy Death of John Bruen of Bruen-Stapleford, in the County of Chester, Esquire*. London, England: R. B. for Philemon Stephens, and Christopher Meredith, 1641.

Hodges, Thomas. *The Creatures Goodness as they Came out of God's Hand* ... London, England: Tho. Parkhurst, 1675.

Jamieson, Robert, Andrew Faussett and David Brown. *Commentary Critical an Explanatory on the Whole Bible*. Grand Rapids, MI: Zondervan, 1999.

Luther, Martin. *Works*. Edited by Jaroslav Pelikan. Saint Louis, MO: Concordia, 1958.

More, Hannah. "History of Tom White, the Postboy." In *The Complete Works of Hannah More, vol.1*: 224–231. New York, NY: Harper, 1840.

Owen, John. "The Strength of Faith: Sermon II." In *Works, vol. 9*. New York, NY: Robert Carter, 1851.

Ray, John. *Three Physico-Theological Discourses*. London, England: William and John Innys, 1721.

Sampson, Philip. *Six Modern Myths*. Downer Grove, IL: IVP, 2001.

Sampson, Philip. "Lord of Creation or Animal Among Animals?" In *The Routledge Handbook of Animal Ethics*, edited by Andrew Linzey and Clair Linzey. Routledge, 2018.

Singer, Peter. *Animal Liberation*. Wellingborough, England: Thorns, 1976.

Spurgeon, C. H. "A Word for Brutes against Brutes." In *The Sword & Trowel, vol. 3*, June1873. Pasadena, CA: Pilgrim, 1975.

Spurgeon, C. H. *The Treasury of David*. Easton, PA: Niche, 2011.

Spurgeon, C. H. "The Blood." In *The Complete Works, vol. 5*. Harrington, DE: Delmarva, 2013.

Thomas, Keith. *Man and the Natural World*. London, England: Penguin, 1984.

Trapp, John. *Commentary on the Old and New Testaments*. London, England: Dickinson, 1867.

Venning, Ralph. *Sin, the Plague of Plagues*. London, England: John Hancock and T. Parkhurst, 1669.

Walker, George. *God Made Visible in His Works*. London, England: G. M. for John Bartlet, 1641.

Wesley, John. "The General Deliverance." In *Sermons on Several Occasions, vol. II*. Sermon 60. Grand Rapids, MI: Christian Classics Ethereal Library, 1872.

Wesley, John. "Letter to Dr. Gibson, Bishop of London, London, June 11, 1747." In *The Letters of John Wesley*. Edited by John Telford. London, England: Epworth Press, 1931.

Westminster Assembly (1643–1652). *The Humble Advice of the Assembly of Divines*. London, England: Evan Tyler, 1647.

Willet, Andrew. *Hexapla in Exodus*. London, England: John Haviland, 1608.

1.8

MYTH AND MEAT

C. S. Lewis sidesteps Genesis 1:29–30

Michael J. Gilmour

A palimpsest is a recycled writing surface with traces of an effaced text remaining. An eraser does not remove pencil markings entirely from paper. Scraping papyrus or vellum for reuse is equally imperfect. The original writing is still there, showing through any new words placed over top. For literary theorist Linda Hutcheon, this is a useful image when contemplating literary influence and artistic adaptations. For her, 'palimpsestuous' works are those haunted by precursors. "If we know that prior text," she writes, "we always feel its presence shadowing the one we are experiencing directly."[1] We "experience adaptations (*as adaptations*) as palimpsests through our memory of other works that resonate through repetition with variation."[2]

The opening chapters of Genesis are an important textual surface on which C. S. Lewis writes, and those familiar with the first pages of the Bible experience that textual haunting Hutcheon describes. He draws heavily on the mythical content of Genesis 1–11 but is equivocal and oddly circumspect when treating the plant-based diet of 1:29–30. He never translates myth into menu.

Lewis loved animals and abhorred cruelty yet never wrote a major work on the topic. The poet and novelist Charles Williams proposed he and Lewis collaborate on a book of "animal stories from the Bible, told by the animals concerned."[3] That amounts to a cruel tease for those curious about Lewis's animal theology, but what we have from his pen is still provocative: A book chapter on animal suffering; an essay condemning vivisection; diary entries and personal correspondence replete with anecdotes about his own encounters with companion animals and free living animals; poems and prose exploring interactions between humans and other creatures. Animals are everywhere in his work, and by all accounts they loomed large in his personal life. One recent biography describes his household as an "eccentric Noah's ark."[4]

So, what do the creation narratives contribute to Lewis's animal-friendly Christianity? This is a difficult question because even though he thought deeply about the fate of mice in the laboratory, the fate of cattle in the abattoir seems a nonissue. He was not a vegetarian and Genesis 1:29–30 left him with a theological Gordian knot, as though a spiritually meaningful vegetarianism were as improbable to his mind as an adequate response to Jesus's charge, "go and sin no more." What does this ambivalence to the primordial vegetarianism of the Priestly creation story look like?

A companion bear, pigs and bacon

A throwaway remark in the 1945 novel *That Hideous Strength* gets us to the marrow of the issue. A bear named Mr. Bultitude lives in a house with the story's heroes. He is a lovable creature but his presence in the home raises an important question for one of its residents: "'The bear … ,' said MacPhee, 'is kept in the house and pampered. The pigs are kept in a stye and killed for bacon. I would be interested to know the philosophical *rationale* of the distinction.'"[5] I think Lewis would too, though suspect the matter confounded him.

Readers of Lewis's fiction recognize Andrew MacPhee (sometimes spelled McPhee) from *Perelandra* and the novel fragment posthumously published as *The Dark Tower*. The template for this character is William Thompson Kirkpatrick (1848–1921), Lewis's tutor who prepared him for Oxford entrance exams in the pre-war years.[6] "If ever a man came near to being a purely logical entity," Lewis writes, "that man was Kirk."[7] In *Surprised by Joy*, Lewis stresses the man's religious skepticism, which informed Lewis's atheism during his teens and twenties: "The reader will remember that my own Atheism and Pessimism were fully formed before I went to Bookham [to study with Kirkpatrick]. What I got there was merely fresh ammunition for the defense of a position already chosen."[8]

The affectionate portrait of his old teacher in *That Hideous Strength* plays a minor part in the novel but an important one for our topic. Who better to express theological reservations, ethical conundrums, or the sometimes-irrational practices of the religiously inclined? MacPhee observes that the actions of his pious friends are inconsistent, even hypocritical. Yes, they recognize the goodness of animals and live in peaceful coexistence with some of them just like Adam in the Garden, and yet others they kill for food. The ideal of biblical vegetarianism is thus a shadowy palimpsest in this episode. There but not there. A religious skeptic inadvertently draws attention to it, making it conspicuous in the breach by querying his God-fearing friends who maintain Genesis is authoritative Scripture. *But what about 1:29–30?* he seems to ask.

MacPhee is not a man of faith so not a mouthpiece for Lewis's views in all respects. He is thus a perfect surrogate for Lewis, allowing him to articulate and explore a moral conundrum without making any theological commitment. MacPhee asks an obvious question no one else knows how to answer. The question just sits there on the page, and it does so rather awkwardly given Elwin Ransom also remains silent on the matter. Ransom is the Christian hero of the space trilogy who is closest to the spirit world and has the best understanding of spiritual realities. Ransom is himself vegetarian at the time MacPhee asks the question, but even he offers no explanation why.

Lewis here attempts to unravel an artistic knot. *That Hideous Strength* combines realism with fantasy. The setting is mid-twentieth century England but there are elements of the supernatural as well. The novel also explicitly echoes biblical myth, evident not least in the book's title and epigraph, which refer to the Tower of Babel.[9] Other themes drawn from Genesis 1–11 include animals in communion with an Adam-like hero; an Adam and Eve, reflected in Jane and Mark Studdock's renewed marriage; Babel-like hubris complete with a confusion of languages; a flood story; and hints of humans and animals multiplying and filling the earth. The vegetarianism of Genesis 1 is there too but Lewis relegates it to the realm of fantasy, not realism. Ransom stops eating meat but he is hardly an everyman. He travels to distant planets and speaks with supernatural beings. By associating vegetarianism with him and him alone, Lewis effectively quarantines the concept, relegating it to the realm of mystics and non-terrestrials. By contrast, not one of the novel's ordinary, everyday, feet-on-the-ground, flesh and blood characters is vegetarian, something noticed by a visitor to Ransom's home who observes, "*Your* people eat dry and tasteless flesh"—not "You people"—thus distinguishing Ransom from the larger community.[10] Ransom's friends eat roast goose and oysters when celebrating the triumph of good over evil.[11] *That* is what normal people do in this story.

Lewis seems to acknowledge this evasion of Genesis 1:29–30 by allowing MacPhee to beg the question. Yes, the pious residents of St. Anne's live peacefully, Adam-like in the company of Mr. Bultitude in their re-enacted Garden of Eden but the pigs in the sty are merely a food source. They have no place in the community; they are literally outside the home and outside the fellowship. What is the difference between a bear and a pig, MacPhee wonders? Why eat one and not the other? If replicating Edenic peacefulness is good, why not replicate it in all respects, eliminating violence against all creatures? The nonbeliever MacPhee raises a question that the Christian Lewis does not resolve.

The gustatory evolution of Prof. Elwin Ransom

As said, Ransom does not eat meat in *That Hideous Strength* but that is at the end of his story. He becomes vegetarian only gradually, as part of his slow transformation into a twentieth-century Adam. Consider that when he travels to Mars and Venus in the first two books of the trilogy, the journeys are both spatial and spiritual. He leaves a fallen world for un-fallen ones, exiting the wilderness that is Earth and entering the Gardens of Eden that are Mars and Venus. While on Perelandra (= Venus), he even passes "an angel with a flaming sword" (alluding to Genesis 3:24) but this figure does not prevent him entering the sacred garden, as was the case for Adam and Eve.[12] God bids him enter that Paradise. Ransom is a long way from home.[13]

One of the ways Lewis signals this journey back to prelapsarian conditions is through various references to the food his protagonist eats. While still on the spaceship *en route* to Malacandra (= Mars), which is to say part way between a fallen and an un-fallen world, his meal is "tinned meat, biscuit, butter and coffee." Again, immediately after landing on Malacandra, he eats more tinned meat but this time, the meal is conspicuously out of place.[14] We discern a tension in the text because the Martian landscape Lewis describes owes much to the biblical Eden. Genesis is again a palimpsest, showing through into Lewis's fantasy, a shadowy, haunting textual presence. There is no meat in the biblical Eden, however. In the novel, terrestrial travelers introduce it, namely the evil men who kidnapped Ransom. They are serpents in the Garden.

Also out of place is Ransom's ambivalence toward animals. For his part, Ransom has a high view of animals compared to the kidnappers Weston and Devine but he is not consistent. One moment he says experimental use of living animals is reprehensible but only a few sentences later he eats that tinned beef.[15] He embraces a theologically informed animal ethic *to a degree* but it does not extend to food. The proximity of these two episodes—his rejection of vivisection and his consumption of a cow—is every bit as awkward as pampering a bear while eating a pig. Eventually, Ransom changes. By the third novel, he not only opposes an evil organization that vivisects, he is also vegetarian.

Ransom escapes his captors soon after arriving on Malacandra. He no longer has access to their tinned meat so securing other food is a priority. His first attempt involves cutting a piece out of a tall plant but—I think significantly—he finds it "quite unswallowable."[16] This is symptomatic of Ransom's circumstances. His appetites, ways of thinking, and very nature are those of corrupted Earth, not those of uncorrupted Mars. Adaptation from one world to another—particularly the transition from the spiritual conditions of one to the other—is not easy. Like the shadow people visiting the foothills of Heaven in Lewis's *The Great Divorce* who find the solid grass painful to step on, adjusting to new spiritual realities is necessary but difficult for Ransom. He finds vegetarianism "unswallowable," at least initially.

On the journey back to Earth, Ransom longs for meat and beer. Before setting out for Venus in the second novel, he eats meat again.[17] But then, a change. After returning from this second visit to a sinless world, there is an unmistakable shift. While on Perelandra, he is rapturous about the satisfying fruit he eats: "It was like the discovery of a totally new genus of pleasures, something unheard of among men."[18] When back on Earth, he tells the friends offering him a traditional breakfast, "'No, thanks, I don't somehow feel like bacon or eggs or anything of that kind. No fruit, you say? Oh well, no matter. Bread or porridge or something.'"[19] By the third book, as mentioned, he eats only bread and wine.

What is Lewis saying by this transformation of a carnivore into a vegetarian in these Genesis-inspired novels? Perhaps he indicates that a mere glimpse into Paradise is enough to alter one forever. See real harmony between species just once, and an appetite for blood diminishes. Unfortunately—and here is the rub—Ransom's experiences are inimitable. It is a utopian vision. No other human being saw both Malacandra and Perelandra and lived to tell the tale, and so no other human being underwent the gustatory conversion the experiences afforded. And so it is a bear back on Earth sits comfortably in the kitchen while its human friends eat a pig. Lewis suggests that enacting the peaceful conditions of Eden is always only partial. We stay the hand and coexist with some animals but never all. (It brings to mind our love of companion animals and the conundrum many face; keeping cats and dogs in the home involves dependence on a meat industry we deplore). Lewis does not allow for the possibility that vegetarianism in the here-and-now is a viable, spiritually enriching choice. He keeps that idea at bay, attaching it *only* to the eccentric and otherworldly-minded Elwin Ransom, and even then, Lewis hedges. Yes, Ransom's diet consists only of bread and wine, but the obvious Eucharistic overtones distract readers from Genesis, suggesting Ransom is some kind of Christ figure.[20] The Garden of Eden gives way, subtly and suddenly, to the Upper Room.

Feasting in Narnia

The Chronicles of Narnia also involve interaction with and evasion of Genesis 1:29–30. There are many references to meat in these books. In *Prince Caspian*, for one, the defeat of Miraz and his forces is cause for celebration complete with a feast. And what a feast it is. Bacchus, Silenus, and the Maenads dance what the narrator tells us is "not merely a dance for fun and beauty … but a magic dance of plenty, and where their hands touched, and where their feet fell, the feast came into existence." First on the long list of foods mentioned are "sides of roasted meat that filled the grove with delicious smell."[21]

Where does this meat come from? Readers of the Chronicles recall there are two kinds of animals in this world, those that talk and those that do not. In *The Magician's Nephew*, Lewis relates the story of Narnia's creation including a memorable account of the emergence of all animals from the dirt, in keeping with Genesis 2:19. After this, Aslan mingles among the assembled creatures and "every now and then he would go up to two of them (always two at a time) and touch their noses with his." To those animals, Aslan says awake, love, think, and speak but those he does not touch simply wander off, presumably to fulfill the biblical mandate to multiply and fill the earth (cf. Genesis 1:22; 8:17).[22]

There are no dietary restrictions preventing the consumption of the non-speaking animals. Talking animals, on the other hand, are "free subjects" and loyal Narnians recoil at the thought of eating them.[23] When Puddleglum discovers his meal of venison came from a talking stag in *The Silver Chair*, he "was sick and faint, and felt as you would feel if you found you had eaten a baby." Readers have no doubt about Puddleglum's horror as he contemplates his actions:

> "We've brought the anger of Aslan on us," he said… "We're under a curse, I expect. If it was allowed, it would be the best thing we could do, to take these knives and drive them into our own hearts."[24]

The Chronicles of Narnia have the biblical creation stories clearly in view and Lewis includes Genesis 1:29–30, after a fashion, by limiting consumption of *some* animal flesh. At the same time, this retelling of Genesis 1 is equally an evasion of those particular verses.

We encounter the distinction between talking and non-talking animals throughout these books. Just a few pages before the *Prince Caspian* feast mentioned earlier there is a lovely scene with Aslan on the stage, and the routing of the forces of evil underway. Lucy awakes to find "everyone was laughing, flutes were playing, cymbals clashing. Animals, not Talking Animals, were crowding in upon them from every direction."[25] These non-speaking animals are part of the dance, part of that enchanted moment celebrating deliverance from tyranny but in the end, they do not join the feast. They become the feast.

To repeat, all creation participates in the celebration. Even the trees dance with those non-speaking animals.[26] As Bacchus leads the festive parade, at

> every farm animals came out to join them. Sad old donkeys who had never known joy grew suddenly young again; chained dogs broke their chains; horses kicked their carts to pieces and came trotting along with them—clop-clop—kicking up the mud and whinnying.

Lewis's stories here and elsewhere reveal a real sensitivity to brutalities against animals, whether vivisection in a lab in the space trilogy or sad donkeys, dogs, and horses who experience human-caused deprivations and pain in the Chronicles of Narnia. There is reprieve for a few of them, but not all. Some of those farmed animals coming out to celebrate the victory with Aslan obviously become the "sides of roasted meat" consumed after the dance.[27]

The Magician's Nephew is important for appreciating Lewis's views on animals and human responsibility toward them. Here he interprets humanity's "dominion" over creation, a concept taken from Genesis 1, in very generous terms. Aslan instructs King Frank, the first human ruler of the new world, to work the soil and name the animals, alluding to Adam in Genesis 2:15 and 2:19 respectively. He also instructs the King to treat the talking animals "kindly and fairly" and all rational species of Narnia to treat non-speaking animals "gently" and to "cherish them."[28] But there is also evasion, a conspicuous sidestepping of the Genesis 1 diet. The taboo about eating rational creatures seems to be a loose equivalent, Lewis's way of incorporating that part of the biblical story without addressing directly thorny ethical questions about food. If you meet a talking hare like Moonwood from *The Last Battle*, he seems to say, you must not eat it, but any other rabbit caught in your snare is fair game. He even ridicules vegetarianism on the one occasion it appears in the Narnia books. Eustace Scrubbs's parents do not smoke, drink, or eat meat, which makes them ridiculous.[29]

Biblical myth lurks behind many of Lewis's stories but like a palimpsest, that precursor is at times faint and undecipherable. The shadowy engagements with

Genesis 1:29–30 are recognizable but subtle, present but not present. Lewis looks to those verses for a moment, and then turns quickly away.

There is little to gain by speculating why, but it is an irresistible thought experiment. Lewis's historical moment likely has much to do with it, as a quick comparison with near contemporary Albert Schweitzer (1875–1965) suggests. Schweitzer's philosophical and theological work also resulted in a high view of animals, a theory of reverence for all life that inspired dramatic changes in the way he lived. These changes, however, were no easy matter. "One should not overlook both the development of thought and the development of practice in Schweitzer's life," writes Ara Paul Barsam.

Schweitzer became deeply critical of animal experimentation and opposed all forms of hunting for sport. It follows that he was even more rigorous with his dietary habits at ninety than he was at thirty, when he first determined: "[I will make] my life my argument." Not incidentally, we should note in passing how grueling such dietary restrictions would have been not only in the 1950s but especially when confronted with the restraints of life in an equatorial forest. Instead of criticizing Schweitzer's early use of meat and fish products, it is critical to grasp how deeply counter-cultural and practically burdensome vegetarianism was at that time.[30]

In various ways, Lewis also made his life his argument with respect to animals. We see it in his joy in nature and abhorrence at its senseless destruction. We see it in his vigorous opposition to animal experimentation and his assertion that theology must be inclusive of the nonhuman. We see it in the "eccentric Noah's ark" of his home life. Lewis loved animals but accepted meat eating as an inescapable, post-Genesis 9 reality. Had he lived to 90 like Schweitzer … who knows.

We can only speculate about reasons why vegetarianism did not emerge as a viable option allowing at least a symbolic, partial enactment of the peaceful co-existence of biblical myth. Certainly, it was not normative in his cultural and family contexts. The challenge of wartime rationing is perhaps a factor for a few of his years, as also the absence of close, respected peers who modelled a plant-based diet. As it was for Elwin Ransom early on, so for Lewis—vegetarianism is a beautiful Edenic image but an idea beyond the capacity of fallen humanity, an "unswallowable" prospect. Yet Lewis goes farther than most in recognizing the significance of Genesis 1 and 2 for giving shape to a theological worldview that takes all creation seriously. Whereas Western theology has a long tradition of de-emphasizing the pre-fall biblical stories, Lewis puts them to the forefront and gestures poetically toward a vision of environmental responsibility and animal care that is most welcome.[31]

Notes

1 Linda Hutcheon, with Siobhan O'Flynn, *A Theory of Adaptation*, 2nd ed. (New York, NY: Routledge, 2013), 6. She introduces the term with reference to Elizabeth Deeds Ermath, "Agency in the Discursive Condition," *History and Theory 40* (2001): 34–58.
2 Hutcheon, with O'Flynn, *Theory of Adaption*, 8.

3 C. S. Lewis, "Preface," in *Essays Presented to Charles Williams*, ed. C. S. Lewis (Grand Rapids, MI: Eerdmans, 1966), xii.
4 Philip Zaleski and Carol Zaleski, *The Fellowship: The Literary Lives of the Inklings* (New York, NY: Farrar, Straus and Giroux, 2015), 284. For a list of animals on the property where he lived, see 283.
5 C. S. Lewis, *That Hideous Strength: A Modern Fairy-Tale for Grown-Ups* (London, England: HarperCollins, 2005), 360. Italics original.
6 See e.g., Zaleski and Zaleski, *The Fellowship*, 53–55.
7 C. S. Lewis, *Surprised by Joy: The Shape of My Early Life* (Boston, MA: Mariner, 2012), 135.
8 Lewis, *Surprised by Joy*, 139–40.
9 Ransom tells Merlin, "'the Hideous Strength confronts us and it is as in the days when Nimrod built a tower to reach heaven'" (Lewis, *That Hideous Strength*, 398).
10 Lewis, *That Hideous Strength*, 396. Italics added.
11 Lewis, *That Hideous Strength*, 507, 512.
12 C. S. Lewis, *Perelandra* (London, England: HarperCollins, 2005), 244.
13 Lewis accentuates the contrast between the Edenic peacefulness of Perelandra and Ransom's home planet by noting the hero's wartime experiences (World War I), and the battles raging on Earth concurrently with events related in the story (World War II).
14 C. S. Lewis, *Out of the Silent Planet* (London, England: HarperCollins, 2005), 30, 50.
15 Lewis, *Out of the Silent Planet*, 28, 30.
16 Lewis, *Out of the Silent Planet*, 60.
17 In order, Lewis, *Out of the Silent Planet*, 191; Lewis, *Perelandra*, 22.
18 Lewis, *Perelandra*, 46.
19 Lewis, *Perelandra*, 30.
20 Lewis, *That Hideous Strength*, 197.
21 C. S. Lewis, *Prince Caspian: The Return to Narnia* (New York, NY: HarperCollins, 1994), 211.
22 C. S. Lewis, *The Magician's Nephew* (New York, NY: HarperCollins, 1994), 122, 124, 126.
23 Lewis, *Magician's Nephew*, 151.
24 C. S. Lewis, *The Silver Chair* (New York, NY: HarperCollins, 1994), 129.
25 Lewis, *Prince Caspian*, 197.
26 Lewis, *Prince Caspian*, 211.
27 Lewis, *Prince Caspian*, 201, 211.
28 Lewis, *Magician's Nephew*, 151, 128.
29 C. S. Lewis, *The Voyage of the 'Dawn Treader'* (New York, NY: HarperCollins, 1994), 3.
30 Ara Paul Barsam, *Reverence for Life: Albert Schweitzer's Great Contribution to Ethical Thought* (Oxford, England: Oxford University Press, 2008), 152.
31 For another version of the argument put forward here, with pertinent biographical and literary contexts, see Michael J. Gilmour, *Animals in the Writings of C. S. Lewis*, The Palgrave Macmillan Animal Ethics Series (Basingstoke, England: Palgrave Macmillan, 2017), esp. 157–61.

Bibliography

Barsam, Ara Paul. *Reverence for Life: Albert Schweitzer's Great Contribution to Ethical Thought*. Oxford, England: Oxford University Press, 2008.

Ermath, Elizabeth Deeds. "Agency in the Discursive Condition." *History and Theory* 40 (2001): 34–58.

Gilmour, Michael J. *Animals in the Writings of C. S. Lewis*. The Palgrave Macmillan Animal Ethics Series. Basingstoke, England: Palgrave Macmillan, 2017.

Hutcheon, Linda. *A Theory of Adaptation*. New York, NY: Routledge, 2006.

Lewis, C. S. *A Preface to Paradise Lost*. New York, NY: Oxford University Press, 1961.

Lewis, C. S., ed. *Essays Presented to Charles Williams*. Grand Rapids, MI: Eerdmans, 1966.

Lewis, C. S. *Prince Caspian: The Return to Narnia*. New York, NY: HarperCollins, 1994.

Lewis, C. S. *The Magician's Nephew*. New York, NY: HarperCollins, 1994.
Lewis, C. S. *The Silver Chair*. New York, NY: HarperCollins, 1994.
Lewis, C. S. *The Voyage of the 'Dawn Treader'*. 1952. New York: HarperCollins, 1994.
Lewis, C. S. *Out of the Silent Planet*. London, England: HarperCollins, 2005.
Lewis, C. S. *Perelandra*. London, England: HarperCollins, 2005.
Lewis, C. S. *That Hideous Strength: A Modern Fairy-Tale for Grown-Ups*. London, England: HarperCollins, 2005.
Lewis, C. S. *Surprised by Joy: The Shape of My Early Life*. Boston, MA: Mariner, 2012.
Zaleski, Philip, and Carol Zaleski, *The Fellowship: The Literary Lives of the Inklings*New York, NY: Farrar, Straus and Giroux, 2015.

1.9

THE MORAL POVERTY OF PESCETARIANISM

Max Elder

Introduction

Pescetarianism has roots that extend deep into history, though the *Oxford English Dictionary* dates the word "pescetarian" back only to 1991 and defines it as "A person who eats fish but avoids eating meat."[1] There are similar terms that involve restricted meat consumption such as "reducetarian" (someone who reduces meat consumption) and "flexitarian" (someone who occasionally eats meat), the latter of which won The American Dialect Society's most useful word of the year in 2003.[2] These terms are important because they not only define our consumption behavior, but also guide it.

Little quantitative research exists on the number of pescetarians today. One piece of evidence that the trend is relatively significant comes from a mobile technology company that polled their users with dietary restrictions (a sample size of 22,400) and found 2% self-identified as pescetarian.[3] More important than a snapshot of pescetarianism in the present is a view of pescetarianism in the future. Globally, demand and supply for fish is only expected to grow. As the FAO reports, world per capita fish supply reached a record 20 kg in 2014, and "world per capita apparent fish consumption increased from an average of 9.9 kg in the 1960s to 14.4 kg in the 1990s and 19.7 kg in 2013, with preliminary estimates for 2015 indicating further growth, exceeding 20 kg."[4] The growth rate for global fish supply for human consumption has doubled the rate of global population growth over the past five decades with 3.2% growth from 1961 to 2013 (compared to 1.6% growth in population over that same time period).[5] Our future looks fishy.

A future of increased fish consumption is not without reason. The 2015 USDA health guidelines advocate an increase in seafood consumption across all age and sex groups.[6] The USDA actually promotes a transition toward greater seafood consumption, stating that "Shifts are needed within the protein foods group to increase seafood intake."[7] The Royal Institute of International Affairs, also known as Chatham House, performed public opinion polling and found that fish was seen by many respondents to be a substitute for meat with fewer negative impacts on health and wellbeing.[8] This public opinion polling is furthered by Google's trend analysis: a pescetarian diet was one of the top ten most searched diets in 2013. That same year, fish accounted for about 17% of animal protein consumed globally—of course, not limited to pescetarians—and that number will likely increase.[9]

When global diets, increased income, concentrated urbanization, and the industrialization of food production is all combined, the 2050 global-average per capita income-dependent diet is estimated to have 82% more fish and seafood.[10] Couple that with the heightened consumer concern about the health and environmental impacts of red meat and it will be no surprise to see many more pescetarians at the dinner table in the future.

The difficulty of dialogue

In 1923, nutritionist Victor Lindlahr popularized the following adage in an advertisement for meat: You are what you eat. The quip has stood the test of time because it rings true, and in two important ways. First, from a nutritional standpoint, Lindlahr is right that our bodies get their necessary nutrients from our food. Second, partially due to the scholarship of author and activist Frances Moore Lappé, many also agree that our diet is inextricably linked to our identity politics. In fact, in 1826 the French gastronome Jean Anthelme Brillat-Savarin wrote: "Tell me what you eat, and I will tell you what you are."[11] Similar to the ballots we cast and the clothes we wear, our diets are statements about the values we hold and the world in which we want to live.

Because one's diet has profound implications for identity formulation and stabilization,[12] conversations about diet are implicitly conversations about politics, philosophy, and ethics.[13] These conversations are more challenging than they often appear on the surface, often easily provoking a sense of alienated or of being attacked.[14] It also leads to a lot of confusion: "When a conversation involves an aspect of a person's social identity and group allegiance and is between people with varied social identities, the potential for misunderstanding increases."[15] You might not think of lunch as a lifestyle (what about brunch?), but "food and lifestyles are intertwined and a critique of certain habits likens to a critique of certain lifestyles."[16] This makes talking about food more challenging than it appears.

In addition to the difficulties of talking about diet in general, the conversation regarding the consumption of fish in particular is thorny. There are over 33,000 species of fish.[17] There are more species of fish than all mammals, amphibians, reptiles, and birds combined. We must be cautious when talking about fish broadly, to avoid erroneously extrapolating our understanding of a few species that have been researched to tens of thousands of others.[18] In terms of diet, some might ethically object to the consumption of certain species of fish that are threatened or endangered (e.g. bluefin tuna) but have no moral qualm with consuming farmed fish. These species-differences are not present in conversations about the consumption of farmed land animals, at least to the same extent.

To further complicate the aquatic picture, there are multiple forms of acquiring fish used for food: Trawling and non-trawling fisheries, recirculating and non-recirculating aquaculture, and even angling, to name but a few. While some might find the method irrelevant to the overall ethical arguments for or against eating fish, many eaters find some methods of capture or production more objectionable than others. For example, the angler who eats her catch might find the consumer who buys her fish from a trawling fishery on tenuous moral ground, yet both could self-identify as pescetarian.

Finally, there are tradeoffs between some of the values we hold concerning our food systems (e.g. sustainability, food security, animal welfare, and many others).[19] There is no perfect system. One challenging tradeoff is between sustainability and animal welfare. What is best for our planet is often worst for our animals. Chatham House provides a paradigmatic example: "Emissions from intensively reared beef tend to be lower than from pasture-fed beef, but the practice raises other problems relating to animal welfare, inefficient use of crops for feed, water pollution and antimicrobial resistance from overuse of antibiotics."[20] In an effort to improve animal welfare on most farms, animals are given more space and feed which, in turn, often increases the environmental impacts those animals have on our planet. Similarly, improving food security in developing countries where the agricultural sector is a vital component of anti-poverty measures, while limiting environmental impacts, is a serious difficulty. These tradeoffs demonstrate why conversations about food are hard. Critical discourse about fish is even harder.

Moral pescetarianism: valid or void?

All else equal, is it morally justified to only eat fish while abstaining from the consumption of other meat? One argument for reducing, or abstaining from, meat consumption comes from a concern for animal welfare. The basic argument is that farmed animals suffer unnecessarily for our gustatory preferences and we ought to work to minimize, if not completely abolish, that unnecessary suffering. The other oft-cited moral argument for the reduction of meat comes out of a concern for the increasingly fragile environment in which we live. The argument is that meat produced from animal agriculture has a large carbon footprint compared to other foodstuffs, and thus we ought to reduce or abolish animal meat in order to minimize our own impact on the environmental.

Animal welfare motivations

Food taboos that engender dietary change can play an important role in animal welfare protection.[21, 22] In fact, a Harris Interactive poll of over 5,000 Americans conducted for *The Vegetarian Times* in 2008 found that the most frequently cited motivating factor for vegetarianism and veganism was animal welfare (54%).[23] The

Humane Research Council, now known as Faunalytics, found in a study of over 11,000 respondents that almost 70% of current vegetarians and vegans are motivated by animal protection.[24] It seems safe to assume, then, that some large percentage of pescetarians are motivated to reduce animal suffering.

A pescetarian motivated by a concern for animal welfare likely believes that either fish do not suffer at all or that fish do not suffer in comparable ways to other farmed animals. Why might pescetarians hold these views?

Most people do not consider fish to be cute. This may be because fish lack certain neotenous characteristics that many people perceive to carry moral weight.[25, 26] This may also be because we simply do not engage with aquatic animals as often or as intimately as we do with terrestrial animals, if only because we cannot breathe underwater. Beyond their physical traits and the foreign aquatic environments in which they live, fish are probably most often determined to lack moral standing because they aren't viewed as conscious beings both alive and living a life. Emotive states like fear, distress, desire, and regret are some of the human-like characteristics that turn, in our minds, animals into morally relevant beings.[27] Studies have shown that carnivores engage in a cognitive dissonance that motivates them to consume animals by ascribing to those animals diminished mental capacities,[28] and nowhere is this more the case than fish—for carnivores and pescetarians alike.

One of the most important traits that turns a being into one that carries moral weight is the capacity to suffer.[29] Brian Key presents the most popular argument against fish suffering in his 2016 article *Why Fish Do Not Feel Pain*, which concludes "that fish lack the necessary neurocytoarchitecture, microcircuitry, and structural connectivity for the neural processing required for feeling pain."[30] In other words, Key makes an anatomical argument that fish do not have the brain structure necessary for the capacity to feel pain. Fish respond to noxious stimuli in their environment; if they did not do so they would be ill-equipped to survive in this world. Key's argument is that there are structures in the brain that are required for those noxious stimuli to rise to a level of consciousness that results in *feeling* the feeling of pain (i.e. suffering). This general line of reasoning has been made by many others and represents one the last arguments still waged against fish suffering.[31, 32, 33, 34]

Key's argument has been heavily criticized. If we deny fish the ability to suffer because they lack a neocortex, we must also deny all non-mammals the ability to suffer as only mammals have a neocortex.[35, 36] It is entirely unclear why conscious pain would be limited to only mammalian neural structures.[37] An evolutionary perspective on brain development suggests that newer brain structures, such as the neocortex, take on functions for which older brain structures used to be responsible. Mammalian characteristics have been seen functionally present in anatomically different brain areas in non-mammals.[38, 39] Similar arguments in other contexts have been disproved in the past, such as the analogy to vision where many animals clearly can see but have very different brain anatomy.[40, 41] Finally, it has not been shown that mammalian cortical pain regions enable any behaviors in mammals that fish do not also demonstrate.[42] Key's argument has difficulty explaining away many fish behaviors that suggest the capacity to suffer. But does this matter?

Fish's capacity to suffer matters more than ever now that we raise a majority of them in aquatic factory farms. Aquaculture has overtaken the oceans as the world's primary source of fish for consumption. Factory farms are characterized by extreme confinement, a lack of individual veterinary care,[43] and an overall industrialization of the farming process that prioritizes efficiency and profit over all else. A majority of aquaculture today meets those criteria, and thus the issue of fish welfare on aquatic factory farms is a pressing concern that rivals the challenging welfare state of terrestrial factory farms.

Environmental motivations

The environmental impact of animal agriculture is increasingly lamented by environmentalists.[44,45] Rightfully so, given that a report from the Worldwatch Institute estimates that over half of our global annual GHG emissions are attributed to animal agriculture products.[46] Beef is particularly feared by many conservationists as the serpent in Eden, to be avoided at all costs. You are most likely to have a net-positive impact on the environment by cutting beef out of your diet as it is the most GHG intensive animal-based food product consumed today.[47] Dietary prohibitions often function as crucial social mechanisms in both species and ecosystem conservation as well as climate change mitigation.[48, 49, 50, 51] So, is there an environmental justification to remove all animal meats from one's diet except fish, and therefore adhere to a sustainability-motivated pescetarian diet?

Tilman and Clark provide a GHG emission lifecycle analysis of 555 food production systems for 22 different food types.[52] For fish food production systems specifically, they segment their analysis into non-trawling fisheries, trawling fisheries, non-recirculating aquaculture, and recirculating aquaculture. GHG emissions can vary significantly depending on the production method, with trawling fisheries and recirculating aquaculture being the most intensive. Whether you look at per kilocalorie, per United States Department of Agriculture-defined serving, or per gram of protein, the GHG emissions for almost every[53] production method is higher for fish than both poultry and pork.

The Environmental Assessment Agency of the Netherlands produced a report on the consumption and production of meat, dairy, and fish in the European Union. They found that energy requirements (electricity, feed production, etc.) for fish farming "are of the same order of magnitude as those of livestock farming,"[54] and, in general, "farmed fish is more or less comparable to poultry in environmental impact."[55] But there are other environmental implications beyond GHG emissions that warrant consideration.

Fish consumption is driving a dangerous reduction in fish population and threatening marine biodiversity. About 90% of marine fish are either fully exploited or overexploited, the latter amounting to about one third of all fish populations in our oceans.[56] A 2006 meta-analysis published in *Science* predicted a global collapse of all species fished, which could result in a fishless ocean by 2048.[57] Not only are fish species declining in absolute numbers, but fish sizes are also getting smaller as we pressure fish populations. As fish dwindle in both population size and physical size, catching them becomes more energy intensive and less sustainable; fishermen are traveling farther and fishing deeper to achieve their quotas.

Mortality due to bycatch (accidental catch) of non-target species while fishing also is a major threat to biodiversity and has been documented to endanger many species of sharks, rays, dolphins, seabirds and sea turtles.[58, 59, 60] Some research estimates that up to 40% of global catch is bycatch, but the real numbers are incredibly difficult to deduce given the available data.[61] In some fisheries, like bottom trawling in the North Sea for sole, bycatch can be as large as 90%.[62]

Discards are also a major environmental problem. Discards are the portion of a catch, and the portion of the bycatch, that is thrown back into the water. There are many reasons why fish may be thrown back: They may not be the target species, they may not meet size requirements, they may have no commercial value, or they may be a species for which that specific fishery does not have a quota. Researching discard mortality is extremely difficult, but it is safe to assume that a non-zero percentage of discards die due to the trauma induced during the capture itself or the increased likelihood of predation since energy levels are so compromised post-capture.

The biodiversity problems of fish-consuming diets are not limited to fisheries. Aquaculture pressures free living fish populations greatly. One simple example of how this happens is through feed. Over 70% of global aquaculture depends upon the supply of external feed, whether that feed is plant-based or fish-based.[63] A 2008 study estimated that about one fifth of free-caught fish is used in aquaculture in the form of fish feed for carnivorous fish.[64] The use of fishmeal and oil in feed has remained relatively steady due to sustained demand since 2000, although feed from all sources has increased significantly, causing the proportion of feed from fish to decline.[65] Plant-based feed is no panacea, as one review article explains: "Aquaculture's environmental footprint may now include nutrient and pesticide runoff from industrial crop production, and depending on where and how feed crops are produced, could be indirectly linked to associated negative health outcomes."[66] Whether direct or indirect, aquaculture feed has an impact on our oceans, our land, and even our health.

The validity of pescetarianism motivated by an environmental ethic is hard to defend given the vast environmental impacts of fish consumption. In terms of the carbon footprint alone, it seems to be arbitrary to abstain from consuming other meat, like chicken, while eating fish, given that the latter has at least an equivalent—if not much larger—footprint across production methods. Half of all fish consumed today come from the ocean and place significant pressure on aquatic biodiversity. The feed used in aquaculture has impacts that are just beginning to be fully appreciated. Across environmental metrics, a truly sustainable meal would appear to leave fish off the plate.

Objection—pragmatic pescetarianism

While pescetarianism may not be morally perfect, some might still believe that pescetarianism is at least better than omnivory. They might think that the reduction of eating meat in general ought to be applauded, and maybe even that the pescetarian's impact on animal welfare and the environment is net-positive. Pescetarians may not be moral saints, but that doesn't make them moral sinners.

This logic is faulty on multiple levels. First, and perhaps most important, the decision to eat only aquatic animals only further entrenches what I call "a bias of warm blood." Aquatic animals have been neglected compared to terrestrial animals. When it comes to animal welfare, all else equal, a diet that involves the consumption of anything *but* fish may have a more positive moral ripple effect than pescetarianism. Such a diet would raise ethical concern for a large group of species who have historically been neglected, and partially because it would combat the growth of the aquaculture industry which is still in its nascence and posed to grow significantly over the coming years. Aquaculture is the fastest growing food production sector in the world,[67] and efforts to retard its growth now will likely have a larger impact than efforts to slow the already established behemoth of terrestrial animal agriculture.

Second, replacing terrestrial meat with aquatic meat, even if one eats meat less frequently as a result, likely kills more individual animals. The average American carnivore kills 25 land animals a year and between 206–417 aquatic animals a year.[68] In other words, fish account for about 90–95% of the individual animals killed for a carnivorous diet. Now imagine an animal-loving carnivore decides to adopt a pescetarian diet, thinking such a diet would have a net-positive impact on animals compared to omnivory. That pescetarian would save 25 terrestrial animal lives over the course of one year (or minimize the number of individual animals killed by 5–10%). However, that assumes this pescetarian doesn't increase fish consumption while eliminating land-based animals from her plate. It seems at least plausible that a pescetarian actually kills *more* individual animals a year than a carnivore when one accounts for the increase in fish consumption. On the other hand, what if that carnivore simply stopped eating fish? The opposite of a pescetarian diet—one where you eat any animal *except* fish—likely saves many more individual lives.

Finally, if the moral objective is to minimize the carbon footprint or biodiversity impact of one's plate, the appropriate diet should focus on the GHG life cycle analysis or the biodiversity impacts, not the specific species of animal. The production method of a farmed animal is a major source of its environmental impact. Diets that are drawn around sustainability lines as opposed to species lines would embody environmental pragmatism better than diets that include fish regardless of their sourcing.

Conclusion

The morally motivated pescetarian appears to be an oxymoron. Perhaps this should not be surprising, as pescetarianism is likely not motivated by an environmental or animal welfare ethic but instead by the unfortunate product of the fact that, as Carl

Safina so eloquently writes, "we have yet to extend our sense of community below the high-tide line."[69] For many people, fish exist outside of the moral concern extended to terrestrial farmed animals. When it comes to the ethics of our dinner table, the question we ought to ask ourselves is not which animals can be consumed but why we need to consume animals at all.

Notes

1 "Pescatarian," *Oxford English Dictionary Online*, accessed May 17, 2016. The OED also defines vegetarian as "A person who abstains from eating animal food and lives principally or wholly on a plant-based diet; esp. a person who avoids meat and *often fish* but who will consume dairy products and eggs in addition to vegetable foods" (italics mine). In this definition, fish are separated from meat, highlighting a linguistic bias that further separates aquatic animals from terrestrial animals.
2 "2003 Words of the Year," American Dialectic Society, last modified January 13, 2004, www.americandialect.org/2003_words_of_the_year.
3 Bijal Shah and Steven Parisi, "Wondering What Dietary Preferences Are Most Common?" *Medium*, March 23, 2005, https://medium.com/ibotta-engine/wondering-what-dietary-preferences-are-most-common-34eeeac69e40#.xb8uzcla1. This research did not utilize a random sample and has other methodological shortcomings, but it is nevertheless more than anecdotal evidence that pescetarianism is a dietary trend on par with trends like veganism, and thus warrants investigation.
4 Food and Agriculture Organization of the United Nations [FAO], "The State of World Fisheries and Aquaculture 2016: Contributing to Food Security and Nutrition for All," Rome, 2016, www.fao.org/3/a-i5555e.pdf.
5 FOA, "The State of World Fisheries and Aquaculture 2016."
6 U.S. Department of Health and Human Services and U.S. Department of Agriculture [USHHS], "2015–2020 Dietary Guidelines for Americans," 8th Edition, December 2015, http://health.gov/dietaryguidelines/2015/guidelines.
7 USHHS, "2015–2020 Dietary Guidelines for Americans," chapter two.
8 Laura Wellesley, Catherine Happer and Anthony Froggatt, "Changing Climate, Changing Diets: Pathways to Lower Meat Consumption," Chatham House, November 2015.
9 FOA, "The State of World Fisheries and Aquaculture 2016."
10 David Tilman and Michael Clark, "Global Diets Link Environmental Sustainability and Human Health," *Nature* 515 (2014): 518–22, doi: 10.1038/nature13959.
11 Jean Anthelme Brillat-Savarin, *The Physiology of Taste: Or, Meditations on Transcendental Gastronomy*, Translated and edited by M.F.K. Fisher. (New York, NY: Knopf, 2009), 15.
12 Nick Fox and Katie Ward, "You Are What You Eat? Vegetarianism, Health and Identity," *Social Science & Medicine* 66, no. 12 (2008): 2585–2595, doi: 10.1016/j.socscimed.2008.02.011.
13 Jessica Greenebaum, "Veganism, Identity, and the Quest for Authenticity," *Food, Culture & Society: An Interdisciplinary Journal of Multidisciplinary Research* 15, no. 1 (2012): 129–144, doi: 10.2752/175174412X13190510222101.
14 Douglas Stone, Bruce Patton and Sheila Heen, *Difficult Conversations: How to Discuss What Matters Most*, (New York, NY: Viking, 1999).
15 Joshua Miller, Susan Donner and Edith Fraser, "Talking When Talking is Tough: Taking on Conversations About Race, Sexual Orientation, Gender, Class and Other Aspects of Social Identity," *Smith College Studies in Social Work* 74, no. 2 (2004): 377–92, doi: 10.1080/00377310409517722.
16 Henk Westhoek et al., "The Protein Puzzle: The Consumption and Production of Meat, Diary, and Fish in the European Union" (The Hague, NL: PBL Netherlands Environmental Assessment Agency, 2011), forward.

17 FishBase, last modified February 2018, www.fishbase.org.
18 Colin Allen, "Fish Cognition and Consciousness," *Journal of Agricultural and Environmental Ethics* 26, no. 1 (2013): 25–39, doi: 10.1007/s10806–011–9364–9.
19 Sometimes these trade-offs are actually synergies, as is the case with the environmental and welfare benefits of a plant-based diet.
20 Wellesley, Happer and Froggatt, "Changing Climate, Changing Diets," viii.
21 Fox and Ward, "You Are What You Eat?"
22 Jennifer Jabs, Carol Devine and Jeffery Sobal, "Model of the Process of Adopting Vegetarian Diets: Health Vegetarians and Ethical Vegetarians," *Journal of Nutrition Education* 30, no. 4 (1998): 196–202, doi: 10.1016/S0022–3182(98)70319-X.
23 Editorial, "Study Shows 7.3 Million Americans are Vegetarians," *Vegetarian Times*, April 16, 2008, www.vegetariantimes.com/article/vegetarianism-in-america. Fifty-three percent of respondents cited a concern for overall health, 47 percent cited environmental concerns, 31 percent cited food-safety concerns, and 25 percent cited weight loss, among others.
24 "Study of Current and Former Vegetarians and Vegans Initial Findings," Humane Research Council, December 2014, https://faunalytics.org/wp-content/uploads/2015/06/Faunalytics_Current-Former-Vegetarians_Full-Report.pdf.
25 Mark Estren, "The Neoteny Barrier: Seeking Respect for the Non-Cute," *Journal of Animal Ethics* 2, no. 1 (2012): 6–11, doi: 10.5406/janimalethics.2.1.0006.
26 Elizabeth Lawrence, "Neoteny in American Perceptions of Animals," *Journal of Psychoanalytic Anthropology* 9, no. 1 (1986): 41–54, http://psycnet.apa.org/record/1987-25122-001.
27 Nicholas Epley, Adam Waytz, Scott Akalis and John Cacioppo, "When We Need A Human. Motivational Determinants of Anthropomorphism," *Social Cognition* 26, no. 2 (2008): 143–155, doi: 10.1521/soco.2008.26.2.143.
28 Brock Bastian, Steve Loughnan, Nick Haslam and Helena Radke, "Don't Mind Meat? The Denial of Mind to Animals Used for Human Consumption," *Personality and Social Psychology Bulletin* 38, no. 2 (2012): 247–256, doi: 10.1177/0146167211424291.
29 Strictly speaking, pain and suffering are two very different phenomena. For the purposes of this paper, the terms are used somewhat interchangeably, partially because the literature often uses them interchangeably and partially for sake of brevity. For a detailed discussion of the important difference, see Max Elder, "The Fish Pain Debate: Broadening Humanity's Moral Horizon," *Journal of Animal Ethics* 4, no. 2 (2014): 16–29, doi: 10.5406/janimalethics.4.2.0016.
30 Brian Key, "Why Fish Do Not Feel Pain," *Animal Sentience* 1, no. 3 (2016), http://animalstudiesrepository.org/animsent/vol1/iss3/1/.
31 James Rose, "The Neurobehavioral Nature of Fishes and the Question of Awareness and Pain," *Reviews in Fisheries Science* 10 (2002): 1–38, doi: 10.1080/20026491051668.
32 James Rose, "Anthropomorphism and 'Mental Welfare' of Fishes," *Diseases of Aquatic Organisms* 75 (2007): 139–154, doi: 10.3354/dao075139.
33 James Rose, "Pain in Fish: Weighing the Evidence." *Animal Sentience* 1, no. 3 (2016), http://animalstudiesrepository.org/animsent/vol1/iss3/25/.
34 James Rose, et al., "Can Fish Really Feel Pain?" *Fish and Fisheries* 15 (2014): 97–133, doi: 10.1111/faf.12010.
35 Jonathan Balcombe, "Cognitive Evidence of Fish Sentience," *Animal Sentience* 1, no. 3 (2016), http://animalstudiesrepository.org/animsent/vol1/iss3/2/.
36 The Cambridge Declaration of Consciousness signed by an international group of renowned scientists, states "The absence of a neocortex does not appear to preclude an organism from experiencing affective states." Philip Low, "Cambridge Declaration of Consciousness," Declaration Presented at the Francis Crick Memorial Conference on Consciousness in Human and Non-Human Animals, at Churchill College, University of Cambridge, July 7, 2012, http://fcmconference.org/img/CambridgeDeclarationOnConsciousness.pdf.
37 Victoria Braithwaite and Paula Droege, "Why Human Pain Can't Tell Us Whether Fish Feel Pain," *Animal Sentience* 1, no. 3 (2016), http://animalstudiesrepository.org/animsent/vol1/iss3/3/.

38 Donald Broom, "Fish Brains and Behaviour Indicate Capacity for Feeling Pain," *Animal Sentience* 1, no. 3 (2016), http://animalstudiesrepository.org/animsent/vol1/iss3/4/.
39 Culum Brown, "Comparative Evolutionary Approach to Pain Perception in Fishes," *Animal Sentience* 1, no. 3 (2016), http://animalstudiesrepository.org/animsent/vol1/iss3/5/.
40 Lynne Sneddon and Matthew Leach, "Anthropomorphic Denial of Fish Pain," *Animal Sentience* 1, no. 3 (2016), http://animalstudiesrepository.org/animsent/vol1/iss3/28/.
41 Robert Elwood, "A Single Strand of Argument with Unfounded Conclusion," *Animal Sentience* 1, no. 3 (2016), http://animalstudiesrepository.org/animsent/vol1/iss3/19/.
42 Adam Shriver, "Cortex Necessary for Pain – But Not in the Sense That Matters," *Animal Sentience* 1, no. 3 (2016), http://animalstudiesrepository.org/animsent/vol1/iss3/27/.
43 Veterinary care primarily consists of largely unregulated antibiotics, often administered prophylactically, to such a rampant extent that many drug-resistant bacterial strains have been found in aquaculture. A 2015 review of 650+ papers found that 76% of antibiotics commonly used in aquaculture and agriculture are important for human medicine and that "resistant bacteria isolated from both aquaculture and agriculture share the same resistance mechanisms, indicating that aquaculture is contributing to the same resistance issues established by terrestrial agriculture" (see Hansa Done, Arjun Venkatesan and Rolf Halden, "Does the Recent Growth of Aquaculture Create Antibiotic Resistance Threats Different from those Associated with Land Animal Production in Agriculture?" *The AAPS Journal* 17, no. 3 (2015): 513–24, doi: 10.1208/s12248–015–9722-z).
44 Marieke Vries and Imke Boer, "Comparing Environmental Impacts for Livestock Products: A Review of Life Cycle Assessments," *Livestock Science* 128, no. 1–3 (2010): 1–11, doi: 10.1016/j.livsci.2009.11.007.
45 Tilman and Clark, "Global Diets."
46 Robert Goodland and Jeff Anhang, "Livestock and Climate Change: What if the Key Actors in Climate Change Are… Cows, Pigs, and Chickens?" *World Watch* 22, no. 6 (2009), www.worldwatch.org/node/6294.
47 Janet Ranganathan et al., "Shifting Diets for a Sustainable Food Future Creating a Sustainable Food Future, Installment Eleven," World Resources Institute, April 2016, http://www.wri.org/publication/shifting-diets.
48 Robert Johannes, "Traditional Marine Conservation Methods in Oceania and Their Demise," *Annual Review of Ecology and Systematics* 9 (1978): 349–64, doi: 10.1146/annurev.es.09.110178.002025.
49 Margaret Chapman, "Environmental Influences on the Development of Traditional Conservation in the South Pacific Region," *Environmental Conservation* 12 (1985): 217–230, doi: 10.1017/S0376892900015952.
50 Marlene Lingard et al., "The Role of Local Taboos in Conservation and Management of Species: The Radiated Tortoise in Southern Madagascar," *Conservation and Society* 1, no. 2 (2003): 223–46, doi: 10.1111/j.1523–1739.2008.00970.x.
51 Elke Stehfest et al., "Climate Benefits of Changing Diet," *Climate Change* 95 (2009): 83–102, doi: 10.1007/s10584–008–9534–6.
52 Tilman and Clark, "Global Diets" (see Table 1).
53 The only instance where GHG emissions are about the same or slightly less is non-trawling fisheries.
54 Westhoek et al., "The Protein Puzzle," 59.
55 Westhoek et al., "The Protein Puzzle," 161.
56 FAO, "The State of World Fisheries and Aquaculture 2016."
57 Boris Worm et al., "Impacts of Biodiversity Loss on Ocean Ecosystem Services," *Science* 3, no. 314 (2006): 787–790.
58 Jean-Christophe Vié, Craig Hilton-Taylor and Simon Stuart, eds. "Wildlife in a Changing World – An Analysis of the 2008 IUCN Red List of Threatened Species" (Gland, Switzerland: IUCN, 2009), https://portals.iucn.org/library/sites/library/files/documents/RL-2009-001.pdf.

59 Graham Raby, Alison Colotelo, Gabriel Blouin-Demers and Steven Cooke, "Freshwater Commercial Bycatch: An Understated Conservation Problem," *BioScience* 61, no. 4 (2011): 271–280, doi: 10.1525/bio.2011.61.4.7.
60 C. Josh Donlan and Chris Wilcox, "Integrating Invasive Mammal Eradications and Biodiversity Offsets for Fisheries Bycatch: Conservation Opportunities and Challenges for Seabirds and Sea Turtles," *Biological Invasions* 10, no. 7 (2008): 1053–1060, doi: 10.1007/s10530–007–9183–0.
61 Amanda Keledjian et al., "Wasted Catch: Unsolved Problems in U.S. Fisheries," *Oceana*, March 2014, http://oceana.org/sites/default/files/reports/Bycatch_Report_FINAL.pdf.
62 Thomas Catchpole, Chris Leslie John Frid and Tim Stuart Gray, "Discards in North Sea Fisheries: Causes, Consequences and Solutions," *Marine Policy* 29 (2005): 421–430, doi: 10.1016/j.marpol.2004.07.001.
63 Albert Tacon and Marc Metian, "Feed Matters: Satisfying the Feed Demand of Aquaculture," *Reviews in Fisheries Science and Aquaculture* 23, no. 1, (2015): 1--10, doi: 10.1080/23308249.2014.987209.
64 Albert Tacon and Marc Metian, "Global Overview on the Use of Fish Meal and Fish Oil in Industrially Compounded Aquafeeds: Trends and Future Prospects," *Aquaculture* 285 (2008): 146–158, doi: 10.1016/j.aquaculture.2008.08.015.
65 Jillian Fry et al., "Environmental Health Impacts of Feeding Crops to Farmed Fish," *Environment International* 91 (2016): 201–214, doi: 10.1016/j.envint.2016.02.022.
66 Jillian Fry et al., "Environmental Health Impacts of Feeding Crops to Farmed Fish," 201.
67 FAO, "The State of World Fisheries and Aquaculture 2014."
68 These numbers exclude shellfish but include bycatch which, in addition to finfish, often contains crustaceans and other aquatic animals. If we exclude bycatch, Americans kill between 159.7–311.3 finfish annually. Harish Sethu, "How Many Animals Does a Vegetarian Save?" *Counting Animals,* February 6, 2012, www.countinganimals.com/how-many-animals-does-a-vegetarian-save.
69 Carl Safina, *Song for the Blue Ocean* (New York, NY: Henry Holt and Company, 1998), 439.

Bibliography

Allen, Colin. "Fish Cognition and Consciousness." *Journal of Agricultural and Environmental Ethics* 26, 1(2013): 25–39, doi: 10.1007/s10806–10011–9364–9369.

American Dialectic Society. "2003 Words of the Year." Last Modified January 13, 2004. www.americandialect.org/2003_words_of_the_year.

Balcombe, Jonathan. "Cognitive Evidence of Fish Sentience." *Animal Sentience* 1, 3(2016). http://animalstudiesrepository.org/animsent/vol1/iss3/2/.

Bastian, Brock, Steve Loughnan, Nick Haslam and Helena Radke. "Don't Mind Meat? The Denial of Mind to Animals Used for Human Consumption." *Personality and Social Psychology Bulletin* 38, 2(2012): 247–256, doi: 10.1177/0146167211424291.

Braithwaite, Victoria and Paula Droege. "Why Human Pain Can't Tell Us Whether Fish Feel Pain." *Animal Sentience* 1, 3(2016). http://animalstudiesrepository.org/animsent/vol1/iss3/3/.

Brillat-Savarin, Jean Anthelme. *The Physiology of Taste: Or, Meditations on Transcendental Gastronomy*. Translated and edited by M.F.K. Fisher. New York, NY: Knopf, 2009.

Broom, Donald. "Fish Brains and Behaviour Indicate Capacity for Feeling Pain." *Animal Sentience* 1, 3(2016). http://animalstudiesrepository.org/animsent/vol1/iss3/4/.

Brown, Culum. "Comparative Evolutionary Approach to Pain Perception in Fishes." *Animal Sentience* 1, 3(2016). http://animalstudiesrepository.org/animsent/vol1/iss3/5/.

Catchpole, Thomas, Chris LeslieJohn Frid and Tim Stuart Gray. "Discards in North Sea Fisheries: Causes, Consequences and Solutions." *Marine Policy* 29(2005): 421–430, doi: 10.1016/j.marpol.2004.07.001.

Chapman, Margaret. "Environmental Influences on the Development of Traditional Conservation in the South Pacific Region." *Environmental Conservation* 12(1985): 217–230, doi: 10.1017/S0376892900015952.

Done, Hansa, Arjun Venkatesan and Rolf Halden. "Does the Recent Growth of Aquaculture Create Antibiotic Resistance Threats Different from those Associated with Land Animal Production in Agriculture?" *The AAPS Journal* 17, 3(2015): 513–524, doi: 10.1208/s12248–12015–9722-z.

Donlan, C. Josh and Chris Wilcox. "Integrating Invasive Mammal Eradications and Biodiversity Offsets for Fisheries Bycatch: Conservation Opportunities and Challenges for Seabirds and Sea Turtles." *Biological Invasions* 10, 7(2008): 1053–1060, doi: 10.1007/s10530–10007–9183–0.

Editorial. "Study Shows 7.3 Million Americans are Vegetarians." *Vegetarian Times*. April 16, 2008. www.vegetariantimes.com/article/vegetarianism-in-america.

Elder, Max. "The Fish Pain Debate: Broadening Humanity's Moral Horizon." *Journal of Animal Ethics* 4, 2(2014): 16–29, doi: 10.5406/janimalethics.4.2.0016.

Elwood, Robert. "A Single Strand of Argument with Unfounded Conclusion." *Animal Sentience* 1, 3(2016). http://animalstudiesrepository.org/animsent/vol1/iss3/19/.

Epley, Nicholas, Adam Waytz, Scott Akalis and John Cacioppo. "When We Need A Human. Motivational Determinants of Anthropomorphism." *Social Cognition* 26, 2(2008): 143–155, doi: 10.1521/soco.2008.26.2.143.

Estren, Mark. "The Neoteny Barrier: Seeking Respect for the Non-Cute." *Journal of Animal Ethics* 2, 1(2012): 6–11, doi: 10.5406/janimalethics.2.1.0006.

FishBase. Last modified February2018. www.fishbase.org.

Food and Agriculture Organization of the United Nations. "The State of World Fisheries and Aquaculture 2014: Opportunities and Challenges." Rome, 2014. www.fao.org/3/a-i3720e.pdf.

Food and Agriculture Organization of the United Nations. "The State of World Fisheries and Aquaculture 2016: Contributing to Food Security and Nutrition for All." Rome, 2016. www.fao.org/3/a-i5555e.pdf.

Fox, Nick and Katie Ward. "You Are What You Eat? Vegetarianism, Health and Identity." *Social Science & Medicine* 66, 12(2008): 2585–2595, doi: 10.1016/j.socscimed.2008.02.011.

Fox, Nick and Katie Ward. "Health, Ethics and Environment: A Qualitative Study of Vegetarian Motivations." *Appetite* 50, 2–3(2009): 422–429, doi: 10.1016/j.appet.2007.09.007.

Fry, Jillian, David Love, Graham MacDonald, Paul West, Peder Engstrom, Keeve Nachman and Robert Lawrence. "Environmental Health Impacts of Feeding Crops to Farmed Fish." *Environment International* 91(2016): 201–214, doi: 10.1016/j.envint.2016.02.022.

Goodland, Robert and Jeff Anhang. "Livestock and Climate Change: What if the Key Actors in Climate Change Are… Cows, Pigs, and Chickens?" *World Watch* 22, 6(2009). www.worldwatch.org/node/6294.

Greenebaum, Jessica. "Veganism, Identity, and the Quest for Authenticity." *Food, Culture & Society: An Interdisciplinary Journal of Multidisciplinary Research* 15, 1(2012): 129–144, doi: 10.2752/175174412X13190510222101.

Humane Research Council. "Study of Current and Former Vegetarians and Vegans Initial Findings." December2014. https://faunalytics.org/wp-content/uploads/2015/06/Faunalytics_Current-Former-Vegetarians_Full-Report.pdf.

Jabs, Jennifer, Carol Devine and Jeffery Sobal. "Model of the Process of Adopting Vegetarian Diets: Health Vegetarians and Ethical Vegetarians." *Journal of Nutrition Education* 30, 4 (1998): 196–202, doi: 10.1016/S0022–3182(98)70319-X.

Johannes, Robert. "Traditional Marine Conservation Methods in Oceania and Their Demise." *Annual Review of Ecology and Systematics* 9(1978): 349–364, doi: 10.1146/annurev.es.09.110178.002025.

Keledjian, Amanda, Gib Brogan, Beth Lowell, Jon Warrenchuk, Ben Enticknap, Geoff Shester, Michael Hirshfield and Dominique Cano-Stocco. "Wasted Catch: Unsolved Problems in U.S. Fisheries." *Oceana*. March2014. http://oceana.org/sites/default/files/reports/Bycatch_Report_FINAL.pdf.

Key, Brian. "Why Fish Do Not Feel Pain." *Animal Sentience* 1, 3(2016). http://animalstudiesrepository.org/animsent/vol1/iss3/1/.

Lawrence, Elizabeth. "Neoteny in American Perceptions of Animals." *Journal of Psychoanalytic Anthropology* 9, 1(1986): 41–54. http://psycnet.apa.org/record/1987-25122-001.

Lingard, Marlene, Nivo Raharison, Elisabeth Rabakonandrianina, Jean-Aime Rakotoarisoa and Thomas Elmqvist. "The Role of Local Taboos in Conservation and Management of Species: The Radiated Tortoise in Southern Madagascar." *Conservation & Society* 1, 2 (2003): 223–246, doi: 10.1111/j.1523–1739.2008.00970.x.

Low, Philip. "Cambridge Declaration of Consciousness." Declaration Presented at the Francis Crick Memorial Conference on Consciousness in Human and Non-Human Animals, at Churchill College, University of Cambridge. July 7, 2012. http://fcmconference.org/img/CambridgeDeclarationOnConsciousness.pdf.

Miller, Joshua, Susan Donner and Edith Fraser. "Talking When Talking is Tough: Taking on Conversations About Race, Sexual Orientation, Gender, Class and Other Aspects of Social Identity." *Smith College Studies in Social Work* 74, 2(2004): 377–392, doi: 10.1080/00377310409517722.

Oxford English Dictionary Online. "Pescatarian." Accessed May 17, 2016.

Raby, Graham, Alison Colotelo, Gabriel Blouin-Demers and Steven Cooke. "Freshwater Commercial Bycatch: An Understated Conservation Problem." *BioScience* 61, 4(2011): 271–280, doi: 10.1525/bio.2011.61.4.7.

Ranganathan, Janet, Daniel Vennard, Richard Waite, Brian Lipinski, Tim Searchinger, Patrice Dumas, Agneta Forslund, Hervé Guyomard, Stéphane Manceron, Elodie Marajo-Petitzon, Chantal Le Mouël, Petr Havlik, Mario Herrero, Xin Zhang, Stefan Wirsenius, Fabien Ramos, Xiaoyuan Yan, Michael Phillips and Rattanawan Mungkung. "Shifting Diets for a Sustainable Food Future Creating a Sustainable Food Future, Installment Eleven." *World Resources Institute*. April2016. www.wri.org/publication/shifting-diets.

Rose, James. "The Neurobehavioral Nature of Fishes and the Question of Awareness and Pain." *Reviews in Fisheries Science* 10(2002): 1–38, doi: 10.1080/20026491051668.

Rose, James. "Anthropomorphism and 'Mental Welfare' of Fishes." *Diseases of Aquatic Organisms* 75(2007): 139–154, doi: 10.3354/dao075139.

Rose, James. "Pain in Fish: Weighing the Evidence." *Animal Sentience* 1, 3(2016). http://animalstudiesrepository.org/animsent/vol1/iss3/25/.

Rose, James, Robert Arlinghaus, Steven Cooke, Ben Diggles, William Sawynok, Don Stevens and Clive Wynne. "Can Fish Really Feel Pain?" *Fish and Fisheries* 15(2014): 97–133, doi: 10.1111/faf.12010.

Safina, Carl. *Song for the Blue Ocean*. New York, NY: Henry Holt and Company, 1998.

Sethu, Harish. "How Many Animals Does a Vegetarian Save?" *Counting Animals*. February 6, 2012. www.countinganimals.com/how-many-animals-does-a-vegetarian-save.

Shah, Bijal and Steven Parisi. "Wondering What Dietary Preferences Are Most Common?" *Medium*. March 23, 2005. https://medium.com/ibotta-engine/wondering-what-dietary-preferences-are-most-common-34eeeac69e40#.xb8uzcla1.

Shriver, Adam. "Cortex Necessary for Pain – But Not in the Sense That Matters." *Animal Sentience* 1, 3(2016). http://animalstudiesrepository.org/animsent/vol1/iss3/27/.

Sneddon, Lynne and Matthew Leach. "Anthropomorphic Denial of Fish Pain." *Animal Sentience* 1, 3(2016). http://animalstudiesrepository.org/animsent/vol1/iss3/28/.

Stehfest, Elke, Lex Bouwman, Detlef van Vuuren, Michel den Elzen, Bas Eickhout and Pavel Kabat. "Climate Benefits of Changing Diet." *Climate Change* 95(2009): 83–102, doi: 10.1007/s10584–10008–9534–9536.

Stone, Douglas, Bruce Patton and Sheila Heen. *Difficult Conversations: How to Discuss What Matters Most.* New York, NY: Viking, 1999.

Tacon, Albert and Marc Metian. "Global Overview on the Use of Fish Meal and Fish Oil in Industrially Compounded Aquafeeds: Trends and Future Prospects." *Aquaculture* 285 (2008): 146–158, doi: 10.1016/j.aquaculture.2008.08.015.

Tacon, Albert and Marc Metian. "Feed Matters: Satisfying the Feed Demand of Aquaculture." *Reviews in Fisheries Science & Aquaculture* 23, 1(2015): 1–10, doi: 10.1080/23308249.2014.987209.

Tilman, David and Michael Clark. "Global Diets Link Environmental Sustainability and Human Health." *Nature* 515(2014): 518–522, doi: 10.1038/nature13959.

Vié, Jean-Christophe, Craig Hilton-Taylor and Simon Stuart, eds. "Wildlife in a Changing World – An Analysis of the 2008 IUCN Red List of Threatened Species." Gland, Switzerland: IUCN, 2009. https://portals.iucn.org/library/sites/library/files/documents/RL-2009-001.pdf.

Vries, Marieke and Imke Boer. "Comparing Environmental Impacts for Livestock Products: A Review of Life Cycle Assessments." *Livestock Science* 128, 1–3(2010): 1–11, doi: 10.1016/j.livsci.2009.11.007.

Watson, Reg and Daniel Pauly. "Systematic Distortions in World Fisheries Catch Trends." *Nature* 414(2001): 534–536, doi: 10.1038/35107050.

Wellesley, Laura, Catherine Happer and Anthony Froggatt. "Changing Climate, Changing Diets: Pathways to Lower Meat Consumption." Chatham House, November2015.

Westhoek, Henk, Trudy Rood, Maurits van de Berg, Jan Janse, Durk Nijdam, Melchert Reudink and Elke Stehfest. "The Protein Puzzle: The Consumption and Production of Meat, Diary, and Fish in the European Union." The Hague, NL: PBL Netherlands Environmental Assessment Agency, 2011.

Worm, Boris, Edward Barbier, Nicola Beaumont, *et al.* "Impacts of Biodiversity Loss on Ocean Ecosystem Services." *Science* 3, 314(2006): 787–790.

1.10

THERE IS SOMETHING FISHY ABOUT EATING FISH, EVEN ON FRIDAYS

On Christian abstinence from meat, piscine sentience, and a fish called Jesus

Kurt Remele

Since the dawn of Christianity the faithful have practiced fasting and abstinence. A brief anonymous treatise called the *Didache* or *The Teachings of the Twelve Apostles,*[1] dated by most contemporary scholars to the first century, called early Christians to fast on Wednesdays and Fridays. Even today, orthodox Christians are required to abstain, in principle, from all vertebrates including fishes[2] and dairy products, eggs and olive oil, wine and other alcoholic beverages on both Wednesdays (in remembrance of Judas' betrayal of Christ) and Fridays (in memory of Christ's crucifixion and death).[3]

Fish on Friday: A long-time, yet one-time mark of Catholic identity

Wednesday eventually vanished as a day of abstinence in most of Christianity and the list of forbidden foods was reduced to mammals and birds. Yet Friday has always remained a Christian day of penance. In the ninth century Pope Nicholas I explicitly decreed that on a Friday every Christian was obliged to practice mortification of his or her own flesh by abstinence from eating meat, albeit fish was tolerated.[4] Fish on Friday gradually became a part of the Catholic culture. In their 1966 statement on penance and abstinence, the United States Conference of Catholic Bishops declared:

> Catholic peoples from time immemorial have set apart Friday for special penitential observance by which they gladly suffer with Christ that they may one day be glorified with Him. This is the heart of the tradition of abstinence from meat on Friday.[5]

Although "abstinence" might be regarded as a form of fasting and both practices often coincide, strictly speaking there is a difference between them: "fasting" in the

Christian context is understood as "the complete or partial abstention from food,"[6] i.e. as eating nothing at all or eating less. "Abstinence", however, signifies "the abstention from the eating of meat [and in some cases] of certain meat products",[7] a behaviour that in traditional Catholic jargon is called "Fish on Friday" and in a more contemporary terminology might be named "Friday pescetarianism". According to the late Andrew Greeley, an internationally renowned sociologist of religion and a Catholic priest, Fish on Friday for many centuries was "one of the great definitions of Catholic identity"[8] and this despite the fact that, strictly speaking, no Catholic was obliged to eat fish on Fridays. Every Catholic had, of course, countless other meat-free food choices but, as it seems, "Baked Beans on Friday" never became a serious competitor.

In February 1966, Pope Paul VI enacted a decree or apostolic constitution named *Paenitemini* (Latin for *Do Penance*) with the intention to "reorganize penitential discipline with practices more suited to our times".[9] The pope authorised national episcopal conferences to "substitute abstinence and fast wholly or in part with other forms of penitence and especially works of charity and the exercises of piety".[10] To put it differently: The pope abolished the universal obligation of the Catholic faithful to abstain from meat on Friday. Other forms of penance like almsgiving, renouncing cigarettes or alcohol or watching television are deemed equally suitable and permitted, should the bishops of a country decide it. Only Good Friday and Ash Wednesday have remained days of both abstinence and fasting in the entire Catholic Church. To ward off misunderstandings: in Catholic/Christian penitential discipline pity on and compassion for animals hardly played a role: "There does not seem to be any spiritual desire to value animals' lives for their own sake. Abstinence from flesh was about acquiring virtue for humans, not cultivating mercy for animals."[11]

In the question and answer section about Lent, which the website of the United States Conference of Catholic Bishops provides, someone posed an inquiry with regard to the rules of abstinence and fasting. He or she asked whether chicken and dairy products are classified as meat. The answer the episcopal conference provides is an excellent example of moral subtleness or casuistry: "Abstinence laws", it explains:

> consider that meat comes only from animals such as chickens, cows, sheep or pigs – all of which live on land. Birds are also considered meat. Abstinence does not include meat juices and liquid foods made from meat. Thus, such foods as chicken broth, consommé, soups cooked or flavored with meat, meat gravies or sauces, as well as seasonings or condiments made from animal fat are technically not forbidden. However, moral theologians have traditionally taught that we should abstain from all animal-derived products (except foods such as gelatin, butter, cheese and eggs, which do not have any meat taste). Fish are a different category of animal. Salt and freshwater species of fish, amphibians, reptiles … and shellfish are permitted.[12]

Although this is a meticulous statement, it still does not tell us whether a 1951 declaration by the Vatican, which ruled that "underwater creatures – even if they come up for air … [generally] count as fish",[13] is still valid.

The Catholic church's giving up the traditional requirement to give up meat on Fridays had some severe economic repercussions: according to Frederick W. Bell, a former chief of economic research at the Bureau of Commercial Fisheries, it most probably "had a negative influence on fish prices and therefore industry revenues [and] create[d] economic problems for many small communities along the coastal United States".[14]

On their website the Catholic bishops of the USA maintain that "fish are a different category of animal".[15] But what exactly is so different with fishes? And why were and partly are these alleged differences relevant with regard to the question of whether Catholics are permitted to eat them on Fridays and other days of abstinence and penance or not?

Caught and cooked: Fishes according to Christian tradition

The answers to the question of why eating fish is permitted on days and during periods of fasting and abstinence, while eating mammals and birds is not, are diverse. A popular one is that, contrary to mammals and birds, who are warm-blooded species, fishes are cold-blooded animals, i.e. their body temperature changes according to the temperature outside. The Hebrew Bible, the Christians' First or Old Testament, relates that from the day Noah departed from the ark, fishes and all the other animals were approved by God for human consumption (Gen. 9.2–3). Of course, at that time the restrictions by the dietary laws of Israel still applied: Edible sea/river/lake creatures had to have both scales and fins (Lev. 11. 9–12).

In the New Testament, fishes are known as a basic staple, as food that is more affordable and therefore more often consumed than meat. Jesus frequently taught along the shores of Lake Galilee. His first disciples were fishermen who had worked there. Because of their new vocation of proclaiming the Gospel, Jesus called them "fishers of humans" (Mark 1.17). Apparently, there were no professional butchers among the disciples. It is somehow remarkable, but to the present day, so-called quiet recreational fishing has a superior reputation compared to noisy so-called leisure hunting, for the most part quite unjustifiably.

In the New Testament there are several reports about the multiplication of fishes in the Gospels: The feeding of the five thousand and the feeding of the four thousand, where Jesus multiplied loaves and fishes, and two reports on the miraculous catch or draft of fish. In the Gospel of Luke (5. 1–11), this miracle occurs before Jesus' crucifixion and resurrection, in John's Gospel (21. 1–14) it is described in the context of Jesus' last appearance. In both versions Jesus tells the disciples, who have not caught a single fish by then, to cast out their nets once more. Accordingly, the disciples cast their net, "and were not able to pull it in because of the number of fish" (John 21. 7). Finally, they succeeded in dragging the net with the fishes to the shore. The exact number of fishes was 153. "When they climbed out on the shore, they saw a charcoal fire with fishes on it and bread. Jesus said to them, 'Bring some of the fish you just caught … Come, have breakfast'" (John 21. 9–10. 12).

Did Jesus eat fish? In his seminal study *Animal Theology*, Andrew Linzey does not mince matters, when he writes: "Jesus was – as far as we know – no crusading vegetarian. While there are no precise biblical accounts of him eating meat, the canonical Gospels leave us in no doubt that he ate fish."[16] I am somewhat more agnostic and sceptical about this, because the eating of fish in John's gospel happened after the resurrection. I doubt that Jesus ate anything at all in this new state of being. I agree with Andrew Linzey,[17] though, that, should Jesus actually have eaten fish, this historical fact or at least possibility does not per se represent a conclusive argument against a contemporary vegetarian or vegan diet. For neither can the demands of contemporary Christian discipleship be met simply by the naïve imitation of the Jesus of first-century Palestine nor must one ignore the necessary particularity of the incarnation with its implied local and temporal cognitive limitations. The eminent Dutch theologian Edward Schillebeeckx put it well when he wrote:

> Jesus, too, not only reveals God, but he also obscures him, since he appears in a non-divine, creaturely humanity. And thus, as a human being, he is a historical, contingent or limited creature that in no way is able to represent the whole abundance of God.[18]

There are two more reasons why the consumption of fishes has been regarded as permissible on days of fasting and abstinence in Christian history:

> It is unclear in biblical interpretation whether fishes were rendered extinct by the Flood. If they were not subjected to the same punishment of drowning by inundation by which all land animals and birds perished, they were by implication free from sin.[19]

Furthermore, fishes were viewed as symbols of sexual and spiritual purity. People commonly believed that fishes did not reproduce by means of eggs and sperm. Instead, they were convinced that, as Basil of Caesarea put it in the fourth century, the "water receives the egg that falls and makes an animal of it".[20]

Caught and released: Fishes according to Pythagoras and Shri Swaminarayan

In the Bible catching fish is widely regarded as something completely normal and ethically unproblematic. It was different for the Greek philosopher Pythagoras and his disciples.[21]

Pythagoras, who was born about half a century before Christ, was a mathematician and philosopher. According to an account recorded by later biographers, one day Pythagoras arrived near the town of Crotone in Southern Italy. There he came upon some fishermen who were drawing up their nets, which were filled with fishes. Pythagoras told the fishermen that he would be able to predict the exact number of fishes they had caught. The fishermen promised to do everything he demanded should the number be correct. Pythagoras was totally accurate in his estimate. He therefore ordered the fishermen to return the fishes, all of whom had survived both the catching and counting, to the sea. It is sometimes claimed that the specific number of fishes that were caught is mentioned in the story and that it was 153, exactly the same number of fishes as in John's report on the miraculous catch of fishes.

There is another story about a wise and holy man and the catching of fishes, which I learned about when I visited the BAPS Shri Swaminarayan Mandir at Neasden in North West London in 2008. This building is Europe's first traditional Hindu temple and the largest Hindu temple outside of India. BAPS stands for Bochasanwasi Shri Akshar Purushottam Swaminarayan Sanstha, a Hindu movement that goes back to Bhagwan or Shri Swaminarayan,[22] a Hindu guru who lived from 1781 to 1830. Nowadays, he is revered as a deity. In his moral code of conduct, he opposed the customary Indian practices of sati (the forced burning of widows), female infanticide, and the sacrificial killing of animals. He prescribed the observance of, among other things, vegetarianism for all his devotees. In the "Understanding Hinduism" exhibition that was offered at the temple I came across a legendary story from the life of Shri Swaminarayan[23]: The young Bhagwan Swaminarayan, who was called Ghanshyam in his childhood years, and his friends went for a swim in the village lake, Meen Sarovar. There, Ghanshyam saw a fisherman emptying his catch into a wicker basket. The sight of the dead fishes saddened the holy boy and his heart bled for the innocent fishes. He resolved there and then to revive the fishes. He looked at them, brought them back to life, and one by one they jumped back into the water. At first, the fisherman was very angry and stormed towards Ghanshyam and his friends. But then Lord Yam, the god of death, appeared to the fisherman and he realized his wrongful act. He prayed to Ghanshyam for forgiveness and pledged never again to kill fishes or any other animal. He bowed in reverence to Ghanshyam.

These three narratives of catching, cooking or releasing fishes suggest that the Pythagorean religious concept of transmigration, in which an immortal soul of a departed being transmigrates into another body including that of an animal and the related Hindu concept of reincarnation together with the Hindu virtues of non-harming (ahimsa) and compassion (karuna) produce more benign consequences for fishes than the biblical admonition to "have dominion over the fish of the sea" (Gen.1. 28). Christians therefore should not shun the insight of comparative religious ethics and moreover be willing to engage in a creation- and animal-friendly re-lecture of the Bible and their other literary sources.

Pain and fear: Fishes as sentient beings with intrinsic value

Fishes were once considered as being insentient and unintelligent by both zoologists and the general public. Although this claim has proven to be remarkably fishy, it is still widespread. Many humans still think of "fish" as an anonymous part of a mass, as things, not as individual living beings. Yet recent research findings have demonstrated that fishes not only feel pain but also experience fear. In many areas, such as memory, individual recognition and self-awareness, the cognitive powers of fish match or exceed those of some non-aquatic vertebrates. Their nervous systems are similar to those of birds and mammals. Although the fish brain has no neocortex, some of the functions performed in the neocortex of mammals are effectively taken care of by other regions of the brain.[24] Maximilian Padden Elder states that "fish seem to have interests just like mammals and birds do" and therefore concludes that "not eating animals while continuing to eat fish has a tenuous moral foundation".[25] The false assumption that fishes do not feel pain has led to a total disregard for the welfare of individual fishes: Fishes caught in shallow waters are brought to the surface alive and die on the decks of boats and ships; fishes brought up from great depths are usually dead when they reach the surface because of the rapidly changing pressure. The number of fishes killed each year by humans is between 1 and 2.7 trillion.[26]

What about fishes and sex? No sex, please, we are fishes? The hard truth is: all the Christian theologians and saints, hermits and virgins, all the Manichaeans and Cathars, who praised fishes as symbols of sexual purity, were dead wrong. Fishes do have sex and they have it in multiple varieties: "There are promiscuous fishes, polygamous fishes, and monogamous ones, including fishes that mate for life."[27] Numerous fish species release their eggs and sperm into the water for so called external fertilization. Most fishes mate without penetration, but there are many exceptions. Males of the family that includes guppies, mollies, platys and swordtails, e.g., all possess a so-called gonopodium which they use for penetrative sex. The great majority of fishes are either male or female throughout life. But there are also sex-changing fishes (sequential hermaphrodites), among them the popular clownfishes of *Finding Nemo* fame, and others (simultaneous hermaphrodites), mainly deep sea fishes who produce both eggs and sperm at the same time and are therefore able to fertilize themselves.[28]

Fishes do have sex and they do have sentience. Moreover, as Jonathan Balcombe put it "a fish has a biography, not just a biology".[29]

A fish called Jesus: A plea for the definitive abandonment of fish on Friday

According to Brian Fagan, an emeritus professor of anthropology at the University of California, Santa Barbara, the acronym *Ichthys* "is as old as Christianity itself … No one knows where it first appeared, perhaps in the bustling streets of Roman

Alexandria, as a protest against the rule of pagan emperors".[30] *Ichthys*, the ancient Greek word for fish, stands for *Iesous Christos Theou Yios Soter* – Jesus [is] Christ, Son of God, [and is] Saviour. In his book *On Baptism* (*De Baptismo*) the second-century church father Tertullian of Carthage wrote: "We being little fishes, as Jesus Christ is our great Fish, begin our life in the water and only while we abide in the water are we safe and sound."[31] The water Tertullian is referring to, of course, is the water of baptism.

A fish called Jesus: The symbolic identification of Christ with the fish enables us to theologically put the torments and pains of fishes on a level with the suffering of Christ. Such an approach refers us back to the famous sermon John Henry Newman held at St Mary's University Church, Oxford, on Good Friday in 1842. Ill-treatment of cattle by barbarous "owners" and the beginnings of vivisection at Oxford University caused Newman to speak out against these practices:

> I mean, consider how very horrible it is to read the accounts which sometimes meet us of cruelties exercised on brute animals ... There is something so very dreadful, so satanic in tormenting those who have never harmed us, and who cannot defend themselves ... that none but very hardened persons can endure the thought of it.[32]

Newman did not only mention the cruel treatment of animals and protest against it, he also placed the animals' pains within a Christological context. According to his sermon the cruelty inflicted upon animals was "the very cruelty inflicted upon our Lord" and thereby Newman "posited nothing less than a moral equivalence between the suffering of animals and the suffering of Christ himself".[33]

Three years after this sermon Newman converted from Anglicanism to Catholicism, at the age of 78 he was made a cardinal and in September 2010 he was beatified by Pope Benedict XVI. One year later, in September 2011, the Catholic bishops of England and Wales re-established Friday abstention from meat and thereby implicitly and inevitably Fish on Friday as a communal and obligatory penitential practice for the Catholic faithful of their dioceses. They declared that penance on Fridays "should be fulfilled simply by abstaining from meat and by uniting this to prayer".[34]

I do not know whether Catholics in England and Wales have taken the re-introduction of meatless Fridays seriously and followed the bishops' directive. Maybe some Catholics have started to honour Friday as a day on which no animals are eaten at all. That would be a good thing in itself and maybe even a first step towards vegetarianism or veganism. Even so, communal practices play a role in religious organisations, the mere reintroduction of meatless Fridays as a penitential practice is nostalgic at best. At a time when sentience of fish has been established and unsustainable overfishing endangers individual fishes and fish species from an ethical point of view, the initiative of the bishops is extremely outdated and insensitive. It is fishy indeed.

Notes

1 "The Lord's teaching to the heathen by the Twelve Apostles," The Didache, ch. 8, www.thedidache.com.
2 Jonathan Balcombe, *What a Fish Knows: The Inner Lives of Our Underwater Cousins* (New York, NY: Scientific America / Farrar, Straus and Giroux, 2016), 6: "We traditionally refer to anything from two to a trillion fish by the singular term 'fish', which lumps them together like rows of corn. I have come to favor the plural 'fishes,' in recognition of the fact that these animals are individuals with personalities and relationships." In this chapter, I follow Balcombe and use the plural "fishes".
3 Orthodox Church in America, "Orthodox Fasting: Questions and Answers," https://oca.org/questions/dailylife/orthodox-fasting. The webpage "The Fasting Rule of the Orthodox Church" admits: "The Church's traditional teaching on fasting is not widely known or followed in our day."
4 Brian Fagan, *Fish on Friday: Feasting, Fasting, and the Discovery of the New World* (New York, NY: Basic Books, 2006), 56.
5 "Pastoral Statement on Penance and Abstinence," United States Conference of Catholic Bishops, November 18, 1966, no.18, www.usccb.org/prayer-and-worship/liturgical-year/lent/us-bishops-pastoral-statement-on-penance-and-abstinence.cfm.
6 Catholic University of America, "Fast and Abstinence," in *New Catholic Encyclopedia*, vol. 5. (Detroit, MI: Gale, 2003), 632.
7 Catholic University of America, "Fast and Abstinence."
8 Andrew Greeley, *The Catholic Revolution: New Wine, Old Wineskins, and the Second Vatican Council* (Berkeley and Los Angeles, CA: University of California Press, 2004), 137.
9 Paul VI, *Paenitemini. Apostolic Constitution on Fast and Abstinence*, ch. 3, February 17, 1966, http://w2.vatican.va/content/paul-vi/en/apost_constitutions/documents/hf_p-vi_apc_19660217_paenitemini.html.
10 Paul VI, *Paenitemini*, ch. 3.
11 Carl Frayne, "On Imitating the Regimen of Immortality or Facing the Diet of Mortal Reality: A Brief History of Abstinence from Flesh-Eating in Christianity," *Journal of Animal Ethics*, vol. 6, no. 5 (Fall 2016): 200.
12 "Questions and Answers about Lent and Lenten Practices," United States Conference of Catholic Bishops, www.usccb.org/prayer-and-worship/liturgical-year/lent/questions-and-answers-about-lent.cfm.
13 *Time*, "Religion: Whale Meat on Friday," *Time*, November 26, 1951, http://content.time.com/time/magazine/article/0,9171,821907,00.html.
14 Frederick, W. Bell, "The Pope and the Price of Fish," *The American Economic Review*, vol. 58, no. 5 (December 1968), 1350.
15 United States Conference of Catholic Bishops, "Questions and Answers."
16 Andrew Linzey, *Animal Theology* (London, England: SCM Press, 1994), 86.
17 Linzey, *Animal Theology*, 86–87.
18 Edward Schillebeeckx, *Menschen: Eine Geschichte von Gott* (Freiburg, Germany: Herder, 1990), 31–32 (my translation).
19 David Grumett and Rachel Muers, *Theology on the Menu: Ascetism, Meat and Christian Diet* (London, England: Routledge, 2010), 85.
20 Basil the Great, *Homily* 7.2 cited in Grumett and Muers, *Theology on the Menu*, 85.
21 Eugen Drewermann, *Der tödliche Fortschritt: Von der Zerstörung der Erde und des Menschen im Erbe des Christentums* (Regensburg, Germany: Pustet, 1981), 204.
22 BAPS Swamynarayan Sanstha, accessed October 1, 2016, www.swaminarayan.org.
23 Sadhu Vivekjivandas, *Baghwan Swaminarayan: Life and Work* (Gujarat, India: Swaminarayan Aksharpith, 2005), 5–7.
24 Sidney J. Holt, "Sea Fishes and Commercial Fishing," in *The Global Guide to Animal Protection*, ed. Andrew Linzey (Urbana, IL: University of Illinois Press, 2013), 47.
25 Maximilian Padden Elder, "The Fish Pain Debate: Broadening Humanity's Moral Horizon," *Journal of Animal Ethics*, vol. 4, no. 2 (Fall 2014), 25.

26 Balcombe, *What a Fish Knows*, 7.
27 Balcombe, *What a Fish Knows*, 181.
28 Balcombe, *What a Fish Knows*, 181–183.
29 Jonathan Balcombe, "Fishes have feelings, too," *International New York Times*, May 17, 2016, 8. See also Peter Singer, "Fish: The Forgotten Victims on our Plate," *The Guardian*, September 14, 2010, www.theguardian.com/commentisfree/cif-green/2010/sep/14/fish-forgotten-victims.
30 Fagan, *Fish on Friday*, 3. See also Mary Ann Beavis and Michael J. Gilmour eds., "Fish, Fisher, Fishing," in *Dictionary of the Bible and Western Culture* (Sheffield, England: Sheffield Phoenix Press, 2012),164.
31 Tertullian, "Homily on Baptism," no. 1, edited with an introduction, translation and commentary by Ernest Evans, www.tertullian.org/articles/evans_bapt/evans_bapt_index.htm.
32 John Henry Newman, *Sermons 1824–1843*, vol. 5 cited in Andrew Linzey, *Why Animal Suffering Matters* (New York, NY: Oxford University Press, 2009), 38–39.
33 John Henry Newman, *Sermons*, 38.
34 "Catholic Witness – Friday Penance," Catholic Bishops' Conference of England and Wales, September 16, 2011, https://rcdow.org.uk/att/files/education/downloads/headteachers/cbcewcatholicwitnessfridaypenanceqa.pdf.

Bibliography

Balcombe, Jonathan. "Fishes have Feelings, too." *International New York Times*, May 17, 2016, 8.

Balcombe, Jonathan. *What a Fish Knows: The Inner Lives of Our Underwater Cousins*. New York, NY: Scientific America / Farrar, Straus and Giroux, 2016.

BAPS Swamynarayan Sanstha. Accessed October 1, 2016. www.swaminarayan.org.

Beavis, Mary Ann and Michael J. Gilmour eds. "Fish, Fisher, Fishing." In *Dictionary of the Bible and Western Culture*. Sheffield, England: Sheffield Phoenix Press, 2012.

Bell, Frederick W. "The Pope and the Price of Fish." *The American Economic Review*, 58, 5 (December 1968): 1346–1350.

Catholic Bishops' Conference of England and Wales. "Catholic Witness – Friday Penance." September 16, 2011. https://rcdow.org.uk/att/files/education/downloads/headteachers/cbcewcatholicwitnessfridaypenanceqa.pdf.

Catholic University of America. "Fast and Abstinence." In *New Catholic Encyclopedia, Vol. 5*. Detroit, MI: Gale, 2003. 632–635.

Drewermann, Eugen. *Der tödliche Fortschritt: Von der Zerstörung der Erde und des Menschen im Erbe des Christentums*. Regensburg, Germany: Pustet, 1981.

Elder, Maximilian Padden. "The Fish Pain Debate: Broadening Humanity's Moral Horizon." *Journal of Animal Ethics*, 4, 2 (Fall 2014): 16–29.

Fagan, Brian. *Fish on Friday: Feasting, Fasting, and the Discovery of the New World*. New York, NY: Basic Books, 2006.

Frayne, Carl. "On Imitating the Regimen of Immortality or Facing the Diet of Mortal Reality: A Brief History of Abstinence from Flesh-Eating in Christianity." *Journal of Animal Ethics*, 6, 5 (Fall 2016): 188–212.

Greeley, Andrew. *The Catholic Revolution: New Wine, Old Wineskins, and the Second Vatican Council*. Berkeley and Los Angeles, CA: University of California Press, 2004.

Grumett, David, and Rachel Muers. *Theology on the Menu: Ascetism, Meat and Christian Diet*. London, England: Routledge, 2010.

Holt, Sidney J. "Sea Fishes and Commercial Fishing." In *The Global Guide to Animal Protection*, edited by Andrew Linzey, 47–50. Urbana, IL: University of Illinois Press, 2013.

Linzey, Andrew. *Animal Theology*. London: SCM Press, 1994.

Linzey, Andrew. *Why Animal Suffering Matters*. New York, NY: Oxford University Press, 2009.

Orthodox Church in America. "Orthodox Fasting: Questions and Answers." https://oca.org/questions/dailylife/orthodox-fasting.

Orthodox Resources. "The Fasting Rule of the Orthodox Church." Accessed October 1, 2016. www.abbamoses.com/fasting.html.

Paul VI. *Paenitemini. Apostolic Constitution on Fast and Abstinence*, February 17, 1966. http://w2.vatican.va/content/paul-vi/en/apost_constitutions/documents/hf_pvi_apc_19660217_paenitemini.html

Schillebeeckx, Edward. *Menschen: Eine Geschichte von Gott*. Freiburg, Germany: Herder, 1990.

Singer, Peter. "Fish: The Forgotten Victims on our Plate." *The Guardian*, September 14, 2010. www.theguardian.com/commentisfree/cif-green/2010/sep/14/fish-forgotten-victims.

Tertullian. "Homily on Baptism." Edited with an Introduction, Translation and Commentary by Ernest Evans. www.tertullian.org/articles/evans_bapt/evans_bapt_index.htm.

The Didache. "The Lord's Teaching to the Heathen by the Twelve Apostles." Chapter 8. www.thedidache.com.

Time. "Religion: Whale Meat on Friday." *Time*. November 26, 1951. http://content.time.com/time/magazine/article/0,9171,821907,00.html.

United States Conference of Catholic Bishops. "Pastoral Statement on Penance and Abstinence." November 18, 1966. www.usccb.org/prayer-and-worship/liturgical-year/lent/us-bishops-pastoral-statement-on-penance-andabstinence.cfm.

United States Conference of Catholic Bishops. "Questions and Answers about Lent and Lenten Practices." www.usccb.org/prayer-and-worship/liturgical-year/lent/questions-and-answers-about-lent.cfm.

Vivekjivandas, Sadhu. *Baghwan Swaminarayan: Life and Work*. Gujarat, India: Swaminarayan Aksharpith, 2005.

PART II

The harms or cruelty involved in institutionalized killing

2.1

"THE COST OF CRUELTY"

Henry Bergh and the abattoirs

Robyn Hederman

In May 1866, Henry Bergh patrolled the streets of New York City looking for violations of his new anti-cruelty law. Mr. Manz, a butcher from Brooklyn, carried a load of calves in his wagon; their legs were bound with cords, and their heads hung over the edges of the cart. One calf's head jolted against a sharp stick that threatened to gauge its eye from the socket. Manz fled when he saw Bergh. Mr. Bergh, in his top hat with his coat tails flying, chased the butcher from Broadway to the Williamsburg Ferry. Catching up with the butcher before he boarded the ferry boat, Bergh brought him before Justice Dowling of the Tombs Court in New York City. After a brief hearing, Mr. Manz was fined $10.00.[1] This was the first conviction under Bergh's new anti-cruelty law.

Henry Bergh galvanized the nineteenth-century animal protection movement. Yet, his crusade to prevent cruelty to animals destined to become food is unrecognized today. Bergh was not a vegetarian, but he was repulsed by how animals were transported, housed, and slaughtered. Bergh exposed the slaughterhouses to the public, while he demonstrated the link between animal cruelty and public health. By examining Berghs' campaigns against the butchers and the swill milk men, we gain insight into the city's changing social, political, economic, and legal system in the years after the American Civil War.[2] His battles also reveal New Yorkers' anxiety about disease, poverty, adulterated food, and sanitary issues affecting the nineteenth-century American city.

American cities expanded after the end of the Civil War. Tenements proliferated, and middle-class New Yorkers feared the influx of immigrants arriving in the city. According to Charles E. Rosenberg in the *Cholera Years*, New Yorkers' growing awareness of the physical dangers of urban living gave birth to the public health movement.[3] In February 1866, the Metropolitan Board of Health was formed in anticipation of New York City's third major cholera epidemic since 1832. The Board was responsible for cleaning the city streets to subdue disease. As noted by Rosenberg, contemporary observers surmised cholera would not be kind to a city "that harbored such filth-and in which pigs still helped to clean the streets."[4]

In this environment, Bergh sought to form the first animal protection society in the United States. New York City residents supported the creation of a private organization to address animal-related problems, believing animals were a factor in the city's sanitary issues.[5] New Yorkers filed complaints with the newly created Board of Health after witnessing screeching cattle prodded on route to the abattoirs. Entrails and other body parts enveloped the streets surrounding slaughter houses and related industries.[6] In August 1866, in response to numerous complaints, the Board of Health ordered the slaughter of animals to be conducted at 110 Street—considered to be outside the city limits.[7]

Yet, beyond the sanitary concerns, New Yorkers were repelled by pitiless scenes they observed every day. In 1864, a *New York Times* editorial wrote that "in this *well governed* City, a dying horse has lain since yesterday morning." The author laments "Why is there no law to punish the owner of the animal? Will someone answer these questions for the sake of Humanity?"[8]

America had not enacted any state or federal animal protection laws at the outset of the nineteenth century, and animal cruelty actions arose from common law.[9] Nineteenth-century legal commentator Joel Prentiss Bishop wrote: "[M]an has always held in subjection the lower animals, to be used or destroyed at will, for his advantage or pleasure." He continued: "Therefore the common law recognizes as indictable no wrong, and punishes no act of cruelty, which they may suffer, however wanton or unnecessary."[10] Common law did not penalize an "owner" from abusing his own animal unless he did the act in public and disturbed the public.[11] Animal cruelty cases were prosecuted under alternative legal theories such as destruction of property, malicious mischief, public nuisance, and commercial deceit, while the animal's pain was not an essential element of these offenses.[12]

In 1829, New York enacted its first anti-cruelty statute, which protected horses, cattle, sheep, and other commercially valuable animals. Although the statute prohibited an "owner" from "maliciously and cruelly" beating or torturing his own animal, any other criminal acts were solely proscribed when these actions were committed against an animal "belonging to another."[13]

Henry Bergh and the ASPCA

New Yorkers expressed their frustration with the 1829 anti-cruelty statute. In 1866, a *New York Times* editorial questioned why the United States had "no society to watch for and procure the punishment of such brutalities." The author noted that "[t]he formation of societies for the prevention of cruelties to animals is not new in the Old World, and should long ago have been effected in the new."[14] In the "Quality of Mercy," Bernard Oreste Unti concluded that the public demand for "action on a range of public matters that involved animals directly or indirectly gave the ASPCA far-reaching influence on the affairs of New York City."[15]

On April 10, 1866, an act of incorporation was passed, and a charter was granted to the American Society for the Prevention of Cruelty to Animals. Bergh was elected as its first president and remained in that position until 1888.[16] The charter granted the Society quasi-police powers in New York State. The ASPCA was given the power to make arrests and enforce all animal protection laws with the aid of the New York City Police Department.[17]

The members of the Society were interested in prosecuting offenders, but they also wished to engage in outreach and education. The Society was to enforce all laws enacted for the protection of animals; "and to secure by lawful means the arrest, conviction, and punishment of all persons violating such laws." The Society further sought "to instruct the people to be kind to animals by the dissemination of humane literature and other effective methods."[18]

Bergh submitted a bill to the New York Legislature entitled "An Act Better to Prevent Cruelty to Animals." The legislation, which replaced the 1829 law, was passed on April 16, 1866. "Ownership" was no longer a defense to animal cruelty, and perpetrators could be prosecuted for negligent acts. The statute still required the prosecutor to prove that the perpetrator acted "maliciously"; this legal element hindered Bergh's early prosecutions.[19] The abandonment of animals was prohibited to prevent the practice of leaving work animals in the street to die when they were no longer useful.[20]

Bergh strove to interest the public in the prevention of animal cruelty, once stating to a colleague "[m]y practice, and recommendation is to keep agitating, and keep continually in the newspapers with our cause."[21] Bergh, nicknamed "The Great Meddler," wrote: "[B]efore undertaking this labor, I took a careful survey of all the consequences to me personally—and I recognized the fact that I should be much abused, and ridiculed, and hence it was necessary for me to forget myself completely."[22]

Abattoirs and transport of animals

Bergh used the media to inform the public about the link between animal cruelty and adulterated meat. He asserted farmed animals should be protected under the animal-cruelty statute. By prosecuting Mr. Manz in 1866, Bergh focused on the brutal methods of transporting farmed animals. Some 15 to 20 live calves were layered on top of each other and transported by wagon. They were unable to move because their legs were tied. The calves were mangled by the time they made it to their destination. The Board of Health prohibited this practice in August 1866, after a public outcry.[23]

Bergh sought to protect all creatures under the anti-cruelty law. Shortly after the Society was established, Bergh arrested butchers for plucking and boiling live chickens. At their criminal trial, Bergh testified that when he arrived at the establishment, he "heard tremendous screaming and screeching of fowls." One bird screamed as two men twisted its wings and picked feathers from its breast. Another man "would pinch the fowl's neck, or stick a blade in it, or throw it into a barrel."

Bergh testified that the fowl were alive during this process. The birds "had not been stabbed or pinched before this tearing out the feathers and bending of the wings back …" Nevertheless the judge acquitted the defendants after finding that the birds had been stabbed in the head before being plucked. The judge determined Bergh failed to prove willful or intentional cruelty.[24]

In a landmark case, Bergh sought to prosecute the captain and several members of the crew for their treatment of live turtles being shipped from Florida to New York. When Bergh entered the ship, he observed the turtles splayed on their backs; their flippers were pierced and tied together. The turtles had remained in this position for three weeks without food or water. Bergh called Louis Agassiz, a professor of zoology and geology at Harvard University, as an expert witness at the trial. Professor Agassiz testified that turtles experience extreme pressure on their organs when they are placed on their backs for an extended time. This pressure can result in their death. In a published letter, Professor Agassiz wrote: "I need not tell you that men had always have excuses enough to justify their wrongdoings." But, after a ten-day trial, the judge acquitted the defendants finding that the turtles were unable to suffer.[25]

The media ridiculed Bergh for prosecuting the turtle case. Yet, in his book *Humane Leaders of the United States*, Sydney Coleman concluded that "[t]he final outcome of the turtle case was to greatly increase the number of supporters and friends of the new society."[26] Subsequently, the ship captain brought a civil suit against Bergh for malicious prosecution; however, the suit was dismissed. Bergh emphasized:

> that so intelligent a judge, has ruled, that there is no evidence of malice, for a citizen witnessing an act of cruelty to a dumb animal—to cause the arrest of the offender. Otherwise, persons disposed to protect the brute, might be deterred by fear of prosecution.[27]

Bergh was not satisfied with the limitations of the 1866 statute, and he drafted new legislation. On April 12, 1867, the New York Legislature passed Bergh's statute, which served as an example for anti-cruelty statutes throughout the country.[28] The 1867 statute expanded the definition of cruelty and pertained to "any living creature." The prosecutor no longer had to prove the malicious intent of the actor.[29] In their article "The Development of Anti-Cruelty Laws," David Favre and Vivien Tsang found that this legislation changed the focus "from the mindset of the individual to objective evidence of what happened to the animal."[30] The statute provided full enforcement powers to the Society. Now, Bergh and his agents had the sole power to arrest violators of the law.

Bergh continued to protect farmed animals under his new animal-cruelty statute. Bergh investigated the hog slaughterhouse of Davis, Atwood, and Crane located on West Thirty-Ninth Street. Bergh observed the butchers break the hogs' legs by hoisting them 40 feet into the air, and plunging them into boiling water while still alive. The workers had not first rendered the hogs insensible to pain. Bergh arrested several employees and ordered the managers to stop the practice. In response, the business brought an injunction restraining the Society from interfering in their operations. In 1878, the case was decided by New York State's highest court in *Davis vs. American Society for the Prevention of Cruelty to Animals*. Here, the Court of Appeals defined the parameters of the Society's police powers by deciding that pursuant to statute, the Society had the right to enter a slaughterhouse to prevent unnecessary cruelty, and had the authority to arrest offenders of the law.[31]

Bergh's early focus on small-scale butchers evolved into a clash with the large factory-style abattoirs springing up in New York and New Jersey. In October 1866, a large-scale slaughter facility was opened in Communipaw, New Jersey, by the New Jersey Yard and Market Company. The public initially believed this facility would decrease the spread of disease from the cattle.[32] The media praised the efficiency and productivity of the new abattoir, while commending its "humane method" of slaughter.[33] The Board of Health claimed that these facilities would "conduce to the health and well-being of the citizens of New York, and remove the greatest sources of disease in the city." The transport and slaughter of cattle now took place outside the sight of the consumer.[34]

Bergh described a different scenario when he visited Communipaw. He observed a "nauseating spectacle," and "a crime in its most odious aspect." He depicted cattle with large ulcers, making them "unfit for human consumption." Bergh pointed out the ulcerated meat to the manager who "shifted the blame off on the deceased animal" and claimed that the "sores upon the flesh and its general emaciated appearance were the results of its own bad behavior."[35]

In 1869, residents living near Communipaw filed injunctions seeking legal protection from the nuisance caused by the facility's "poisonous gasses." The residents further claimed that the cruel conditions at the abattoir constituted a public hazard. An article in the *Times*, dated August 10, 1869 expressed horror at the "sickening sight" of 250 cattle offered for sale at the Communipaw Cattle Yard. These cattle were transported from Florida having received "almost no food or water since they left their native shore." These "miserable objects" were in the "last agonies of starvation and utterly unable to rise." The newspaper emphasized that these cattle were being offered as food for "a civilized and refined people."[36] Yet, according to historian Bernard Unti, as late as 1876, few reforms took place at Communipaw.[37]

Bergh pushed for legislation regulating the railroad transportation of farmed animals. Animals were transported by trains from the West to New York for over 80 hours without food or water. In 1867, a bill was passed limiting the time that animals could be carried on New York railroads without unloading for rest, food, and water. Railroad and farmed animal companies included numerous exemptions to the application of the bill. Nevertheless, the first federal animal protection statute, the Twenty-Eight Hour Law, was passed in 1873.[38] Congress substantially weakened the proposed law by not including criminal sanctions or any substantial enforcement provisions. But, the humane and public health concerns escalated after the cattle epidemics of Chicago's meatpacking industry. In Chicago, the Illinois

Society for the Prevention of Cruelty to Animals was created as a response to the public furor.[39]

Swill milk

Bergh continued to bring the concept of animal-related health issues to the public's attention by focusing on the "swill milk" operators who sold diseased milk produced by cows fed upon distillery slops and garbage. By exposing the unsanitary conditions that produced swill milk and diseased meat, Bergh also revealed the industry's brutal treatment of these animals.

Cows were housed in filthy conditions in Brooklyn. These ulcerated and "stump-tailed" animals were milked up to their death.[40] As many as one thousand animals were chained together in unsanitary stables. Each cow occupied a stall about two and a half to three feet wide. In 1858, Frank Leslie of *Frank Leslie's Illustrated Newspaper* conducted an investigation of the swill milk facilities after receiving complaints from a Brooklyn resident. Leslie wrote a series of articles exposing the plight of the mistreated cows. Drawings of sick cows, held up by slings and milked until death, inflamed the public. Nevertheless Brooklyn officials refused to take any action.[41]

Bergh entered the campaign in 1867. He arrested Morris Phelan, a Brooklyn swill milk operator, and charged him with animal cruelty. A grand jury indictment was issued against Phelan, but the Brooklyn District Attorney delayed the case and refused to prosecute him.[42] The Brooklyn Board of Health released a statement insisting the dairies "are cleaned four times a day, and are in better condition than hundreds of houses in the City of New York in which human beings are compelled to live."[43]

New Yorkers were alarmed about what they were putting in their coffee every morning, and the Board of Health was finally shamed into taking action. In 1873, New York City prohibited the sale of swill milk, or any other adulterated milk.[44] But, in spite of an 1874 public health report claiming that swill milk was the chief cause of infant and child mortality, the Brooklyn authorities allowed the dealers to operate. In a lecture given to the Farmer's Club in 1878, Bergh concluded that although a man is sent to prison "for uttering a spurious coin or bank note," no punishment is imposed where "health is destroyed, lingering and distressing diseases are induced, and life itself destroyed with impunity, to say nothing of the torments inflicted on dumb animals."[45] As late as the 1890s, the Society continued raiding swill milk facilities and making arrests. But, swill milk related illnesses and deaths continued until the advent of refrigeration and sterilization. In his article, "The Cost of Cruelty," Bergh addressed the link between animal cruelty and public health. He asked whether "the alarming increase of cancers, tumors, and scrofulous diseases [can] be traced to this disregard of the simplest principles of humanity."[46]

Conclusion

Early in his career, Bergh became disturbed by the mistreatment of cattle in a New York City drove-yard. He wrote to the "owners" protesting their unconscionable treatment of cattle who "stood huddled together unsheltered from the sleet and wintery blasts." He reproached the disregard of the "owners" for such animals, an attitude that persists today:

> It seems to be regarded that a creature which has been condemned to death, in order to nourish and sustain human life, loses at once all claim upon the merciful consideration of its slayers, and is treated as though it was so much a stone, iron or other sensible merchandise.

Today, the food industry and agribusiness corporations continue to conceal repugnant food production practices from consumers. Several U.S. states have passed legislation criminalizing the methods used by animal activists and food safety advocates to expose the realities of factory farming. Perhaps one approach to our own ethical problems of eating animals lies in reexamining Bergh's tactics. By revealing the unsanitary conditions that produced diseased meat and milk, Bergh also unmasked the industry's brutal treatment of these animals. Analyzing Bergh's campaigns may reveal how we can now transform our attitudes towards sentient beings condemned to be processed into food.

Notes

1 Ten dollars in 1866 is worth approximately 154 dollars in 2013. "Seven Ways to Compute the Relative Value of a U.S. Dollar Amount—1774 to Present," Measuring Worth, accessed September 27, 2016, www.measuringworth.com/calculators/uscompare/.
2 Swill milk was produced from cows fed on distillery waste.
3 Charles E. Rosenberg, *The Cholera Years: The United States in 1832, 1849, and 1866* (Chicago, IL: University of Chicago Press, 1962), 175–177.
4 Rosenberg, *The Cholera Years*, 191.
5 Bernard Oreste Unti, "The Quality of Mercy: Organized Animal Protection in the United States, 1866–1930" (Ph.D. diss., American University, 2002), 86–87; Rosenberg, *The Cholera Years*, 191, 202–3, 232.
6 "The Code of Health," *New York Times*, May 20, 1866, 4; "Kitchen Refuse," *New York Times*, December 18, 1870, 8.
7 Unti, "The Quality of Mercy," 91.
8 "Inhumane Cruelty," *New York Times*, July 29, 1864, 2; "Cruelty to Animals: Letters from the People," *New York Times*, January 10, 1864, 5.
9 For a discussion of common law in the United States see, Herbert Pope, "The English Common Law in the United States," *Harvard Law Review* 24, no. 1 (November 1910): 6–30. See also, *Western Union Telegraph Company v. Call Publishing Company* 181 U.S. 92 (1901). See, David Favre and Vivien Tsang, "The Development of Anti-Cruelty Laws During the 1800s," *Detroit College of Law Review* 1 (Spring 1993):1–35, citing William Blackstone, *Commentaries on the Laws of England* (Oxford, England: Clarendon Press, 1765).
10 Joel Prentiss Bishop, *Commentaries on the Criminal Law*, Sixth Edition, revised and greatly enlarged, Volume I (Boston, MA: Little, Brown, and Company, 1877), 335.
11 See, *People v. Smith*, 5 Cow. 258 (N.Y. 1825) where a man was convicted of killing another man's cow. Here citing *Republica v Teischer*, the court found that the offence stated in the indictment was a proper subject of criminal prosecution finding such an action "dangerous to society." The court described this action "an evil example of the most pernicious tendency." *People v. Smith*, at 2.
12 Favre and Tsang, "Anti-Cruelty Laws," 5–6. In *State v. Briggs*, 1 Aik. 226 (1826), a man in Vermont was indicted for malicious mischief by "wounding and torturing a living

animal, not with force and arms, but with wicked and malicious motives and intention..." In 1788, a Pennsylvania man was convicted upon an indictment for "maliciously, willfully, and wickedly killing a horse." On appeal, Chief Justice M'Kean upheld the indictment claiming that this action constituted a "public wrong." *Respublica v Teischer*, 1 U.S. 335 (1788).

13 NY. Rev. Tit. 6, Sec 26 (1829) provides "Every person who shall maliciously kill, maim or wound any horse, ox, or other cattle, or any sheep, belonging to another, or shall maliciously and cruelly beat or torture such animals, whether belonging to himself or another, shall upon conviction, be adjudged guilty of a misdemeanor."

14 "Work for the Idle," *New York Times*, January 10, 1866, 4.

15 Unti, "The Quality of Mercy," 76.

16 "American Society for the Prevention of Cruelty to Animals," *New York Times*, April 24, 1866, 4.

17 Roswell Cheney McCrea, *The Humane Movement: A Descriptive Survey, Prepared on the Henry Bergh Foundation for the Promotion of Humane Education in Columbia University* (New York, NY: The Columbia University Press, 1910), Appendix VI, Charter of the American Society for the Prevention of Cruelty to Animals, 200.

18 McCrea, *The Humane Movement*, Appendix VII, By-Laws of The American Society for the Prevention of Cruelty to Animals, Ch. II pg. 203.

19 Favre and Tsang, "The Development of Anti-Cruelty Laws," 14–15. There was no statutory definition of the term "maliciously." See, *State v. Avery*, 44 N.H. 392 (1862), which demonstrated that "as long as some excuse could be presented to the court, it was difficult to prove malice."

20 N.Y. Rev. Stat. Ch. Sec. 682.2 (1866); Animal Welfare Institute (AWI), *Animals and Their Legal Rights* (Washington, DC: AWI, 1990), 5.

21 Bernard Oreste Unti, "The Quality of Mercy," 107.

22 Marion S. Lane and Stephen L. Zawistowski, *Heritage of Care: The American Society for the Prevention of Cruelty to Animals* (Westport, CT.: Praeger Publishers, 2008), 11.

23 "Cruelty to Animals—First Case of Punishment under the New York Law," *New York Times* April 13, 1866, 8; "Cruelty to Animals," *New York Tribune*, August 3, 1866, 5; "Cruelty to a Calf," *New York Times*, October 7, 1866, 6; "Cattle Driving in the Streets," *Frank Leslie's Illustrated Newspaper*, April 28, 1866, 81; "Society for the Prevention of Cruelty to Animals," *Frank Leslie's Illustrated Newspaper*, May 12, 1866.

24 "The Chicken-Plucking Case," *New York Times*, June 8, 1866, 2; "Chicken-Butchers and the Anti-Cruelty Society," *New York Times*, June 9, 1866, 4; Lane and Zawistowsky, *Heritage of Care*, 22.

25 "The Turtle Case—Letter from Professor Agassiz," *New York Tribune*, October 2, 1866; "Professor Agassiz and Cruelty to Animals," *New York Times*, October 2, 1866, 4.

26 Sydney Coleman, *Humane Society Leaders in America, with a Sketch of the Early History of the Humane Movement in England* (Albany, NY: American Humane Association, 1924), p. 43

27 Unti, "The Quality of Mercy," 110; "The Reversed Turtle," *New York Times*, June 7, 1876, 4; "Score one for Bergh," *New York Herald*, June 12, 1876, 6.

28 AWI, *Animals and Their Legal Rights*, 5–7.

29 N.Y. REV. STAT. Sec 375. 2–9. See, Favre and Tsang, "Anti-Cruelty," 16; Appendix A: The 1867 New York Anti-Cruelty Law.

30 Favre and Tsang, 16.

31 "Hog-Slaughtering: Application to Restrain Mr. Bergh from Interfering with the Trade—How the Animals are Tortured," *New York Times*, March 4, 1874; *Davis vs. American Society for the Prevention of Cruelty to Animals*, 75 N.Y. 362 (1878). The Court determined that pursuant to section 8 of the 1867 statute, agents of the ASPCA were designated by the sheriff to arrest offenders of the law. The Court described Bergh as "clothed with authority to execute the law." The Court of Appeals affirmed the Society's right to enter a slaughterhouse to prevent unnecessary cruelty and found that it was "therefore unnecessary to determine, in this case, whether plaintiffs were, as a matter

of fact, guilty of violating the law." "Another Victory for Mr. Bergh," *New York Times*, June 4, 1874, 5.

32 "Abattoirs," *Frank Leslie's Illustrated Newspaper*, February 16, 1867, 81; "The Communipaw Abattoirs," *New York Times*, June 29, 1866; "Transport of Cattle and other Animals to Market," *New York Times*, September 28, 1868, 5; "Cattle Disease: Two Pamphlets on the Disease of Cattle and Their Causes—How Cattle Suffer the Transportation-Patent Truck and Water Troughs," *New York Times*, October 16, 1868, 5.

33 "The Communipaw Abattoir: A Description of the Mammoth Slaughterhouse—How the Animals are Prepared for Market," *The New York Times*, November 12, 1868, 8.

34 Unti, "The Quality of Mercy," 113; "The Mode of Slaughtering in the New Abattoir in Communipaw, New Jersey," *Frank Leslie's Illustrated Newspaper*, February 16, 1867, 344.

35 "Mr. Bergh's Observation at the Communipaw Abattoir," *New York Times*, December 16, 1868, 2. Bergh concluded his letter stating: "Thus ended an inquiry, which, however much I may be abused for making, was nevertheless performed at a cost to my sensibilities, beyond my powers of expression, and strictly in the interests of the public, and those inferior creatures, whose claims on man's mercy we are striving—not wisely perhaps for our own peace to enforce." "Cattle Disease," *New York Times,* August 12, 1868, 8.

36 "The Communipaw Abattoir—Application for an Injunction," *New York Times*, August 14, 1869, 2; "Injunction Against the Communipaw Abattoir," *New York Times*, August 15, 1869, 8; "Unsound Beef on Sale," *New York Times*, August 10, 1869, 2.

37 Unti, "The Quality of Mercy," 115.

38 The Twenty-Eight Hour Law was revised in 1906. U.S. Revised Statutes, 1906. Ch. 3, 594. To see this statute, see McCrea, *The Humane Society*, 226. Present law can be found at 49 U.S.C. § 80502.

39 For a further discussion of George T. Angel who traveled to Chicago in 1870 to assist in the founding of the Society, see, Coleman, *Humane Society Leaders in America*, 89–114.

40 Cows fed on distillery waste sickened. Teeth, hooves, and tails broke from their bodies.

41 Alvin F. Harlow, *Henry Bergh: Founder of the ASPCA* (New York, NY: Julian Messanger, Inc., 1957), 84–86; "The Startling Exposure of the Milk Trade of New York and Brooklyn," *Frank Leslie's Illustrated Newspaper*, May 8, 1858; "Our Exposure of the Milk Trade of New York and Brooklyn," *Frank Leslie's Illustrated Newspaper*, May 15, 1858. See Edward Buffet, "Bergh's War on Vested Cruelty," Vol. IV, Unpublished Manuscript. (New York, NY: ASPCA Archives, Undated).

42 Buffet, "Bergh's War on Vested Cruelty," Vol. IV; Zulma Steele, *Angel in Top Hat* (New York, NY: Harpers and Brothers Publishers, 1942), 111; Unti, "The Quality of Mercy," 118–122; "Cruelty to Animals," *New York Herald*, March 2, 1867; "Cruelty to Animals," *New York Times,* March 8, 1867, 2; "The Cow-Stable Business," *New York Times*, March 15, 1867, 8; "The Swill Milk Trade, Mr. Bergh on the Condition of the Swill Milk Manufactures of Brooklyn," *New York Times*, January 14, 1869.

43 Harlow, *Henry Bergh*, 87.

44 Steele, *Angel in Top Hat*, 119.

45 "A Lecture by Henry Bergh: He tells the Farmers' Club What He Knows About Milk and Swill Milk," *New York Times*, March 27, 1878, 10; "Henry Bergh and the Brooklyn Board of Health," *New York Times*, April 27, 1871, 8. A Code of Health Ordinances was passed by the Board of Health prohibiting, among other so called nuisances, swill milk establishments, however, enforcement was ineffectual. See, "The Code of Health—Practical Requirements of the Sanitary Law," *New York Times*, May 20, 1866.

46 Henry Bergh, "The Cost of Cruelty," *The North American Review* 133, No. 296 (July, 1881): 75–81, 79.

Bibliography

Animal Welfare Institute (AWI), *Animals and Their Legal Rights*. Washington, DC: AWI, 1990.

Bergh, Henry, "The Cost of Cruelty." *The North American Review* (July1881): 75–81.
Bishop, Joel Prentiss. *Commentaries on the Criminal Law*. Sixth Edition. Revised and Greatly Enlarged. Volume I. Boston, MA: Little, Brown, and Company, 1877.
Blackstone, William. *Commentaries on the Laws of England*. Oxford, England: Clarendon Press, 1765.
Buffet, Edward P. "Bergh's War on Vested Cruelty." Unpublished Manuscript. New York, NY: ASPCA Archives, Undated.
Coleman, Sydney, *Humane Society Leaders in America, with a Sketch of the Early History of the Humane Movement in England*. Albany, NY: American Humane Association, 1924.
Davis vs. American Society for the Prevention of Cruelty to Animals, 75 N.Y. 362(1878).
Favre, David and Vivien Tsang. "The Development of Anti-Cruelty Laws during the 1800s." *Detroit Law Review* 1(Spring 1993): 1–35.
Frank Leslie's Illustrated Newspaper. "Our Exposure of the Milk Trade of New York and Brooklyn." May 15, 1858.
Frank Leslie's Illustrated Newspapers. "The Startling Exposure of the Milk Trade of New York and Brooklyn." May 8, 1858.
Frank Leslie's Illustrated Newspaper. "Cattle Driving in the Streets." April 28, 1866.
Frank Leslie's Illustrated Newspaper. "Society for the Prevention of Cruelty to Animals." May 12, 1866.
Frank Leslie's Illustrated Newspaper. "Abattoirs." February 16, 1867.
Frank Leslie's Illustrated Newspaper. "The Mode of Slaughtering in the New Abattoir in Communipaw, New Jersey." February 16, 1867.
Harlow, Alvin F. *Henry Bergh: Founder of the ASPCA*. New York, NY: Julian Messanger, Inc., 1957.
Lane, Marion S. and Stephen L. Zawistowsky. *Heritage of Care: The American Society for the Prevention of Cruelty to Animals*. Westport, CT: Praeger Publishers, 2008.
McCrea, Roswell Cheney. *The Humane Movement: A Descriptive Survey, Prepared on the Henry Bergh Foundation for the Promotion of Humane Education in Columbia University*. New York, NY: The Columbia University Press, 1910.
Measuring Worth. "Seven Ways to Compute the Relative Value of a U.S. Dollar Amount – 1774 to Present." Accessed September 27, 2016. www.measuringworth.com/calculators/uscompare/.
New York Herald. "Cruelty to Animals." March 2, 1867.
New York Times. "Cruelty to Animals-First Case of Punishment under New York Law." April 13, 1866.
New York Times. "The Code of Health-Practical Requirements of the Sanitary Law." May 20, 1866.
New York Times. "Chicken-Butchers and the Anti-Cruelty Society." June 9, 1866.
New York Times. "The Communipaw Abattoirs." June 29, 1866.
New York Times. "Professor Agassiz and Cruelty to Animals." October 2, 1866.
New York Times. "Cruelty to a Calf." October 7, 1866.
New York Times. "Cruelty to Animals." March 8, 1867.
New York Times. "The Cow-Stable Business." March 15, 1867.
New York Times. "American Society for the Prevention of Cruelty to Animals—Letter from Mr. Bergh." March 22, 1868.
New York Times. "Cattle Disease." August 12, 1868.
New York Times. "Transport of Cattle and other Animals to Market." September 28, 1868.
New York Times. "Cattle Disease: Two Pamphlets on the Disease of Cattle and Their Causes-How Cattle suffer the Transportation-Patent Truck and Water Troughs." October 16, 1868.

New York Times. "The Communipaw Abattoir: A Description of the Mammoth Slaughter-house-How the Animals are Prepared for Market." November 12, 1868.
New York Times. "Mr. Bergh's Observation at the Communipaw Abattoir." December 16, 1868.
New York Times. "The Swill Milk Trade, Mr. Bergh on the Condition of the Swill Milk Manufactures of Brooklyn." January 14, 1869.
New York Times. "Unsound Beef on Sale." August 10, 1869.
New York Times. "The Communipaw Abattoir-Application for an Injunction." August 14, 1869.
New York Times. "Injunction Against the Communipaw Abattoir." August 15, 1869.
New York Times. "Henry Bergh and the Brooklyn Board of Health." April 27, 1871.
New York Times. "Another Victory for Mr. Bergh." June 4, 1874.
New York Times. "The Reversed Turtle." June 7, 1876.
New York Times. "A Lecture by Henry Bergh: He tells the Farmers' Club What He Knows About Milk and Swill Milk." March 27, 1878.
New York Tribune. "Cruelty to Animals." August 3, 1866.
New York Tribune. "Turtle Case-Letter from Professor Agassiz." October 2, 1866.
New York Tribune. "Score One for Bergh." June 12, 1876.
N.Y. Rev. Stat. Ch. Sec. 682.2 (1866).
People v. Smith, 5 Cow. 258 (N.Y. 1825).
Pope, Herbert. "The English Common Law in the United States." *Harvard Law Review*24, 1 (November1910): 6–30.
Respublica v Teischer, 1 U.S. 335 (1788).
Rosenberg, Charles E. *The Cholera Years: The United States in 1832, 1849, and 1866*. Chicago, IL: University of Chicago Press, 1962.
State v. Avery, 44 N.H. 392 (1862).
State v. Briggs, 1 Aik. 226 (1826).
Steele, Zulma. *Angel in Top Hat*. New York, NY: Harpers and Brothers Publishers, 1942.
Unti, Bernard Oreste. "The Quality of Mercy: Organized Animal Protection in the United States, 1866–1930." Ph.D. diss., American University, 2002.
U.S. Revised Statutes, 1906. Ch. 3, 594.
Western Union Telegraph Company v. Call Publishing Company, 181 U.S. 92 (1901).

2.2

"ALL CREATION GROANS"

The lives of factory farmed animals in the United States

Lucille Claire Thibodeau

Today, more animals suffer at human hands than at any other time in history. It is therefore not surprising that an intense and controversial debate is taking place over the status of the 60+ billion animals raised and slaughtered for food worldwide every year. To keep up with the high demand for meat, industrialized nations employ modern processes generally referred to as "factory farming." This article focuses on factory farming in the United States because the United States inaugurated this approach to farming, because factory farming is more highly sophisticated here than elsewhere, and because the government agency overseeing it, the Department of Agriculture (USDA), publishes abundant readily available statistics that reveal the astonishing scale of factory farming in this country.[1]

The debate over factory farming is often "complicated and contentious,"[2] with the deepest point of contention arising over the nature, degree, and duration of suffering food animals undergo. "In their numbers and in the duration and depth of the cruelty inflicted upon them," writes Allan Kornberg, MD, former Executive Director of Farm Sanctuary in a 2012 Farm Sanctuary brochure, "factory-farm animals are the most widely abused and most suffering of all creatures on our planet."

Raising the specter of animal suffering inevitably raises the question of animal consciousness and sentience. Jeremy Bentham, the eighteenth-century founder of utilitarianism, focused on sentience as the source of animals' entitlement to equal consideration of interests. Simply put, an interest is anything a being may want to do and strive for. Human and non-human animals have many interests in common, such as living, eating, drinking, sleeping, reproducing, being comfortable, avoiding pain and suffering, having their needs met, and the like. The moral principle of equal consideration of interests holds that one should both include all affected interests when calculating the rightness of an action and weigh those interests equally.[3] In *An Introduction to the Principles of Morals and Legislation*, Bentham, in a long portentous footnote, asserted that "the question is not, Can they *reason*? Nor, Can they *talk*? But, Can they *suffer*?"[4] In one sentence, Bentham challenges all those who would deny rights to animals based on reason, language, or any other morally irrelevant factor. Sentience, the ability to experience pleasure and pain, is the only morally significant factor, because that capacity is "*a prerequisite for having interests at all,*—at the very least an interest in not suffering."[5] Bentham's footnote was a battle cry: Our moral responsibilities to animals are neither different from nor less important than our responsibilities to human beings. His cry, silent for over two hundred years, was echoed by Princeton University professor of bioethics Peter Singer in the book that has become the definitive classic of the animal movement.

Contemporary scientific evidence—supplementing what most people have understood, with plain common sense, since pre-history—shows that mammals and birds, among other species, do possess sentience and self-awareness.[6] The arguments of Bentham, Singer, and Linzey, among others, are supported by a 2008 report published jointly by the Pew Charitable Trusts and the Johns Hopkins Bloomberg School of Public Health. In "Putting Meat on the Table: Industrial Farm Animal Production in America," the Commission found that:

> the present system of producing food animals in the United States is not sustainable and presents an unacceptable level of risk to public health and damage to the environment, as well as unnecessary harm to the animals we raise for food.[7]

The Commission recommends the total phase-out of intensive confinement practices and asserts unequivocally that "the most intensive confinement systems ... constitute inhumane treatment" and thus fall short of current ethical and societal standards.[8]

Sixty-plus years ago, the American countryside was studded with thousands of small, family-owned farms where one crop was grown and several species of farmed animals were raised. The growing demand for meat by a booming post-World War II population led to the gradual replacement of these small farms by large facilities housing 2,000 animals or more. These operations were supported (and still are) by huge government subsidies for certain crops—corn especially—that were found to be cost-effective in feeding large numbers of farmed animals. The aim of these large operations was to grow more animals more quickly in less space by using cost-effective food and by replacing human labor with technology as much as possible. The industry devised a series of discrete production processes connected down to the minutest detail by economic transactions.[9] When the economic logarithms kicked in, farmers and consumers found themselves at the extreme ends of the spectrum of animal "production." "The factory farm ... succeeded by divorcing people from their food, eliminating farmers, and ruling agriculture by corporate fiat."[10]

When profit is the only duty and efficient the only law, ethical standards are bound to fall short. How they fall short leads us to ask: "What happens to animals in such a system?" Are they able to express behaviors natural to them? Are they given food that their digestive systems are designed to metabolize? Are they subjected to physical alterations in the interest of their not harming one another or the humans who handle them? In other words, what is it like to be an animal living in such a system?[11]

On factory farms, 98 percent of the 9 billion chickens (up from 1.6 billion in 1960) raised and slaughtered in a year in the U.S. are egg-laying hens the vast majority of whom live out their entire two-year lives in battery cages—tiny wire cages in which several hens are packed together, each one allotted the industry standard (set by United Egg Producers) of 67 square inches of space—less than a standard sheet of 8½ × 11-inch paper. Such limited space obviously interferes with a hen's ability to express behaviors natural to her such as stretching her wings, taking short flights, and pecking on the ground or in the grass for food.[12] Furthermore, as a tiny chick, she will have her beak cut off without benefit of anesthetic by a specially designed machine so that as she grows, she will not peck at or cannibalize the other hens in the cage. "During beak trimming, workers ... place each chick's beak into a hot-iron guillotine-type machine, and when it snaps the tip [the front third to front half] of the beak off, the chick's face smokes and the chick struggles."[13] Because a hen's laying cycle is linked to light, egg barns are kept dark most of the time. The hens never experience sunlight, except on the day they are shoved into crates and trucked to the slaughterhouse. Moreover, by taking away light and food for as long as 14 days, producers trigger year-round unnatural but lucrative laying cycles. Artificially forced to produce at this rate, and kept relatively disease-free by antibiotics in her filthy, crowded cage until her immune system fails her, a hen will lay over 300 eggs a year, whereas the norm before factory farming was about one-third of that.

A typical battery cage is 12 by 18 inches—the size of a file drawer—and holds no fewer than five hens. The barns that house the battery cages can be more than 450 feet long and 25 feet high. The air in the barns is dense with dander and dust and the smell of chickens and their ammoniac manure. Some cages are 12 feet long, 4 feet wide, and contain more than 75 hens.[14] Excrement drops into pits below the cages, and piles can grow to be six feet high. After a year and a half to two years—their natural life span could reach at least ten years—, the hens are "spent," their egg production wanes, and they are sent to the slaughterhouse where, hooked by one leg (often broken in the process) to a quickly moving conveyor belt line, their throats are slit by a whirring blade, again without benefit of anesthetic. Government estimates suggest that about four million hens come loose from these lines in a year, and fall into a vat of boiling water meant to remove their feathers, while they are still alive.[15]

Hens receive no veterinary care. They suffer from untreated sores, cysts, infections, and uterine prolapses because care costs more than the bird is worth to the producer. Emaciated, featherless, and covered with the feces that drops from above, many hens die in their cages. Left there to decompose over several months, their bodies get pressed into the wire and stepped on by their cage mates who must fight for, literally, every square inch of space in order to survive. Workers rip dead hens, flattened to an inch, off the bottom of the cage, a practice referred to as "carpet pulling."[16] In a medium-size facility housing about two million hens, thousands of dead and moribund hens are carried off by bulldozers to landfills or incinerators. Such are the living conditions the egg industry considers suitable for these lively, curious, intelligent creatures[17] possessed, ironically, of wings, the universal symbol of freedom. The barns, the battery cages, the de-beaking machines, the pits of excrement, the conveyor belts with their whirring, neck-slitting blades, the vats of boiling water, the hen-filled dumpsters—these are the monstrous, filthy things we have constructed, the hell we have created for hens because we eat eggs at a historically unprecedented rate, we want an unlimited supply of them at all times, and we want them cheap.

What about the male chickens? According to the most recent USDA statistics, every year over 250 million male chicks are "rendered" on egg farms the day they are born because they are of no use. Deemed unfit to be fed because they lay no eggs and their flesh is of poor quality, the chicks are macerated alive in high-speed grinders, the slurry of their tiny carcasses disposed of as trash or fed to other chickens and farmed animals.[18] Chickens selectively bred as "broilers" on other farms, especially for breast meat, are fed growth hormones so as to get up as quickly as possible to a market weight of five pounds in seven weeks or less. (Pre-factory-farmed chickens took twice as long to grow up to three pounds.) These chickens become too heavy to stand; their legs break under their weight, and they topple over onto their chests ("flip-over syndrome"). Some of them starve when they can no longer stand up to reach food or water. Or their organs shut down and their hearts stop beating, unable to cope with their body's unnaturally rapid growth. For them, it's about 35 days from birth to death. For the industry, the equation is simple: If the extra weight on birds who make it to slaughter brings in a greater profit than the loss in dead birds, then it's good for business. The same equation applies to laying hens: Increased total egg output outweighs the number of hens who die due to crowding, stress, and illness. A massive production rate presupposes, and factors in, a high attrition rate. In the war we humans wage on the animals we eat, these little chickens are just collateral damage.[19]

What is life like for cows on factory farms? Dairy cows are kept tethered to a stall and artificially inseminated on what the industry calls a "rape rack" in order that the cow will continue to lactate and provide milk that will be denied to her calf, who will be taken away from her two hours after birth. (The milk denied the calf is sold to us humans who don't need to drink cow's milk. The calf, who does need its mother's milk, gets "milk replacer," a cheap, inferior nutrition.) If the calf is female, like her mother she too will be forcibly impregnated and injected with synthetic bovine growth hormone to increase her milk production. She will give birth to four or five calves in a four-to-five-year period, will develop udder infections, and if she hasn't been literally milked to death, will become a "chopper cow," slaughtered at a fraction of her natural life span. On the kill floor, she will be stunned into unconsciousness by a captive bolt gun (if she is among the fortunate few)[20] and hoisted upside down by one leg over rapidly moving conveyor belt lines where, within roughly 45 minutes, she will be bled, skinned, dismembered, and eviscerated until what was a large living animal is "processed" into several shrink-wrapped packages of hamburgers. If her calf is male, he will either be killed on the spot if he is sickly, or chained by the neck and confined for life (about 15 weeks) in a wooden crate in which he is unable to turn around. He will be fed an iron-deficient diet to make him anemic and turn his flesh pale because consumers prefer the taste and color of "veal" that comes from anemic calves.[21]

Pigs also suffer in ways hard to imagine. For 124 days (her gestation period), a sow is confined in a "gestation crate" in which she cannot turn around or roll over or lie down comfortably.[22] After she gives birth in a "farrowing crate," her piglets are torn away from her just a few days later instead of the natural few weeks. She is then artificially re-inseminated and the cycle repeats itself as many as eight times until she dies or is sent to slaughter.[23] Female piglets, when grown, are put into gestation crates in which, like their mothers, they will live out their two-year lifespan completely immobilized. If they die in their crates before slaughter, they are buried in a "dead hole," a mass grave of thousands of sows, each one in an airtight bag, or "rendered"—ground into feed for other sows.[24] When artificially grown to industry standards (about 250 pounds, half their adult weight), male pigs are slaughtered. "Since uniform size is so important to packers, piglets who don't grow quite fast enough … are quickly weeded out. Picked up by their hind legs, thousands are swung and then bashed headfirst onto the concrete floor. This standard practice used by mega-farm workers is called 'thumping.'"[25] Then their intestines are ground up and fed to their mothers.

In the U.S., there are only two federal protections for farmed animals. The first, "the 28-hour law," originally passed in 1873 when trains were the predominant method of animal transport, stipulates that animals being transported to slaughter or for any other reason must be let out of the transport to be fed, watered, and rested for at least five hours before transport is resumed. This law was enacted not for the sake of the animals but to prevent heavy losses before they could be slaughtered for market.[26] The second law, the Humane Methods of Slaughter Act, originally passed in 1958, stipulates that the animal must be rendered insensible to pain, i.e., unconscious, before being slaughtered.[27] The USDA exempts chickens from the protective provisions of these two laws. Chickens, who outnumber pigs and cattle in the U.S. nine to one (nine billion to one billion), have no protections whatsoever. It needs also to be noted that these two laws cover the animals only on the way to, or at, slaughter, where they may spend only minutes before being dispatched. No federal laws exist to protect any farmed animals from living conditions on the farmed or from the horrific practices to which handlers regularly subject them: Beatings, sexual assault, kicking, bashing their heads in with pickaxes and sledgehammers, and other acts of unspeakably sickening cruelty. Furthermore, farmed animals are specifically exempted from the protections of the Animal Welfare Act. As for state laws, the anti-cruelty statutes of most states exempt "accepted," "common," "customary" practices considered to be "industry standards." Fifty states now have animal anti-cruelty statutes carrying felony penalties, but in only seven states do they apply to farmed animals. Thus, no farmed activity can be deemed cruel by either federal or state standards, no matter how painful or unnecessary it is—as long as enough workers are doing it.[28]

Over the past 60 years, people in the U.S. have had less and less access to farmed animals. When they used to be outdoors much of the time, cows, pigs, and chickens could easily be seen by passersby. But that is no longer the case. Our almost complete lack of exposure to farmed animals makes it much easier for us to push aside questions about how our actions might affect their treatment. Hidden away from public view on large factory farms laid out on thousands of acres that have been bought up by big corporations like Cargill and Tyson and ConAgra, farmed animals today have been removed from visibility and access. Ordinary citizens cannot just stroll onto these factory farms as if they are tourist attractions, open doors, and take a look at what actually goes on. That is called trespassing, and it carries fines or imprisonment.[29] The agriculture industry has received much bad press in recent years due to the release of photos and videos taken by undercover investigators, some of whom apply for jobs as farmhands. These investigations have led to plant closures, numerous recalls regarding food safety issues, citations for environmental and labor condition violations, criminal convictions, and civil litigation.

We know what we know today about factory farms because of the brave actions of undercover investigators—whistleblowers, journalists, animal activists, advocates, protesters, sanctuary workers, rescuers.[30] They record what goes on daily in a system that enables and tolerates cruelty that goes unpunished not because of an absence of law enforcement but because of an absence of actual laws to enforce. They do the work that the federal and state governments are not doing. Over the past decade, advocates have tried to chip away at cruel, unsafe, illegal but common practices by promoting legislation prohibiting some of the worst abuses. In some cases, their efforts have met with success. Unfortunately, reforms don't often reach the states where they are needed most, and where the farm industry and its lobby are most powerful.

Not surprisingly, the push for farmed animal protection has led to a backlash from the agriculture industry. In an effort to stifle criticism, the industry has been aggressively pushing to criminalize undercover investigations on factory farms across the country by introducing bills into state legislatures that would make it a criminal offense for anyone to take photos or make videos of animals being treated cruelly or of conditions that are unsafe for animals and workers. At the moment, seven states have these "Ag-Gag" laws: Utah, Wyoming, North Dakota, Missouri, Kansas, Iowa, and North Carolina. In the past year, 14 other states have tried to put Ag-Gag laws on the books, and all but one have failed. In August 2015, a coalition headed by the Animal Legal Defense Fund (ALDF) argued a case before the U.S. District Court in Idaho. In a historic decision, the court threw out Idaho's law banning whistleblowing and undercover investigations on factory farms, declaring it unconstitutional, violating the First Amendment[31] and the equal protection clause of the Fourteenth Amendment.[32] This was the first time a court declared an Ag-Gag statute in violation of the U.S. Constitution. It was a landmark victory for farmed animals.[33] ALDF has since filed lawsuits against Utah's, Wyoming's, and North Carolina's Ag-Gag statutes, challenging them as unconstitutional.

Rationally speaking, factory farming is so obviously unethical, so plainly wrong in so many ways. But what moves us to action are our feelings, the "moral sentiments" of empathy, regret, grief, and love. To allow ourselves to feel for the plight of farmed animals is to push against a culture and against a food system that trains us to believe that non-humans are not fully subjective beings, and that therefore their lives and sufferings and deaths are not worthy of our consideration. The centuries of stories we have told ourselves about our superiority to and separation from animals—the speciesism[34] to which we have subscribed—these centuries of stories have sanctified our anthropocentric, tyrannical self-interest and created a world of barbaric atrocities, poisoned landscapes, and glaring human rights abuses.

The similarities between how Nazis treated people targeted for extermination and how we treat farmed animals are impossible not to recognize.[35] Nazis kept their victims in filthy, crowded camps, fed them as little as possible or not at all, forced them to huddle together naked and defenseless, and struck them with fists and rifle butts to push them along more quickly into the chutes that led to the gas chambers. The Nazis used intimidation, brute force, and speed in order to minimize the resistance of their victims and to quell the scruples of the exterminators, in just the same way that slaughterhouse supervisors keep the lines moving at top speed by intimidating workers and threatening to fire them. And just as producers send to slaughter every day very young animals who have lived only a tiny portion of their natural lives, the Nazis showed no mercy to children, killing them as brutally as they killed adults.

Polish-born Jewish-American writer Isaac Bashevis Singer, 1978 Nobel laureate and the most compassionate champion of animals in modern literature who lost many members of his own family in the Holocaust, wrote in a short story, "In relation to [animals], all people are Nazis; for the animals it is an eternal Treblinka."[36] The philosopher Theodor Adorno (1903–1969), a German Jew who was forced into exile by the Nazis but returned to Germany after the war to a professorship at Frankfurt University, wrote: "Auschwitz begins wherever someone looks at a slaughterhouse and thinks: They're only animals."[37]

"Take sides," Elie Wiesel has said. "Take sides. Neutrality helps the oppressor, never the victim. Silence encourages the tormentor, never the tormented."[38] Are we really bound by justice to respect the right of animals to live a decent life? Are we really going to fall back on the "humans come first" argument, more often used as an excuse to do nothing about the suffering of humans and non-humans alike? If you knew you could help alleviate massive and intense suffering in a way that took no time, no money, and little effort, wouldn't you do it? Is cruelty only the willful causing of unnecessary suffering, or is our indifference also cruelty? "Just how destructive does a culinary preference have to be," asks Jonathan Foer, "before we decide to eat something else? If contributing to the suffering of billions of animals that live miserable lives and … die in horrific ways isn't motivating, what would be? … And if you are tempted to put off these questions of conscience, to say *not now*, then *when*?"[39] What we decide will ultimately test how we respond to the powerless, to the most distant, to the voiceless and defenseless. Our decision will test how we act when no one is forcing us to act one way or another. Our decision will be nothing less than a test of who we truly are. In choosing conscience and care over craving and convenience, compassion over mindless, heartless consumption, we will truly be putting our values where our mouth is. And the animals would thank us if they could.

Notes

1 "Upwards of 99 percent of all animals eaten in this country come from 'factory farms.'" Jonathan Safran Foer, *Eating Animals* (New York, NY: Little, Brown, 2009), 12. Factory farms are also called CAFOs ("confined animal feeding operations") and factory farming is sometimes referred to as "industrialized farm animal production" (IFAP).

2 Paul Waldau, *Animal Rights: What Everyone Needs to Know* (Oxford, England: Oxford University Press, 2011), 34.

3 See Peter Singer, *Animal Liberation* (New York, NY: HarperCollins, updated edition, 2009), 5. Singer bases the argument of his book, and his entire ethical theory, on this moral principle. The principle opposes theories that either exclude some interests from the moral calculus or weigh certain interests differently from others. Where animals have a characteristic equal to humans, such as the ability to feel, for example, one must provide for an equal consideration of interests.

4 Jeremy Bentham, *An Introduction to the Principles of Morals and Legislation* (Mineola, NY: Dover Publications, 2007), 311. This Dover edition is an unabridged republication of the edition published at the Clarendon Press, Oxford, 1907. It in turn was a reprint of "A New Edition, corrected by the Author," published in 1823. The work originally appeared in 1780 and was first published in 1789.

5 Singer, *Animal Liberation*, 7.

6 "The Cambridge Declaration on Consciousness," proclaimed on July 7, 2012, presents the conclusions of a prominent international group of neuroscientists, and validates the findings of numerous prior scientific studies. The declaration, as well as several of the seventeen talks and presentations of the Francis Crick Memorial conference, can be found at www.fcmconference.org. On animal sentience, self-awareness, emotion, and suffering, see also Marc Bekoff, *The Emotional Lives of Animals: A Leading Scientist Explores Animal Joy, Sorrow, and Empathy—And Why They Matter* (Novato, CA: New World Library, 2007); Andrew Linzey, *Why Animal Suffering Matters: Philosophy, Theology, and Practical* Ethics (New York, NY: Oxford University Press, 2009); Carl Safina, *Beyond Words: What Animals Think and Feel* (New York, NY: Henry Holt, 2015); Frans de Waal, *Are We Smart Enough to Know How Smart Animals Are?* (New York, NY: W.W. Norton, 2016); and Jennifer Ackerman, *The Genius of Birds* (New York, NY: Penguin Press, 2016).

7 "Putting Meat on the Table: Industrial Farm Animal Production in America," a report of the Pew Commission on Industrial Farm Animal Production, viii, www.Pewtrusts.org.

8 Pew Commission, "Putting Meat on the Table," 38. A longer version of my chapter examines some of the sources of ethical standards in the Western world regarding the treatment of animals. The positions of Pythagoras, Plutarch, the Hebrew Bible, the New

Testament, the hagiographical literature of medieval monasticism, Thomas Aquinas, the Catechism of the Catholic Church, and Pope Francis are briefly highlighted. The positions of many other philosophers, religious figures, and activists receive extensive treatment in Norm Phelps, *The Longest Struggle: Animal Advocacy from Pythagoras to PETA* (New York, NY: Lantern Books, 2007).

9 In the industry these processes are called "vertical integration." Matthew Scully, *Dominion: The Power of Man, the Suffering of Animals, and the Call to Mercy* (New York, NY: St. Martin's Press, 2002), 250. Scully points out that "As in all forms of tyranny, management intensifies as knowledge [of], interest in, or even curiosity about the subjects passes away," 272.

10 Foer, *Eating Animals*, 237.

11 Jacques Derrida is one of the few contemporary philosophers who has taken on these questions. He writes: "no one can today deny … the *unprecedented* proportions of this subjection of the animal. Such a subjection … can be called violence in the most morally neutral sense of the term … No one can deny seriously anymore, or for very long, that men do all they can to dissimulate this cruelty or to hide it from themselves; in order to organize on a global scale the forgetting or misunderstanding of this violence, which some would compare to the worst cases of genocide … " *The Animal That Therefore I am*, ed. Marie-Louise Mallet, trans. David Wills (New York, NY: Fordham University Press, 2008), 25–26.

12 An adult chicken needs a minimum of 197 square inches to turn around, 138 square inches to stretch, 290 square inches to flap wings, and 172 square inches to preen—all basic biological activities. Karen David, *Prisoned Chickens, Poisoned Eggs: An Inside Look at the Modern Poultry Industry* (Summertown, TN: Book Publishing Company, 1997), 100.

13 Deb Olin Unferth, "Cage Wars: A Visit to the Egg Farm," *Harper's Magazine* (November 2014), 48. Research based on observation of their behaviors has shown that hens suffer chronic acute pain all their lives because of this procedure.

14 Unferth, "Cage Wars," 45.

15 See Foer, *Eating Animals*, 299, note for page 133.

16 Unferth, "Cage Wars," 49.

17 "They have complicated cliques and can recognize more than a hundred other chicken faces, even after months of separation. They recognize human faces too. They have distinct voices and talk among themselves, even before they hatch … Adult chickens have at least thirty different categories of conversation, centered around, to name a few, mating, eating, nesting, rearing, and warning, each with its own web of coos and calls and clucks." Unferth, "Cage Wars," 45.

18 "United Egg Producers—the industry group that represents hatcheries that produce 95 percent of all eggs produced in the United States—announced Thursday that it would end this 'culling' of millions of chicks by 2020, or as soon as it's 'economically feasible' and an alternative is 'commercially available' … In a country that's hardly famous for humane animal farming practices, this is a big deal." Karin Brulliard, "Egg Producers Pledge to Stop Grinding Newborn Male Chickens to Death," *Washington Post*, June 10, 2016, www.washingtonpost.com/news/animalia/wp/2016/06/10/egg-producers.

19 For detailed information on what happens to broilers and egg-laying hens as well as pigs, calves, and cows in the factory farm system, see Peter Singer, *Animal Liberation*, 95–157.

20 Animals in U.S. slaughterhouses are routinely beaten, skinned, dismembered, and scalded while fully conscious. See Gail A. Eisnitz, *Slaughterhouse: The Shocking Story of Greed, Neglect, and Inhumane Treatment Inside the U.S. Meat Industry* (Amherst, NY: Prometheus Books, 2007), 300. Eisnitz has interviewed individuals who, having spent a combined total of more than two million hours on kill floors, speak publicly about what's really taking place behind the closed doors of America's slaughterhouses. No work of investigative journalism on this topic is as comprehensive as her exposé.

21 A word like "veal" helps us forget what we are eating. The words "pork" and "beef" do the same thing. "Even the wooden stalls and neck chains are part of the plan, as these restrictions keep the calf from licking his own urine and feces to satisfy his craving for

iron." Jim Mason and Mary Finelli, "Brave New Farm?" *In Defense of Animals: The Second Wave*, ed. Peter Singer (Oxford, England: Blackwell, 2006), 110.

22 "The pregnant pigs … must lie or step in their excrement to force it through the slatted floor … The system makes good welfare practices more difficult because lame and diseased animals are almost impossible to identify when no animals are allowed to move." Foer, *Eating Animals*, 184, and note for page 184 on page 316. Sows suffer from "sores, tumors, ulcers, pus pockets, lesions, cysts, bruises, torn ears, swollen legs." Scully, *Dominion*, 267.

23 Over 90 percent of large hog farms use artificial insemination. See Foer, *Eating Animals*, 157, and note for page 157 on page 306.

24 Scully, *Dominion*, 261 and 266.

25 Eisnitz, *Slaughterhouse*, 220.

26 *United States Code*, Title 49, Transportation, Subtitle X: Miscellaneous, Chapter 805, Miscellaneous. www.gpo.gov, 1994. See also Michigan State University College of Law. Animal Legal and Historical Center, www.animallaw.info. This web page is more readily accessible than the U.S. Code, and contains a summary as well as the full text of the statute. The 28-hour law was amended in 1994 to cover other methods of animal transport such as trucks. The law is rarely enforced, and the maximum penalty is only $500.

27 The Humane Methods of Slaughter Act (HSA) of 1958, 1978, 2002; Pub. L. 87–765, Aug. 27, 1958, 72 Stat. 862. See *United States Code Annotated*, Title 7, Agriculture, Chapter 48, Humane Methods of Livestock Slaughter. www.gpo.gov. The USDA mandates the enforcement of HSA regulations. But both slaughterhouse workers in their affidavits and USDA meat inspectors, blowing the whistle on their own agency, have stated that due to faster production speeds and industry deregulation, they did not abide by or enforce the HSA. One inspector explains that "there's a specific problem with enforcing the Humane Slaughter Act. That's because these large slaughtering operations are primarily concerned with productivity and profit. They don't care about the effects on the animals." Dave Carney, former USDA meat inspector and chairman of the National Joint Council of Food Inspection Locals, quoted in Eisnitz, *Slaughterhouse*, 188. Enforcement of the law has been so lax that in 2002 Congress passed a resolution entitled Enforcement of the Humane Slaughter Act of 1958. In February 2004, the government's General Accounting Office [GAO] reported that the Act was still not being adequately enforced. See Mason and Finelli, "Brave New Farm?" 120.

28 Common Farming Exemptions (CFEs) make legal any method of raising farmed animals as long as it is common practice in the industry. CFEs are enacted state by state. Undercover investigations have consistently revealed that slaughterhouse workers, laboring under what Human Rights Watch calls "systematic human rights violations," let loose their frustrations on farmed animals or keep the slaughter lines moving at all costs in order to keep their jobs (many of them are very poor or undocumented) because that's what their supervisors demand. "Blood, Sweat, and Fear: Workers' Rights in US Meat and Poultry Plants," Human Rights Watch, January 24, 2005, 2, www.hrw.org/report/2005/01/24/blood-sweat-and-fear/workers-rights-us-meat-and-poultry-plants. Workers' wages are low, union organizing is virtually non-existent, worker injuries are extremely common, and dying on the job is a real possibility. Human beings can be neither human nor humane under the conditions of a factory farm or a slaughterhouse.

29 https://signs.com/blog/state-by-state-guide-to-no-trespassing-laws-signage/.

30 Some of these investigators work under contract for organizations like Mercy for Animals, Compassion over Killing, PETA, and The Humane Society of the United States.

31 Because it suppressed free speech concerning topics of great public importance, namely, the safety of the public food supply, the safety of agricultural workers, the treatment and health of farmed animals, and the impact of business activities on the environment. The entire case of ALDF (Animal Legal Defense Fund) et al. v. Otter is found at www.ALDF.com.

32 Because it was motivated by hostility against animal rights advocates. "The now-defunct Idaho Ag-Gag law was originally written by the Idaho Dairymen's Association after an undercover investigation by an animal rights group revealed workers beating, stomping and sexually abusing cows at an Idaho dairy farm. It was signed into law in February 2014 by Governor C. L. "Butch" Otter, "a rancher." "ALDF Ag-Gag Victory: Hope for Farmed Animals," *The Animals' Advocate* 34(3) (Fall 2015): 5.

33 Senior attorney Matthew Liebman of the ALDF stated: "This is the first step in defeating similar Ag-Gag laws across the country, and should dissuade other states from considering similar laws … People have the right to know how their diet is contributing to suffering—and how common industry practices could be making them sick." "ALDF Ag-Gag Victory," 5.

34 The term was coined by Richard Ryder. Apparently, its first inclusion in a formal publication was in Ryder's contribution to *Animals, Men and Morals: An Enquiry into the Maltreatment of Non-humans*, ed. Stanley Godlovitch, Roslind Godlovitch, and John Harris (London, England: Gollancz, 1971), 81. The term now appears in *The Oxford English Dictionary*, second edition (Oxford, England: Clarendon Press, 1989), s.v. "speciesism." It also appears in *The Concise Oxford English Dictionary* (Oxford, England: Oxford University Press, 2008), 11th edition revised, where its definition is: "the assumption of human superiority over other creatures, leading to the exploitation of animals."

35 See Charles Patterson, *Eternal Treblinka: Our Treatment of Animals and the Holocaust* (New York, NY: Lantern Books, 2002). Treblinka is the name of a Nazi concentration camp. The last three chapters profile people—Jewish and German, perpetrators and survivors—whose animal advocacy has been shaped by the Holocaust.

36 Isaac Bashevis Singer, "The Letter Writer," in *The Collected Stories* (New York, NY: Farrar, Straus and Giroux, 1982), 271.

37 Translation of "*Auschwitz beginnt da, wo jemand auf Schlachthof steht und denkt: Es sind ja nur Tiere.*" Quoted in Christa Blanke, *Da krähte der Hahn: Kirche für Tier? Eine Streitschrift* (Eschbach, Germany: Verlag am Eschbach, 1995), 48.

38 Quoted in Patterson, *Eternal Treblinka*, 137. No source is given.

39 Foer, *Eating Animals*, 243.

Bibliography

Ackerman, Jennifer. *The Genius of Birds*. New York, NY: Penguin Press, 2016.

"ALDF Ag-Gag Victory: Hope for Farmed Animals." *The Animals' Advocate*. 34, 3, Fall 2015.

Bekoff, Marc. *The Emotional Lives of Animals: A Leading Scientist Explores Animal Joy, Sorrow, and Empathy—And Why They Matter*. Novato, CA: New World Library, 2007.

Bentham, Jeremy. *An Introduction to the Principles of Morals and Legislation*. Mineola, NY: Dover Publications, 2007.

Blanke, Christa. *Da kräte der Hahn: Kirche für Tier? Eine Streitschrift*. Eschbach, Germany: Verlag am Eschbach, 1995.

Brulliard, Karin. "Egg Producers Pledge to Stop Grinding Newborn Male Chickens to Death." *Washington Post*, June 10, 2016. www.washingtonpost.com/news/animalia/wp/2016/06/10/egg-producers.

David, Karen. *Prisoned Chickens, Poisoned Eggs: An Inside Look at the Modern Poultry Industry*. Summertown, TN: Book Publishing Company, 1997.

Derrida, Jacques. *The Animal That Therefore I Am*. Edited by Marie-Louise Mallet. Translated by David Wills. New York, NY: Fordham University Press, 2008.

de Waal, Frans. *Are We Smart Enough to Know How Smart Animals Are?*New York, NY: W. W. Norton, 2016.

Eisnitz, Gail A. *Slaughterhouse: The Shocking Story of Greed, Neglect, and Inhumane Treatment Inside the U.S. Meat Industry*. Amherst, NY: Prometheus Books, 2007.

Foer, Jonathan Safran. *Eating Animals*. New York, NY: Little, Brown, 2009.

Human Rights Watch. "Blood, Sweat, and Fear: Workers' Rights in US Meat and Poultry Plants." January 24, 2005. www.hrw.org/report/2005/01/24/blood-sweat-and-fear/workers-rights-us-meat-and-poultry-plants.

Linzey, Andrew. *Why Animal Suffering Matters: Philosophy, Theology, and Practical Ethics*. New York, NY: Oxford University Press, 2009.

Mason, Jim, and Mary Finelli. "Brave New Farm?" In *In Defense of Animals: The Second Wave*, edited by Peter Singer, 104–122. Oxford, England: Blackwell, 2006.

Michigan State University College of Law. Animal Legal and Historical Center. www.animallaw.info [the 28-hour law].

Patterson, Charles. *Eternal Treblinka: Our Treatment of Animals and the Holocaust*. New York, NY: Lantern Books, 2002.

Pew Commission on Industrial Farm Animal Production. "Putting Meat on the Table: Industrial Farm Animal Production in America." A Report. www.Pewtrusts.org.

Phelps, Norm. *The Longest Struggle: Animal Advocacy from Pythagoras to PETA*. New York, NY: Lantern Books, 2007.

Ryder, Richard. "Experiments on Animals." In *Animals, Men and Morals: An Enquiry into the Maltreatment of Non-Humans*. Edited by Stanley Godlovitch, Roslind Godlovitch, and John Harris, 41–82. London, England: Gollancz, 1971.

Safina, Carl. *Beyond Words: What Animals Think and Feel*. New York, NY: Henry Holt, 2015.

Scully, Matthew. *Dominion: The Power of Man, the Suffering of Animals, and the Call to Mercy*. New York, NY: St. Martin's Press, 2002.

Singer, Isaac Bashevis. "The Letter Writer." In *The Collected Stories*, 250–276. New York, NY: Farrar, Straus and Giroux, 1982.

Singer, Peter. *Animal Liberation*. New York, NY: HarperCollins, updated edition, 2009.

Unferth, Deb Olin. "Cage Wars: A Visit to the Egg Farm." *Harper's Magazine*, November 2014.

United States Code Annotated, Title 7, Agriculture, Chapter 48, Humane Methods of Livestock Slaughter. www.gpo.gov.

United States Code, Title 49, Transportation, Subtitle X: Miscellaneous, Chapter 805, Miscellaneous. www.gpo.gov, 1994 [the 28-hour law].

Waldau, Paul. *Animal Rights: What Everyone Needs to Know*. Oxford, England: Oxford University Press, 2011.

2.3

L'ENFER, C'EST NOUS AUTRES

Institutionalized cruelty as standard industry practice in animal agriculture in the United States

Patricia McEachern

In his play *Huis-Clos*, known in English as *No Exit*, Jean-Paul Sartre famously stated: "*L'enfer, c'est les autres*" (hell is other people).[1] Being of a more sanguine nature than Sartre, I long took umbrage with his sweeping assessment of human nature. But the more I have learned about the horrific suffering we inflict on 9.2 billion animals exploited for food every year in the United States alone,[2] the more I have been compelled to reassess my opinion. Human beings, it seems, have conspired to create a living hell for the billions of animals who spend their entire lives in factory farms before they are shipped off to brutal slaughterhouses. From the moment of their births to the moment when they are slaughtered, we terrorize, traumatize, and torture these innocent, sentient individuals. Almost all of their natural behaviors are thwarted. To believe that it is ethical to terrorize and slaughter billions of sentient animals, simply to please our palates, is to believe that no form of animal cruelty is unethical. It is to believe that we should be at ease as "participants" in what J. M. Coetzee's character Elizabeth Costello wondered might be "a crime of stupefying proportions."[3] Eating animals in our times relies on institutionalized cruelty and terror. It is inconceivable that anyone could consider cruelty and terror ethical.

Undercover investigations and the Internet have made it possible for anyone to do for animals what Nobel Peace Prize laureate and Holocaust survivor Elie Wiesel does for victims of the Holocaust, albeit in a modest way. He implores us to understand our "duty to bear witness for the dead and for the living."[4] In this case, we are called to bear witness to the gruesome reality that lurks behind such propaganda campaigns as the "happy cow," an American advertising campaign designed to manipulate consumers into believing that dairy cows enjoy their lives.[5] In a television commercial, viewers see "happy cows" playing soccer in a field. According to the California Milk Advisory Board "Great milk comes from happy cows. Happy cows come from California."[6] In reality, dairy cows are impregnated repeatedly and hooked up to milking machines in crowded conditions before being turned into hamburger when they are no longer able to produce milk. The heartbroken mother cows bellow for their calves who are ripped away from them, often to be turned into veal. In a final act of hellish cruelty, these naturally loving and protective mothers are impregnated one final time before they go to slaughter. Their unborn calves are removed during slaughter for a variety of exploitive purposes. Ethical, caring human beings are called to unmask the myth of the "happy cow," or at a minimum, at least not to support it by buying dairy or meat products. Similarly, we are called to unmask the myth known as "humane" meat, an Orwellian term if ever there were one, which has been foisted upon us of late, simultaneously encouraging us to eat sentient animals and placate our consciences by convincing ourselves that they have been treated humanely.[7]

As people have become increasingly more willing to bear witness, many refuse to remain silent. The animal industrial complex has responded not by changing standard industry practices, but by enacting so-called "ag-gag" laws that punish whistle blowers.[8] These laws essentially gag journalists, activists, undercover investigators, and even animal agriculture employees. Whistle blowers can be convicted of violating the law simply for documenting animal cruelty or safety violations.[9] Ag-gag laws are an implicit confession on the part of the animal industrial complex. They know that when people see the kind of institutionalized cruelty that takes place as standard industry practice, many demand that these practices be stopped. Eight states in the United States have passed "ag-gag" laws[10] and, although they have been unsuccessful, many more states have attempted to do so. We currently have some reason for optimism in the U.S. because the Idaho ag-gag law was ruled unconstitutional in federal court in 2015 under the first and fourteenth amendments.[11, 12] This landmark victory could mark the beginning of the end for other similar state laws.

Former United States Department of Agriculture (USDA) worker, Timothy Walker, reported having firsthand knowledge of seeing cattle skinned while they were still alive at Kaplan Industries in Bartow, Florida in 1989.[13] His reports to the USDA, the Veterans Association, and the U.S. Congress bore no results. As it turned out, his senator, Bob Graham, whom he had contacted, owned a large dairy operation.[14] He finally sought the help of investigator Gail Eisnitz, author of *Slaughterhouse: the Shocking Story of Greed, Neglect, and Inhumane Treatment inside the U.S. Meat Industry*, who at that time worked for a Washington D.C.-based animal protection organization.[15] If Walker had made those same reports of illegal acts by Kaplan in an "ag-gag" state today, and if Eisnitz had investigated, both of them could have been charged and found guilty of breaking the law. According to Walker, "What I saw when I walked into the plant looked like illustrations for Dante's *Inferno*. Hell can't be any worse than what exists at this place."[16] Walker, who had always received excellent performance evaluations, was eventually fired by the USDA. His supervisor wrote in his termination letter: "It is my decision that to continue your employment beyond the probationary period would impede the efficiency of government service."[17]

In recent years, as more people have become aware of the inherent cruelty involved in contemporary factory farms and slaughterhouses, many American farmers have begun to justify their practices with what often appear to be sincerely held religious convictions, the dominion argument from Genesis being the most popular. Although such religious convictions may be genuine, they are utterly bereft of critical theological underpinnings.

I have decided to include in this chapter some of my personal experiences in founding an Animal Studies minor at Drury University, a small liberal arts university that is located in a small Midwestern city in S. W. Missouri, a region that is particularly hostile to animal rights. If we can accomplish what we have in such a hostile region, surely much more is possible elsewhere. Animal agriculture is a major economic force in S. W. Missouri. The Missouri Soybean Merchandising Council[18] recently reported that "For Missouri, the aggregated animal production and processing industries supported 154,268 jobs and paid $6.212 billion in labor income." In addition, S. W. Missouri is known as the puppy mill capital of the United States.[19] Hunte Corporation, the largest international puppy distributor in the world, is located in Springfield.[20] Dickerson Park Zoo, located in Springfield, is infamous having had an Asian elephant by the name of Chai removed by the USDA in 2001 after numerous trainers beat her mercilessly in an effort to make her comply with commands so they could breed her. She lost 1,000 pounds during the time she spent at Dickerson Park Zoo.[21] Recently, a female tiger by the name of Petra had one of her legs amputated up to the shoulder after a keeper left her vulnerable to attack by a male tiger.[22] In 2013, an elephant by the name of Patience killed a keeper.[23] Springfield is also home to Bass Pro Shops, the largest hunting and fishing chain store in the U.S.

When news appeared in 2008 that we would be teaching an interdisciplinary Animal Ethics class at Drury University the Missouri Cattlemen's Association descended upon the president *en masse*. The story was big news because Bob Barker of *Price is Right* fame, a Drury alumnus, had provided the university with a generous endowment to "do something to improve the lives of animals."[24] The cattlemen attempted to muscle and intimidate Drury's president into quashing the class. *The Ozarks Farm and Neighbor*, a local agricultural periodical, quoted Dr. Beth Walker, a local agriculture instructor and producer, as stating that "the people who use this language, they're terrorists" (referring to vivisection, speciesism, and non-human animals).[25] Evidently, someone hacked into my computer and found a draft version of the syllabus that included those terms. Accusations against me of terrorism also appeared in blogs and in an on-line response to a letter to the editor of the *Springfield News-Leader*. Drury's president refused to back down, and, far from intimidating me, they encouraged me enormously: I concluded that if the entire Missouri Cattlemen's Association were afraid of one, small class at Drury University, we must be on to something.

I have long been of the belief that truth withstands, even invites, scrutiny. With that in mind, the second year of the class I invited members of the Missouri Cattlemen's Association to give a presentation in the Animal Ethics class. I wanted students to hear as many voices as possible. Many of our students come from farming families and I wanted their voices to be represented. John Kleiboecker, Executive Director of the Missouri Beef Industry Council, was the speaker. To my astonishment, the presentation was based entirely on a fundamentalist, literal reading of the dominion passage in Genesis 1:26, which he quoted:[26]

> Then God said, "Let us make man in our image, after our likeness. And let them have dominion over the fish of the sea and over the birds of the heavens and over the livestock and over all earth and over every creeping thing that creeps on earth."
>
> *(English Standard Version)*

After the presentation, I asked Kleiboecker privately if he were aware that the prophet Daniel was a vegetarian. He answered in the affirmative, but added that he hoped that no one else present knew that.[27] Then I asked if he were aware of Genesis 1:28–31, in which all living individuals in the creation narrative were vegan, and he appeared to be utterly dumbfounded.[28] That verse had escaped his notice entirely, as it usually does, in my experience, for most Christians who quote Genesis 1:26 as a justification for eating animals.

Apparently, in attempt to appease the cattlemen, an administrator invited cattleman Trent Loos, an "Americana" radio broadcaster and blogger, to speak to a group of freshman students. Like Kleiboecker, Loos based his defense of eating animals on the dominion verse from the creation narrative. According to Loos' website:

> Trent is a sixth generation (sic) United States farmer with a passion for agriculture that started years ago – he had his first pig when he was just 5 years old on the family farm near Quincy, IL. He established his strong rural foundation in farming side-by-side with his father and grandfather. That strong connection between man and his God-given natural resources was the source for all of their farm management decisions. *(H)is real fondness was for the animals* (emphasis added). When anti-agricultural activists threatened the way of life he cherished, it was time to take a stand. Consumers, even individuals in other aspects of agriculture, were listening to and believing the lies being spread by celebrities and vegan zealots.[29]

Loos' colorful talk included tossing out questions such as: "Chickens in cages. Good or bad?"[30] In Loos' world, keeping chickens in cages is very good. His argument, which relied on the fallacy of false equivalence, went something like this: "We all know stories about couples who cannot conceive. Then they go on vacation. They are not stressed. They conceive. It's the same thing with chickens. As long as they are laying eggs, they're not stressed."[31] Ergo, as Loos would have us believe, keeping chickens in cages is good.

Polyface, Inc., the self-proclaimed "farm of many faces," is a family-owned, "beyond organic, local-market" farm located in Virginia's Shenandoah Valley.[32]

One wonders who they see when they look at the "many faces" of the individual animals whom they raise and slaughter for food. Adopting religious terminology, their website informs us: "We are in the redemption business: Healing the land, healing the food, healing the economy, and healing the culture."[33] Redemption sounds like a lofty ideal, but it is of no value to the animals, God's creatures, who are raised and slaughtered at Polyface, Inc. According to Polyface owner Joel Salatin, a self-proclaimed devout Christian, "People have a soul; animals don't. It's a bedrock belief of mine. Unlike us, animals are not created in God's image, so when they die, they just die."[34] One could argue that if you do not have a soul, your brief life on earth would be even more precious to you than if you believe you will live forever in Salatin's version of Heaven. He describes his book, *The Marvelous Pigness of Pigs*, as a "clarion call to readers to honor the animals and the land and produce food based on spiritual principles."[35] The belief that you can exploit the life and death of an individual whose life matters to her and "honor" her at the same time makes me feel as though I have stepped through the proverbial Looking Glass.

The Salatins believe that "the Creator's design is still the best pattern for the biological world."[36] Like John Kleiboecker, Joel Salatin is apparently unaware of Genesis 1:28–31, which is also the Creator's design:

> Then God said, I give you every seed-bearing plant on the face of the whole earth and every tree that has fruit with seed in it. They will be for your food. And to all the beasts of the earth and all the birds in the air and all the creatures that move on the ground—everything that has breath of life in it—I give every green plant for food"
>
> *(New International Version)*

To be fair, the animals destined for the dinner plate who are raised at Polyface have it far better than those raised in factory farms. What is of particular interest to me here is what I believe to be a distinctly American type of religiosity connected to raising animals for food and the preening and self-righteous justifications based on a kind of *sola scriptura* reading of the Bible that conveniently ignores scripture passages that do not support their argument.

Obviously, Sartre did not have the current animal industrial complex in mind when he penned *Huis Clos*, but both the title and his famous line "Hell is other people" came to my mind when I considered the subject the ethics of eating animals. Both the literal and the common English translation of the title could speak to the issue. Literally, *huis clos* is translated "behind closed doors." Institutionalized cruelty depends upon secrecy, upon every action happening behind closed doors; standard industry practices are kept from public view just as carefully as individual acts of sadistic cruelty, which apparently happen on a regular basis. We know this because of evidence provided by many undercover investigators such as Gail Eisnitz, Matt Rossell,[37] Nathan Runkle, and Bryan Monell, and organizations like Mercy for Animals and PETA[38] who regularly document them.

In English, *Huis clos* is usually translated as "No Exit." Such is the case for the billions of animals born in factory farms. Over *nine billion* animals were slaughtered for food in the United States alone in 2015.[39] Most pigs see natural light only when they are transported to "finishing farms" when they weigh around 100 lbs, and then again when they are shipped to slaughterhouses. Mercy for Animals founder Nathan Runkle described a typical scenario.

> The animals had been packed into trucks and transported—often over hundreds of miles—to slaughterhouses. They arrived weak, dehydrated, and coated in urine and feces. Many endured brutal beatings and other abuses before their deaths. During the slaughter process, some of the animals were scalded and dismembered while still alive and fully conscious.[40]

One worker at a hog slaughterhouse reported that when hogs "don't want to go" into the chutes to be killed, they are beaten with pipes.[41] "I've beaten eleven to death in one day. I hope that doesn't sound like bragging, because it's not" he shared.[42] There is no exit for factory farmed animals except through a gruesome and brutal death. Patty Shenker, a tireless American animal rights advocate, made a succinct and poignant argument in defense of farmed animals in a meme she created for social media: "Every being is born to die, but no being should be born to be killed."[43]

In June 2016, many of us were revolted and horrified by the barbarity of the Yulin dog and cat eating festivals in China. The festivals generated an international outcry and protests. A Change.org petition opposing Yulin garnered 11 million signatures.[44] What far too many dog lovers missed was the opportunity to understand that we in the west do equally horrific things in our own factory farms and slaughterhouses.[45] We have no moral high ground to stand on in this regard. Yulin should provide a window on our own moral relativism. It is an opportunity to recognize that we should extend our moral compass to include all sentient beings. Those who eat animals, eggs, and dairy, and are horrified by the Yulin dog and cat eating festival, should consider changing their diets to be consistent with their ethics.

It is painful to read about these atrocities. It is even more painful to witness them in videos and photographs. How much more excruciating must it be for the 9 billion innocent animals in the U.S. who endure them every year simply for the pleasure of our palates?[46] The days when animal sentience was called into question are long gone. The moral question that remains to us is one about which Andrew Linzey has so eloquently written in his book *Why Animal Suffering Matters*. "The moral issue is not whether their suffering is identical in all respects to our own, but rather whether their suffering is *as important to them as ours is to us*."[47]

Dave Carney, a former USDA meat inspector and later the chairman of the National Joint Council of Food Inspection Locals, the federal meat inspectors' union, spoke to former undercover investigator, Gail Eisnitz, with surprising candor. Speaking about large slaughtering operations and the Humane Slaughter Act, Carney said:

> It's as if they're not even killing the animals. They're disassembling them, processing raw materials in a manufacturing operation. To keep that production line moving, quite often uncooperative animals are beaten, they have prods poked in their faces and up their rectums, they have bones broken and eyeballs poked out, suspects (i.e., those suspected of being unfit for consumption) are left unattended for days ... Some even reach various stages of the slaughtering process alive. They have their hooves cut off, parts of their bung and their cavities opened.[48]

The United States has a Humane Slaughter Act (HSA), but, according to Carney, there is no procedure for checking for humane slaughter.[49] "Inspectors are required to enforce humane regulations on paper only."[50] They are not even allowed to go the areas where the slaughtering takes place.[51] "The Humane Slaughter Act is a regulation on paper only. It is not being enforced."[52] Appallingly, the HSA excludes birds, who account for the vast majority of animals killed for human consumption. Chickens and turkeys go into "the scald tanks still alive, breathing and sucking in the water."[53]

Even if we choose to ignore the enormous suffering of animals whom we exploit for food, we are still left with the ethical problem posed by the fact that animal agribusiness is the major source of destruction to the environment. In her book, *Why We Love Dogs, Eat Pigs, and Wear Cows*, Melanie Joy points out how this $125 billion industry in the United States is controlled by a small number of corporations that are exploiting both animals and the environment at inconceivable levels.[54] "Livestock-based agribusiness causes 55 percent of the erosion and sediment produced in the United States. Also, 37 percent of all pesticides and 50 percent of all antibiotics used in this country are used by animal agribusiness."[55] While the World Health Organization warns that antibiotic resistance is one of the greatest threats to public health, and doctors are warned not to overprescribe antibiotics to their human patients,[56] animal agribusiness blithely and arrogantly continues to dispense 50 percent of all antibiotics for use in confined feeding operations.[57] The fact that we are facing a global crisis brought on, in large part, by this immoral industry is reason enough to conclude that eating animals is unethical. We are depleting the oceans, and "sixty to seventy percent of the world's fish catch goes to feed livestock."[58] That is incomprehensible. As Captain Paul Watson of the Sea Shepherd Conservation Society says, "If the oceans die, we all die."[59] How can anyone defend this practice as ethical? While many environmental groups encourage us to ride bicycles and take short showers, they neglect to warn us that "the methane produced by cattle and their manure has a global warming effect equivalent to that of 33 million automobiles."[60] Thirty-three million cars! "Greenhouse gases produced by livestock constitute 37 percent of all methane, 65 percent of nitrous oxide, and 64 percent of ammonia in the atmosphere."[61] This is a human-created environmental and ethical catastrophe of never-before-seen proportions.

It is time to bring the virtue of humility to our relationships with animals. Just as it was appalling arrogance that allowed human beings to convince themselves that they had a right to own other human beings as slaves, it is arrogance that allows us to believe it is ethical to consume animals. The animal rights movement is arguably the most important social justice movement since the abolition of slavery. We are on the right side of history. We are talking about billions of individuals whose lives are precious to them. From our current vantage point, it is obvious that an attempt to reform or improve slavery would have been absurd. Slavery needed to be abolished, not reformed. Similarly, the conditions under which we enslave and slaughter farmed animals do not need to be improved or reformed. We need to abolish the practice entirely.

Notes

1 J. Sartre, *Huis Clos* (Paris, France: Gallimard, 1947), 182.
2 "Farm Animal Statistics: Slaughter Totals," Humane Society of the United States, Last modified June 25, 2015, www.humanesociety.org/news/resources/research/stats_slaughter_totals.html.
3 J. M. Coetzee, *The Lives of Animals* (Princeton, NJ: Princeton University Press, 1999), 69.
4 E. Wiesel, *Night* (New York, NY: Bantam Books, 1982), xv.
5 "About Us," California Milk Advisory Board, accessed October 1, 2016, www.realcaliforniamilk.com/about-us/.
6 California Milk Advisory Board, "About Us."
7 "Our Mission," Humane Farm Animal Care, accessed October 1, 2016, http://certifiedhumane.org/.
8 "Taking Ag-Gag to Court," Animal Legal Defense Fund, accessed October 1, 2016, http://aldf.org/cases-campaigns/features/taking-ag-gag-to-court/.
9 Animal Legal Defense Fund, "Taking Ag-Gag to Court."
10 L. Runyon, "Judge Strikes Down Idaho 'Ag-Gag' Law, Raising Questions for Other States," *National Public Radio*, August 4, 2015, www.npr.org/sections/thesalt/2015/08/04/429345939/idaho-strikes-down- ag-gag-law-raising-questions-for-other-states.
11 G. Prentice, "U. S. Court Judge Strikes Down Idaho Ag-Gag Laws: 'The Remedy is More Speech, Not Enforced Silence,'" *Boise Weekly*, August 3, 2015, www.boiseweekly.com/boise/us-court-judge-rules-idaho-ag-gag-law-unconstitutional-the-remedy-is-more-speech-not-enforced-silence/Content?oid=3553359.
12 Runyon, "Judge Strikes Down."
13 G. Eisnitz, *Slaughterhouse: The Shocking Story of Greed, Neglect, and Inhumane Treatment Inside the U.S. Meat Industry* (Amherst, NY: Prometheus Books, 2007), 30.
14 Eisnitz, *Slaughterhouse*, 30.
15 Eisnitz, *Slaughterhouse*, 17–18.
16 Eisnitz, *Slaughterhouse*, 25.
17 Eisnitz, *Slaughterhouse*, 58.
18 "Economic Contribution of Animal Agriculture to Missouri," Missouri Soybean Merchandising Council, May 9, 2016, https://mosoy.org/wp-content/uploads/2016/05/Animal-Ag-Contribution-Report-5.9.16.pdf.
19 R. Teuscher, "The Puppy Industry in Missouri: A Study of the Buyers, Sellers, Breeders and Enforcement of the Law: Executive Summary," *Missouri Better Business Bureau*, March 2010, http://stlouis.bbb.org/Storage/142/Documents/Puppy%20Mills%20study.pdf.
20 Teuscher, "The Puppy Industry in Missouri."
21 "2008 Ten Worst Zoos for Elephants," In Defense of Animals, last modified December 29, 2016, www.idausa.org/assets/files/campaign/Elephants/2008%20Ten%20Worst%20Zoos%20for%20Elephants%20-%20In%20Defense%20of%20Animals-%20In%20Defense%20of%20Animals.pdf.

22 T. Gounley, "Injured Tiger Not Going Anywhere, Dickerson Park Zoo Says," *Springfield News-Leader*, September 20, 2016, www.news-leader.com/story/news/2016/09/20/injured-tiger-not-going-anywhere-dickerson-park-zoo-says/90731456/.
23 C. Johnston, "Elephant Kills Keeper at Springfield, Missouri Zoo," *CNN*, October 11, 2013, www.cnn.com/2013/10/11/us/missouri-zoo-death/.
24 B. Barker, personal communication, August 2007.
25 M. Fuller, "A Closer Look at Drury's Study of Animal Rights," *Ozarks Farm and Neighbor*, November 30, 2009, www.ozarksfn.com/2009/12/07/a-closer-look-at-drurys-study-of-animal-rights-2/.
26 J. Kleiboecker, Classroom Presentation to the Animal Ethics Class at Drury University, Springfield, MO, October, 2010.
27 J. Kleiboecker, personal communication, October 2010.
28 Kleiboecker, personal communication.
29 "About Trent," Loos' Tales, accessed July 1, 2016, www.loostales.com/abouttrent.html.
30 T. Loos, Presentation at Drury University, Springfield, MO, September 2011.
31 Loos, Presentation.
32 "Polyface: The Farm of Many Faces," Polyface, Inc., accessed October 1, 2016, www.polyfacefarms.com/.
33 Polyface, Inc., "Polyface."
34 Polyface, Inc., "Polyface."
35 Polyface, Inc., "Polyface."
36 Polyface, Inc., "Polyface."
37 M. Rossell, Presentation to the Animal Ethics Class at Drury University, Springfield, MO, October 2015.
38 "Pigs: Intelligent Animals Suffering in Farms and Slaughterhouses," People for the Ethical Treatment of Animals, accessed October 1, 2016, www.peta.org/issues/animals-used-forfood/animals-used-food-factsheets/pigs-intelligent-animals-suffering-factory-farms-slaughterhouses/.
39 Humane Society, "Farm Animal Statistics."
40 N. Runkle, Presentation to the Animal Ethics Class at Drury University, Springfield, MO, November, 2015.
41 Eisnitz, *Slaughterhouse*, 82.
42 Eisnitz, *Slaughterhouse*, 82.
43 P. Shenker, personal communication, June 2016.
44 "Boycott Chinese Goods to Stop Yulin Dog Meat 'Festival'," Change.org., accessed June 29, 2016, www.change.org/p/global-animal-advocates-boycott-chinese-goods-to-stop-yulin-dog-meat-festival?source_location=topic_page.
45 N. Runkle, "Our Outrage Over China's Yulin Dog Meat Festival Exposes a Disgusting Hypocrisy," *New York Daily News*, June 29, 2015, www.nydailynews.com/opinion/runkle-disgusting-hypocrisy-eating-animals-article-1.227512.
46 Humane Society, "Farm Animal Statistics."
47 A. Linzey, *Why Animal Suffering Matters* (New York: NY: Oxford University Press, 2009), 53.
48 Eisnitz, *Slaughterhouse*, 188–189.
49 Eisnitz, *Slaughterhouse*, 189.
50 Eisnitz, *Slaughterhouse*, 190.
51 Eisnitz, *Slaughterhouse*, 190.
52 Eisnitz, *Slaughterhouse*, 191.
53 Eisnitz, *Slaughterhouse*, 194.
54 M. Joy, *Why We Love Dogs, Eat Pigs, and Wear Cows* (San Francisco, CA: Conari Press, 2010), 88.
55 Joy, *Why We Love Dogs*, 86.
56 "Antibiotic resistance," World Health Organization, last modified February 5, 2018, www.who.int/mediacentre/factsheets/antibiotic-resistance/en/.
57 Joy, *Why We Love Dogs*, 86.

58 Joy, *Why We Love Dogs*, 87.
59 P. Watson, "If the Ocean Dies, We All Die!" *Sea Shepherd*, September 29, 2015, https://seashepherd.org/2015/09/29/if-the-ocean-dies-we-all-die/.
60 Joy, *Why We Love Dogs*, 87.
61 Joy, *Why We Love Dogs*, 87.

Bibliography

Animal Legal Defense Fund. "Taking Ag-Gag to Court." Accessed October 1, 2016. http://aldf.org/cases-campaigns/features/taking-ag-gag-to-court/.

California Milk Advisory Board. "About Us." Accessed October 1, 2016. www.realcalifor niamilk.com/about-us/.

Change.org. "Boycott Chinese Goods to Stop Yulin Dog Meat 'Festival'." Accessed June 29, 2016. www.change.org/p/global-animal-advocates-boycott-chinese-goods-to-stop-yulin-dog-meat-festival?source_location=topic_page.

Coetzee, J. M. *The Lives of Animals*. Princeton, NJ: Princeton University Press, 1999.

Eisnitz, G. *Slaughterhouse: The Shocking Story of Greed, Neglect, and Inhumane Treatment Inside the U.S. Meat Industry*. Amherst, NY: Prometheus Books, 2007.

Fuller, M. "A Closer Look at Drury's Study of Animal Rights." *Ozarks Farm and Neighbor*. November 30, 2009. www.ozarksfn.com/2009/12/07/a-closer-look-at-drurys-study-of-animal-rights-2/.

Gounley, T. "Injured Tiger Not Going Anywhere, Dickerson Park Zoo Says." *Springfield News-Leader*. September 20, 2016. www.news-leader.com/story/news/2016/09/20/injured-tiger-not-going-anywhere-dickerson-park-zoo-says/90731456/.

Humane Farm Animal Care. "Our Mission." Accessed October 1, 2016. http://certi fiedhumane.org/.

Humane Society of the United States. "Farm Animal Statistics: Slaughter Totals." Last modified June 25, 2015. www.humanesociety.org/news/resources/research/stats_sla ughter_totals.html.

In Defense of Animals. "2008 Ten Worst Zoos for Elephants." Last modified December 29, 2016. www.idausa.org/assets/files/campaign/Elephants/2008%20Ten%20Worst%20Zoos %20for%20Elephants%20-%20In%20Defense%20of%20Animals-%20In%20Defense%20of %20Animals.pdf.

Johnston, C. "Elephant Kills Keeper at Springfield, Missouri Zoo." *CNN*. October 11, 2013. www.cnn.com/2013/10/11/us/missouri-zoo-death/.

Joy, M. *Why We Love Dogs, Eat Pigs, and Wear Cows*. San Francisco, CA: Conari Press, 2010.

Kleiboecker, J. "Classroom Presentation to the Animal Ethics Class at Drury University, Springfield, MO." October, 2010.

Linzey, A. *Why Animal Suffering Matters*. New York, NY: Oxford University Press, 2009.

Loos, T. "Presentation at Drury University, Springfield, MO." September, 2011.

Loos' Tales, "About Trent." Accessed July 1, 2016. www.loostales.com/abouttrent.html.

Missouri Soybean Merchandising Council. "Economic Contribution of Animal Agriculture to Missouri." May 9, 2016. https://mosoy.org/wp-content/uploads/2016/05/Animal-Ag-Contribution-Report-5.9.16.pdf.

People for the Ethical Treatment of Animals. "Pigs: Intelligent Animals Suffering in Farms and Slaughterhouses." Accessed October 1, 2016. www.peta.org/issues/animals-used-forfood/animals-used-food-factsheets/pigs-intelligent-animals-suffering-factory-farms-sla ughterhouses/.

Polyface, Inc. "Polyface: The Farm of Many Faces." Accessed October 1, 2016. www.polyfa cefarms.com.

Prentice, G. "U.S. Court Judge Strikes Down Idaho Ag-Gag Laws: 'The Remedy is More Speech, Not Enforced Silence.'" *Boise Weekly*. August 3, 2015. www.boiseweekly.com/boise/us-court-judge-rules-idaho-ag-gag-law-unconstitutional-the-remedy-is-more-speech-not-enforced-silence/Content?oid=3553359.

Rossell, M. "Presentation to the Animal Ethics Class at Drury University, Springfield, MO." October2015.

Runkle, N. "Our Outrage Over China's Yulin Dog Meat Festival Exposes a Disgusting Hypocrisy." *New York Daily News*. June 29, 2015. www.nydailynews.com/opinion/runkle-disgusting-hypocrisy-eating-animals-article-1.227512.

Runkle, N. "Presentation to the Animal Ethics Class at Drury University, Springfield, MO." November, 2015.

Runyon, L. "Judge Strikes Down Idaho 'Ag-Gag' Law, Raising Questions for Other States." *National Public Radio*. August 4, 2015. www.npr.org/sections/thesalt/2015/08/04/429345939/idaho-strikes-down-ag-gag-law-raising-questions-for-other-states.

Sartre, J. *Huis Clos*. Paris, France: Gallimard, 1947.

Teuscher, R. "The Puppy Industry in Missouri: A Study of the Buyers, Sellers, Breeders and Enforcement of the Law: Executive Summary." *Missouri Better Business Bureau*. March, 2010. http://stlouis.bbb.org/Storage/142/Documents/Puppy%20Mills%20study.pdf.

Watson, P. "If the Ocean Dies, We All Die!" *Sea Shepherd*. September 29, 2015. https://seashepherd.org/2015/09/29/if-the-ocean-dies-we-all-die/.

Wiesel, E. *Night*. New York, NY: Bantam Books, 1982.

World Health Organization. "Antibiotic resistance." Last modified February 5, 2018. www.who.int/mediacentre/factsheets/antibiotic-resistance/en/.

2.4

WELFARE AND PRODUCTIVITY IN ANIMAL AGRICULTURE

Jeff Johnson

A number of features of our food environment encourage consumers to think that there is nothing terribly problematic in eating animals. The practice is widely accepted, traditions are built around eating animals, we are encouraged to enjoy what we think tastes good to us, etc. Sometimes, though, the practice of eating animals comes under ethical scrutiny. From time to time, for example, undercover videos emerge that raise awareness around what have become standard practices on farms.

These investigations reveal that animals have their beaks, tails, and testicles cut off or seared off without any pain relief. For nearly their entire pregnancy, mother pigs are kept in cages so small that they can't turn around. Chickens who lay eggs are kept in cages so small that they're each afforded less than an iPad's worth of space in which to live out their lives.

When questions are raised about these practices, consumers are often told that they ought not to worry too much, since producers who engage in them have an interest in promoting the welfare of the animals. You can find a range of general remarks about the industry's commitment to animal welfare on industry sites. Here's an example from the United States Farmers and Ranchers Alliance:

> Farmers and ranchers are committed to the safest and most appropriate care for their animals. They care deeply about the health and safety of their animals and take pride in them. They also know that consumers are concerned about animal care … USFRA believes that farmers and ranchers work diligently to keep their farm animals safe, healthy and comfortable.[1]

And here's an example from the Animal Agriculture Alliance:

> Animal care has always been important to livestock and poultry producers … Producers take their ethical obligation to providing the best quality care to

> their animals very seriously. Science-based animal care guidelines are in place across all sectors of the industry to help farmers, ranchers and processors improve the lives of the animals that depend on them.[2]

In these quotes we find claims about how much producers care about their animals and about how seriously they take what they see as their ethical obligation to provide care. We also see an appeal to the idea that the ways producers engage with animals is "science-based," with the implication that the kinds of systems producers favor in production are the result of careful research on how to best "improve the lives" of animals.

These sorts of claims are designed to offer a kind of moral cover, both for consumers and for the industry. While consumers might be surprised by what they learn from undercover investigations, they're meant to be assured that really everything is alright because these practices are being done for the sake of the animals. After all, they're told, the people doing these things to animals care deeply about them. And while consumers might feel challenged by the idea that extreme confinement, for example, could be in the interest of an animal, they are told that the housing systems producers choose for animals are consistent with guidelines grounded in science, backed by the say-so of experts. Unless consumers know something about the science, it's hard to see how they could respond.

If this attempt to offer moral cover is successful, the industry comes out immune from critique and consumers can feel at ease with practices they may have initially found to be troubling.

Whether the attempt to give moral cover succeeds, however, depends on the industry's conception of welfare. If their conception of welfare aligns with that of consumers, then consumers will have reason to feel at ease with what producers do. But if these conceptions diverge, consumers may not be able to take moral cover under the industry's claims. A recent meta-analysis seems to suggest that consumers take humane treatment (which includes concerns both about physiological and psychological well-being) and natural living conditions (which provide space and freedom for animals to engage in natural behaviors) to be central to providing for good animal welfare on farms.[3]

So what does the industry understand by animal welfare?

In order to begin to respond to this question, it will help to attend to a specific case in which the industry works to show that practices which may appear troublesome at first ought to understood as being in keeping with an interest in animal welfare. I focus for the purposes of this chapter on the use of gestation stalls in sow confinement facilities.

Gestation stalls are metal cages used to confine sows during nearly the entire duration of their four-month pregnancy. The dimensions of these stalls (typically two feet by seven feet) are such that the sows confined in them can only take one step forward and one step back. They can't turn around or lie down comfortably. Sows are transferred to other cages with similar dimensions before they give birth, and they are then re-impregnated and returned to gestation stalls just a few weeks later. This cycle continues for the sow's entire life, which on farms is typically between two and three years.

There's no question that on hearing of (or seeing) such extreme confinement for so long, it's easy to think something problematic is afoot. We need only ask what we might think of subjecting our dogs or cats to such confinement to see the issues that arise. Nevertheless, the industry endorses this kind of confinement—indeed, it's the norm on sow confinement facilities here in Minnesota. David Preisler, Executive Director of the Minnesota Pork Producers Association, has indicated that anywhere from 80%–90% of the sows in the state are confined to gestation stalls.[4]

It's important to note that industry sanctioned research acknowledges what are straightforwardly welfare issues with this sort of confinement. The American Veterinary Medical Association's policy statement on pregnant sow housing states that, "[s]tall systems restrict normal behavioral expression."[5] In her piece "Making Difficult Welfare Choices—Housing for Pregnant Sows," the AVMA's Gail Golab is a bit more forthcoming:

> Examples of individual housing include 2 ft by 7 ft gestation stalls (most common by far) … Individual housing … presents challenges when it comes to [promoting good mental health] … sows housed individually have less opportunity to socialize and … the sow's movement may be restricted and she may not be able to behave in ways that would be considered "normal" in a more natural environment—all potential welfare negatives.[6]

Despite these disadvantages, the industry understands the use of gestation stalls as providing for the sow's needs and even as offering welfare advantages over alternatives like group housing systems in which sows can interact with one another.

When it comes to the idea that gestation stalls satisfy the needs of sows, we find some revealing remarks in the "Swine Welfare Fact Sheet":

> A typical gestation stall may have an inside dimension of 22 inches … by 7 feet long … Note that the body of a large sow weighing 660 lb. … will be contained in a stall that is 17 inches wide …, 6.5 feet … long… Thus, a 2 by 7-foot stall … could easily meet the static space needs of a 660 lb. … sow.[7]

Because a 660 lb. sow is roughly 17 inches wide by 6.5 feet long, a cage that's 22 inches wide and 7 feet long meets her "static space needs." So it's in this sense that in confining a sow in a gestation stall producers think of themselves as meeting her needs—gestation stalls offer sows enough space for their stationary bodies to occupy.

The American Veterinary Medical Association's policy statement on pregnant sow housing cites advantages of gestation stalls:

> Gestation stall systems may minimize aggression and injury, reduce competition, and allow individual feeding and nutritional management, assisting in control of body condition.[8]

These advantages largely mirror those identified in the National Pork Board's position statement on gestation stalls.[9] And they seem to be targeted at what we might think of as welfare advantages. Being sure to get enough to eat and avoiding conflict and confrontation certainly seem to be good for the animals. But in order to bring out the reasons the industry sees these considerations as trumping what seem like fairly substantial welfare disadvantages, we need to see more clearly why the industry weighs these considerations so highly.

Let's first consider the question of the importance of individual feeding and nutritional management. Note that it's said to be important as a way to control the body condition of the sow. According to Ronald Bates, a swine researcher at Michigan State University:

> Correctly managing feed resources will optimize pig and sow performance and maintain favorable feed costs per cwt of pork sold. Underfeeding or over-feeding sows can unfavorably impact sow productivity and therefore adversely affect cost of production.[10]

The purpose of controlling body condition in sow facilities, it appears, has to do with maximizing productivity. When sows eat too much or too little or at the wrong times, they may give birth to fewer piglets and they may not produce milk for nursing in the ways the industry sees as optimal for sow performance.

Now consider the question of reducing competition. When sows are raised in group housing and they're fed without the help of individual feeders, sows will compete for food. The winners of the competition will get more food and the losers will go with less. Yuzhi Li, a swine researcher at the University of Minnesota, says of this conflict that it "results in the subordinate young sows becoming fearful of further conflicts while attempting to obtain feed which may lead to inadequate feed intake …"[11] When we take this together with the issue about controlling body condition, it becomes clear that these advantages of gestation stalls are not so much valued because they are advantages for the sow herself. Industry research points to these as important ways to ensure that sows are maximally productive.

When it comes to the issue of the role of gestation stalls in minimizing aggression and injury, it seems we might have come to a consideration that takes the interests of the sows themselves squarely into consideration. Aggression and injury, after all, seem like bad things for a sow to have to endure.

Yuzhi Li notes that when sows are mixed in group housing systems, fights are inevitable—sows will spend the first 24 hours or so fighting with one another in an effort to establish a dominance hierarchy in the herd. After a week, though, stable social groups form.[12]

There is no question that the injuries sows may endure in this brief period of fighting aren't good for the sows. But as we think about whether to count the concern about aggression as robust evidence of a commitment to animal welfare on the part of producers, we need to keep in mind that the alternative to this is nearly four months in gestations stalls with all the welfare disadvantages we might expect (poor mental health, lack of ability to engage in natural behaviors, limited ability to socialize, etc.). You might reasonably think that serious concern for the interests of the sows would lead to the view that spending a day fighting would be worth it if the sows also get to spend nearly four months able to move around in meaningful ways and to interact in stable social groups with other sows.

So why don't producers who decide to keep sows in gestation stalls come to this same conclusion?

We learn from Yuzhi Li and Lee Johnston that "aggression between two and three weeks after breeding can result in loss of embryos, which may cause reproductive failure."[13] Li and Johnston report that the conception rate of sows can decrease by up to 5% as a consequence of this sort of aggression. In a facility that houses thousands of sows, this adds up to a big hit to productivity.

So though it first appeared that a concern about aggression and injury counted primarily as a concern for the interests of the sows themselves, once again we find that this advantage to gestation stalls over other housing systems has to do with maximizing productivity.

The industry counts keeping sows in gestation stalls as a case of looking after the welfare of the sows. A survey of the considerations the industry offers in favor of gestation stalls in sow production, though, helps to show that what it means for a producer to take an interest in animal welfare is quite different from what we might expect.

The Animal Agriculture Alliance assures us that "[f]armers and ranchers know that when animals are well cared for, they will be healthy and productive."[14] In this very common refrain, we see the notion of welfare being connected up with the notion of productivity. What close inspection reveals, however, is that here attention to the welfare of the sows means attention to their productivity. Issues about productivity seem to be decisive in assessing whether gestation stalls provide for good welfare. It's no surprise, then, that productivity is typically seen as an indication of welfare.

We have some reason to think this focus generalizes across the industry. Bernard Rollin, for example, observes:

> by welfare, producers mean the presence of conditions relevant to the purposes for which the animal is raised. The animal receives food, water, shelter, protection from predators, and so on, all of which allow it to thrive in terms of the producer's purpose for the animal—being sold for food or producing products sold for food. Even treatment or prevention of disease enters into this view only in so far as it is relevant to the animal's productivity ...[15]

Productivity, however, has far more to do with the interests of the producers than it does with the interests of the sows themselves. It's thanks to the productivity of the animals that producers are able to make money off of them.

We may think that the collapse of concerns for welfare into concerns about productivity shows that when producers claim that they have an interest in animal welfare they are being deceptive. But this need not be so. Producers may share a conception of well-being with Aristotle.

In *Nicomachean Ethics*, Aristotle says that "the good, i.e. [doing] well … for whatever has a function and [characteristic] action, seems to depend on its function …"[16] A classic example of Aristotle's thought here involves thinking of a cook's knife. The function of a cook's knife is to cut food. Because of this, we can conclude that a bad knife (a knife that's not doing well as a knife) is one that's dull and so cuts poorly. A good knife (a knife that's doing well as a knife) will be the one that is sharp and so cuts well. On such a view, sharpening a cook's knife can be seen as benefiting the knife, because it helps the knife perform its function well.

If we understand producers as seeing the function of animals on their farms to be to produce meat, milk, or eggs, then we may be able to make sense of their claims about how a focus on productivity amounts to looking after the welfare of their animals. While the confinement of sows in gestation stalls, for example, puts pressure on a range of physiological and psychological aspects of the sows, this practice is designed to maximize productivity. Insofar as producers see productivity as the function of animals on farms, in setting the conditions for optimum productivity, producers may think of themselves as helping animals to perform their function well. Since this is thought to be the good for the animals in question, producers may see themselves as benefiting the animals after all.

It's in this light that we may see producers as genuinely understanding themselves to have the welfare of animals in mind when they engage in standard practices on intensive farms. Whether consumers can take moral cover under the claim that producers care about the welfare of their animals will then depend on whether consumers share this way of seeing animals with producers. This way of seeing animals, however, isn't the only way of seeing animals available to us and there are some compelling reasons to challenge it.

It's true that as a matter of fact animals produce meat (a rather odd way of saying that they grow), that they may produce milk (a rather odd way of saying that they may lactate), and that they may produce eggs (a rather odd way of saying that they may lay eggs). But these are not the only things they do.

The only straightforward sense in which the animals in question could be thought of as having as their primary function the production of meat, milk, or eggs, is that producers have an interest in those things the animals are able to do. That is, these capacities are brought into focus because they are the capacities of interest to producers, the capacities the animals have been bred for in the first place.

But the idea that we may benefit an animal by setting the conditions for her to perform well some capacity we happen to have an interest in seems problematic.

Let's suppose that I take in Bubs the cat so that I can watch him grow. That is, let's suppose watching him grow is my reason for bringing him into my life. Because I don't have any special interest in seeing him playing or otherwise bounding around the house and especially because doing those things would expend precious energy that could be directed to the business of growing, let's suppose I create a cage that immobilizes Bubs. It's not so small that he becomes too sick to grow, but it's not so big that he can expend extra energy on things like moving around in meaningful ways (which, after all, would cost me more when it comes to cat food). In doing this to Bubs, I set the conditions for him to perform well the capacity of his that I have an interest in.

The idea that I'm benefiting Bubs in confining him as I do in such a case seems strange. And it would collapse into absurdity if Bubs didn't much like being in such extreme confinement. If you came over and saw Bubs in his cage trying desperately to get free, it wouldn't do for me to claim that I'm benefiting him despite appearances because I'm helping him to fulfill the purpose I had for him when I took him in. The same goes, though, if in his cage Bubs had fallen limp, resigned to a life in confinement.

Apart from the challenges in thinking it makes sense to target productivity as the characteristic function of other animals, there are other concerns with this way of seeing them.

Matthew Scully offers a reminder of a way in which how we see other animals can become corrupted when we only see them through the lens of our own interests:

> when you look at a rabbit and can see only a pest, or vermin, or a meal, or a commodity, or a laboratory subject, you aren't seeing the rabbit anymore. You are seeing only yourself and the schemes and appetites we bring into the world ...[17]

And Stephen R. L. Clark reminds us of the cost of seeing animals so narrowly:

> by stripping away the false perception that such creatures exist for us, we are enabled to see them in their beauty ...[18]

Scully and Clark raise worries that seeing other animals through the lens of our interests keeps us from seeing them for who they are. This set of considerations has a lot of power to bring out the way that a focus on productivity flattens our ability to appreciate our fellow creatures. But it also offers a reason to take seriously some work animal scientists have done around questions of animal welfare. Fraser and Weary, for example, point out that productivity may well be a measure of proper biological functioning of an animal, but it misses altogether concerns about their emotions and feelings and about whether they have opportunities to engage in natural behaviors.[19] To miss those seems to miss what's central to an ethical concern for animal welfare.[20]

We've been exploring the possibility of offering producers the benefit of the doubt when it comes to their conception of animal welfare. Perhaps they think that they offer care for animals in the sense that they set the conditions for animals to do their best in production. But we've also seen what I take to be a range of serious issues with such a view. It turns out, though, that since this view requires producers to have an interest in promoting the productivity of individual animals, there is reason to worry whether producers can even hold this view.

When it comes to housing animals on farms, producers routinely opt for systems that maximize the productivity of a barn as a whole rather than the productivity of individual animals. In egg production, for example, there is a point at which increasing the amount of space hens are afforded will result in diminishing marginal productivity. While the animal's individual productivity will continue to increase a bit, the extra space allowed the hen will displace other hens. The result of this displacement is that the number of eggs produced in the barn as a whole will drop. Agricultural economists Norwood and Lusk offer this assessment of the situation:

> space allotment chosen by farmers does not maximize animal well-being, or even individual animal productivity. When farmers are constrained by land, labor, barn size, or even availability of capital, the economically optimal stocking density will be more crowded than will suit the animals.[21]

A shift from thinking about an interest in individual animal productivity to an interest in productivity of the operation as a whole seems to collapse from a concern that could potentially be seen as a concern with some ethical dimension (however misguided) to an overt concern for profit. Indeed, Norwood and Lusk ask us to keep in mind that on farms "animals are a commodity and they are treated well to the extent that it is profitable."[22]

Whether consumers can take moral cover under the language the industry offers about how much they care about the welfare of their animals depends centrally on sharing with producers their conception of welfare. For my own part, I can't. A conception of animal welfare that's so closely tied to the notion of productivity and, ultimately, profit seems to me to have far more to do with the interests of producers than it does with the interest of the animals themselves.

It is these considerations, then, that lead me to think it's difficult to take moral cover under the notion that producers do what they do out of an interest in animal welfare.

Notes

1 "USFRA's View on Animal Welfare," United States Farmers and Ranchers Alliance, accessed August 26, 2015, www.fooddialogues.com/foodsource/usfras-view-on-animal-welfare.
2 "Animal Care and Modern Food Production," Animal Agriculture Alliance, accessed August 26, 2015, http://animalagalliance.org/images/upload/Animal%20Welfare%20Brochure%202013%20-%20Final.pdf%20with%20bleed.pdf.
3 Beth Clarke et al., "A Systematic Review of Public Attitudes, Perceptions and Behaviours Towards Production Diseases Associated with Farm Animal Welfare," *Journal of Agricultural and Environmental Ethics* 29, no. 3 (2016): 455–478.
4 David Preisler, email, August 23, 2013.
5 "Pregnant Sow Housing," American Veterinary Medical Association, accessed August 26, 2015, www.avma.org/KB/Policies/Pages/Pregnant-Sow-Housing.aspx.
6 Gail Golab, "Making Difficult Welfare Choices: Housing for Pregnant Sows," American Veterinary Medical Association, May 18, 2012, http://atwork.avma.org/2012/05/18/making-difficult-welfare-choices-housing-for-pregnant-sows/.

7 "Swine Welfare Fact Sheet," National Pork Board, accessed August 2015, https://porkcdn.s3.amazonaws.com/sites/all/files/documents/Factsheets/WellBeing/SWINEWELFARsowsandspace.pdf.
8 American Veterinary Medical Association, "Pregnant Sow Housing."
9 "Position Statements on Sow Housing," National Pork Board, accessed August 2015, www.pork.org/position-statements-sow-housing/.
10 Ronald Bates, "Sow Body Condition Influences Productivity and Profitability," Michigan State University Extension, accessed August 26, 2015, http://msue.anr.msu.edu/news/sow_body_condition_influences_productivity_and_profitability.
11 Yuzhi Li, "Controlling Aggression Among Group-Housed Gestating Sows," University of Minnesota Extension, accessed August 10, 2015, www.extension.umn.edu/agriculture/swine/components/pdfs/controlling_aggression_among_group_housed_gestating_sows.pdf.
12 Li, "Controlling Aggression Among Group-Housed Gestating Sows."
13 Yuzhi Li and Lee Johnson, "Tips for Managing Pregnant Sows in Group Housing," *Farm and Dairy*, July 24, 2012, www.farmanddairy.com/news/experts-offer-eleven-tips-for-managing-pregnant-sows-in-group-housing/39482.html.
14 Animal Agriculture Alliance, "Animal Care and Modern Food Production."
15 Bernard Rollin, *Farm Animal Welfare* (Ames, IA: Iowa State Press, 2000), 29.
16 Aristotle, *Nicomachean Ethics*, trans. Terrence Irwin (Indianapolis, IN: Hackett, 2000), 8.
17 Matthew Scully, *Dominion* (New York, NY: St Martin's Press, 2002), 2–3.
18 Stephen R. L. Clark, "Vegetarianism and the Ethics of Virtue," in *Food for Thought: The Debate Over Eating Meat*, edited by Steve Sapontzis (Amherst, NY: Prometheus Books, 2004).
19 David Fraser and Daniel Weary, "Quality of Life for Farm Animals: Linking Science, Ethics, and Animal Welfare," in *The Well-Being of Farm Animals*, edited by John Benson and Bernard Rollin (Ames, IA: Blackwell, 2004), 39–60.
20 Cf. I. J. H. Duncan, "Animal Welfare Defined in Terms of Feelings," *Acta Agriculturae Scandinavica, Section A, Animal Science*, Supplement 27 (1996): 29–35.
21 F. Bailey Norwood and Jayson Lusk, *Compassion by the Pound* (Oxford, England: Oxford University Press, 2011), 102.
22 Norwood and Lusk, *Compassion by the Pound*, 102.

Bibliography

American Veterinary Medical Association. "Pregnant Sow Housing." Accessed August 26, 2015. www.avma.org/KB/Policies/Pages/Pregnant-Sow-Housing.aspx.

Animal Agriculture Alliance. "Animal Care and Modern Food Production." Accessed August 26, 2015. http://animalagalliance.org/images/upload/Animal%20Welfare%20Brochure%202013%20-%20Final.pdf%20with%20bleed.pdf.

Aristotle. *Nicomachean Ethics*. Translated by Terrence Irwin. Indianapolis, IN: Hackett, 2000.

Bates, Ronald. "Sow Body Condition Influences Productivity and Profitability." Michigan State University Extension. Accessed August 26, 2015. http://msue.anr.msu.edu/news/sow_body_condition_influences_productivity_and_profitability.

Clark, Stephen R. L. "Vegetarianism and the Ethics of Virtue." In *Food for Thought: The Debate Over Eating Meat*, edited by Steve Sapontzis. Amherst, NY: Prometheus Books, 2004.

Clarke, Beth, Gavin B. Stewart, Luca A. Panzone, I. Kyriazakis and Lynn J. Frewer. "A Systematic Review of Public Attitudes, Perceptions and Behaviours Towards Production Diseases Associated with Farm Animal Welfare." *Journal of Agricultural and Environmental Ethics* 29, 3(2016): 455–478.

Duncan, I. J. H. "Animal Welfare Defined in Terms of Feelings." *Acta Agriculturae Scandinavica, Section A, Animal Science*, Supplement 27(1996): 29–35.

Fraser, David and Daniel Weary. "Quality of Life for Farm Animals: Linking Science, Ethics, and Animal Welfare." In *The Well-Being of Farm Animals*, edited by John Benson and Bernard Rollin, 39–60. Ames, IA: Blackwell, 2004.

Golab, Gail. "Making Difficult Welfare Choices: Housing for Pregnant Sows." American Veterinary Medical Association. May 18, 2012. http://atwork.avma.org/2012/05/18/making-difficult-welfare-choices-housing-for-pregnant-sows/.

Li, Yuzhi. "Controlling Aggression Among Group-Housed Gestating Sows." University of Minnesota Extension. Accessed August 10, 2015. www.extension.umn.edu/agriculture/swine/components/pdfs/controlling_aggression_among_group_housed_gestating_sows.pdf.

Li, Yuzhi and Lee Johnson. "Tips for Managing Pregnant Sows in Group Housing." *Farm and Dairy*. July 24, 2012. www.farmanddairy.com/news/experts-offer-eleven-tips-for-managing-pregnant-sows-in-group-housing/39482.html.

National Pork Board. "Position Statements on Sow Housing." Accessed August 3, 2015. www.pork.org/position-statements-sow-housing/.

National Pork Board. "Swine Care Handbook." Accessed August 3, 2015. http://porkcdn.s3.amazonaws.com/sites/all/files/documents/AnimalWell-Being/swine%20care%20handbook%202003.pdf.

National Pork Board. "Swine Welfare Fact Sheet." Accessed August 3, 2015. https://porkcdn.s3.amazonaws.com/sites/all/files/documents/Factsheets/WellBeing/SWINEWELFARsowsandspace.pdf.

Norwood, F. Bailey and Jayson Lusk. *Compassion by the Pound*. Oxford, England: Oxford University Press, 2011.

Rollin, Bernard. *Farm Animal Welfare*. Ames, IA: Iowa State Press, 2000.

Scully, Matthew. *Dominion*. New York, NY: St Martin's Press, 2002.

United States Farmers and Ranchers Alliance. "USFRA's view on animal welfare." Accessed August 26, 2015. www.fooddialogues.com/foodsource/usfras-view-on-animal-welfare.

2.5

TAKING ON THE GAZE OF JESUS

Perceiving the factory farm in a sacramental world

Jim Robinson

Introduction

In vivid, eucharistic imagery, Wendell Berry asserts that in order to exist, "we must daily break the body and shed the blood of Creation."[1] When we do so "knowingly, lovingly, skillfully, reverently, it is a sacrament," but when we do so "ignorantly, greedily, clumsily, destructively, it is a desecration."[2] And it is by desecrating the body of creation that we "condemn ourselves to spiritual and moral loneliness," while simultaneously subjecting "others to want."[3] This is our situation: Our sustenance as selves depends, tenuously and intimately, upon our relationship with the broader body of creation, which we must break and eat on a daily basis. Each meal marks our intricate complicity in the breaking and taking of life. The food we consume issues a call to contemplation, prompting us to question whether we are involving ourselves in a loving communion with the broader body of creation, or in destructive desecration. And yet, the rippling ramifications of our daily bread far too often elude our view. The food that we purchase at a restaurant, or in a supermarket is, as Peter Singer observes, all too frequently the "culmination of a long process, of which all but the end product is delicately screened from our eyes."[4]

In this chapter, I attend to factory farms as obscured sites of desecration, within which bodies are mechanically broken and blood is methodically shed. The beings who are moved through these spaces are seen through a desensitizing economic lens, and are consumed as meat, deliberately sanitized of its association with grave desecration and concrete suffering. I lift the factory farm from an economic context, and ground it instead in a dramatically different worldview, which takes the entirety of creation to be a sacrament, suffused with the presence of God. I propose that the sacramental worldview is a particularly vivid resource, developed within the Christian tradition, for promoting a capacity to see and sense the irreplaceable sacredness of the beings who are routinely defiled in factory farms, and for emphasizing the immense moral and spiritual gravity of this defilement. I stress the necessity for sensitive perception that sensing the sacramentality of our world requires. And I argue that an enhanced sensitivity to the sacramental nature of the world might animate a capacity to more deeply perceive and more readily respond to the desecration of creation, especially as manifested in the deliberately concealed spaces of factory farms. I additionally emphasize the extent to which our desecration of creation signifies not only our ruptured and deeply wounded relationship with the broader body of creation, but our estrangement and alienation from God. To desecrate creation is to distort its sacramental quality, thereby obscuring divine presence and obstructing our capacity for relating to God in all. I therefore read the renunciation of meat to be a centrally significant ascetical practice, which interrupts our complicity in desecration, and moves us into a more sensitive, harmonious, and reverent relationship with this sacramental world and with the loving God underlying it.

Receiving a sacramental world

In *Laudato Si': On Care for Our Common Home*, Pope Francis quotes and affirms the remarks of Patriarch Bartholomew, delivered at the inaugural Halki Summit in 2012, in which Bartholomew identifies the world itself as a veritable sacrament of communion.

> As Christians, we are … called "to accept the world as a sacrament of communion, as a way of sharing with God and our neighbours on a global scale. It is our humble conviction that the divine and the human meet in the slightest detail in the seamless garment of God's creation, in the last speck of dust of our planet.[5]

In echoing Bartholomew, Francis urges his readers to receive creation as a sacrament of communion, to recognize that we ceaselessly meet the divine in the myriad contours of materiality, down to and including the specks of dust drifting amongst us, and to relate to the world in light of these insights. Each thread in the garment of creation, within which we, as a species, are inextricably interwoven, announces our interdependence with everyone and everything, as well as our constant capacity for acknowledging and connecting with the subtle, loving presence of God undergirding all. There is, in this vision, both a luminous transparency and a fluid harmony about our world. It is a seamless garment, a relational space within which we might freely mingle and share with each other, meeting the divine in all of our encounters. The cleanly delineated borders, the rigid walls, the oppressive sanitization, the rows of cages, the maze of machinery, the agenda of containing, concealing, and controlling animal bodies that characterizes the atmosphere of a factory farm are utterly foreign to this vision of the world. The non-world of the factory farm, conceived by human minds and constructed by human hands, collides violently with the sacramental vision of creation articulated by Bartholomew and Francis.

Francis declares that the "entire material universe speaks of God's love."[6] As such, the soil, the water, the mountains, and every other aspect of creation serve as a "caress of God."[7] God, in this picture, ceaselessly reaches through matter, affirming every aspect of creation as worthy of love and capable of revelation. Just as God makes intimate contact with our bodies through caresses of soil, water, mountains, sky, we might likewise celebrate our communion with God in and through the textures of creation. To attune ourselves to the touch of the divine in the suchness of every moment is to receive the world, in its shifting actuality, as an endlessly novel sacrament of communion. This option, Francis avers, is definitively on the table. And yet, we far too frequently fail to receive and reciprocate. The earth "cries out" in a muffled and unheeded chorus of voices in response to our violent and violating footprint.[8] The severity of the ecocrisis, which is epitomized in factory farms, signals the degree to which we are so radically *out of touch* with the natural world, through which God goes on caressing us, despite our callousness. To begin to sense the presence of the divine amongst us, and to see the unrepeatable sacredness of each unique being, might compel us to work more fervently to dismantle systems of desecration, which warp our world into so many zones of devastation, oppression, and death.

Opening to ecological conversion

The patterns of violence that exist in our hearts and minds are, for Francis, "reflected in the symptoms of sickness evident in the soil, in the water, in the air and in all forms of life."[9] We have turned our backs upon the earth, even as we inhale the earth's air, even as we drink in the earth's waters, even though our very flesh is an assemblage of earthly elements.[10] To heal the sickness which has infiltrated our hearts, and which has come to bear bitter fruit in the degradation of soil, water and air, will require both an internal and external interruption of the status quo. We are called, in this context, to undergo what Francis terms an "ecological conversion."[11] We are called to re-turn to the earth, to acknowledge our interdependence, and to begin to develop healthy, holy and mutually enriching modes of living into the future.

To encounter the world as a sacrament of communion must aid in stimulating this necessary *metanoia*, for, in the words of Leonardo Boff, "a sacrament without conversion is condemnation."[12] In order to receive the world as a sacrament of communion, in order to relate to reality as a seamless garment, saturated with the divine, we must work to disentangle ourselves from desecrating patterns of being, which fragment the earth community, and lock us all into sickness and suffering. As Aristotle Papanikolaou emphasizes, the "sacramentality of creation" is encountered through the concrete "performance of ascetical practices that opens one up to communing with the life of God that is in and around creation."[13] There is, then, a direct correlation between an enhanced attentiveness to the sacramentality of creation and a commitment to ecological conversion through concrete ascetical practices, such as the renunciation of meat. While our deepening reception of the earth as a sacrament animates a more vivid commitment to ecological conversion, our ascetical practices of disrupting desecration feeds a deepening sensitivity to the sacramentality of the earth.

Sacramental participation as training to be a beholder

Pope Francis opens the section of his encyclical explicitly devoted to sacramental theology, "Sacramental Signs and the Celebration of Rest," with the observation that the "universe unfolds in God, who fills it completely."[14] His sensitivity to the extent to which the unfolding cosmos is saturated in the presence of God leads him to affirm that there is deep mystical meaning to be found in the entire order of creation, from leaves to mountain trails, from glittering dewdrops to the faces of people experiencing poverty.[15] There is, by way of extrapolation, deep mystical meaning to be found in chickens and cows, lambs and lobsters, pigeons and pigs. Francis presents the seven public sacraments of the Church as particularly privileged instances in which nature becomes an efficacious "means of mediating supernatural life."[16] It is in the Eucharist, he asserts, that creation finds its "greatest exaltation."[17] It is here that Christ intertwines with our most "intimate depths" in the form of a humble "fragment of matter."[18] The Eucharist confirms that God "comes not from above, but from within" so that we might "find [God] in this world of ours."[19] As such, the celebration of the Eucharist should enhance our capacity for attending to the presence of God within, where God touches even our most "intimate depths," as well as in the world around us, which God accompanies, loves and indwells.[20] For Francis, the Eucharist is an "act of cosmic love," which "embraces and penetrates all creation" and which should function as a "source of light and motivation for our concerns for the environment."[21]

Fr. Michael Himes proposes that to participate in the life of the sacraments is to engage in a gradual process of "training to be beholders."[22] In this perspective, to enter into the rhythms of liturgy consists of an attunement to the divine mystery, which is always present but seldom sensed, reflected upon and responded to.[23] This sensitivity to divinity, which is uniquely developed in liturgical life, bleeds out into a far broader vision of sacramentality that numbers the sacraments as "virtually infinite."[24] For Himes, there are as many sacraments as there are "things in the universe," as any person or place, any event, sight, sound, any taste, touch, or smell that prompts us to acknowledge the love that undergirds our being and all beings can be received as a sacrament.[25] In this respect, Himes insists that nothing is "by definition profane," as everything possesses the potential to operate sacramentally, by opening us to acknowledge the divine presence, which unfailingly "supports all that exists."[26] We can encounter the presence and gratuitous love of God in all, but we must first learn to properly perceive. Himes's insight resonates closely with Leonardo Boff's assertion that it is only when we are "awake" in the world that we can perceive it as a "sacrament of God."[27] To become awakened beholders, to cultivate a capacity for sacramental seeing, is to assume a stance of radical attentiveness, receptivity, and sensitivity to the world, and to the presence of God penetrating all. In a context of obscuration, the struggle to see might be adequately assessed as a radically transgressive act.

A sacramental worldview refuses a vision of divinity cordoned off from the contours of creation, hovering forever elsewhere. It simultaneously refuses the move to reduce non-human beings to mere objects, stripped of the inherent value indelibly impressed upon them by the presence of God underlying their beaks and eyes, their muscles, tongues and tails, rendering them sacred, revelatory, irreplaceable. A sacramental worldview affirms the inherent depth and significance of every aspect of existence, and directs our attention toward the beings of our world as sites of divine disclosure. In this spirit, Thomas Berry declares that "when we destroy the living forms of this planet," we "destroy modes of divine presence."[28]

John Hart asserts that the universe is sacramental because it is a "revelation of the Spirit's ongoing creativity."[29] To hold the cosmos as sacramental means, for Hart, to acknowledge the fact that the "totality of creation" is "infused with the vision, love, creative presence, and active power" of the Spirit.[30] As such, the cosmos, both in its holistic, expanding scope, as well as in its most delicate detail, can be properly read and related to as the milieu in which we confront divine presence, grace and blessing.[31] It is in and through the textures of creation that we sense the sacred, and experience our intimate relationality with the entire earth community. The fact that the Spirit suffuses all of creation, taking its aspects as its dwelling place, confirms, for Hart, that all space is sacred space.[32] Wendell Berry reminds us that "there are no unsacred places;" instead, there are "only sacred places" and "desecrated places."[33] The factory farm is, definitively, a radically desecrated place. Our training to become beholders, awake and alert to the sacramentality of creation, must aid us in attending to the desecration of our world, which is not only hidden in our midst, but deliberately concealed.

A place of desecration

In a harrowing depiction of contemporary hog farming, Matthew Scully captures the extreme degree to which American consumers are removed from contact, let alone communion, with pigs. He additionally conveys the extent to which these animals are extracted from, and denied access to, the rhythms of the natural world. According to the National Pork Producers Council, Scully notes, 80 million of the 95 million hogs brought to slaughter in the United States on an annual basis are "intensively reared in mass-confinement farms," which translates to a life deprived of soil and sunshine.[34] Scully writes,

> Genetically designed by machines, inseminated by machines, fed by machines, monitored, herded, electrocuted, stabbed, cleaned, cut, and packaged by machines—themselves treated like machines 'from birth to bacon'—these creatures, when eaten, have hardly ever been touched by human hands.[35]

This is the portrait of an obscured site of desecration. Annually, 80 million unique beings, who are loved into being by God, whose bodies are saturated with mystical meaning and revelatory potential, who might communicate to us the divine caress in the textures of their flesh, are ensnared in and eclipsed by the machinery of our factory farms. We cannot see them. We cannot reach them. We cannot sense their suffering. The workers[36] recruited to electrocute, stab, and carve the bodies of these pigs—those who do the dirty work of desecration—wear earplugs in order to muffle the howls.[37] Scully sensitively links the systemic degradation articulated above to a simple impulse: The "hankering for a hot dog."[38] Beyond the space of the factory farm, the remnants of sacramental bodies are related to as bacon, hot dogs and ham, sanitized and scrubbed of the residue of deep and concrete suffering. To give into this "hankering for hot dogs" is to solidify one's complicity in this system, and therefore to involve oneself in desecration. To *give up* this hankering, along with the desires for animal flesh, milk and eggs in all of their carefully repackaged forms, can be adequately assessed as an ascetical act of deep ethical and spiritual significance.

Conclusion: Taking on the gaze of Jesus

In the section of his encyclical entitled "The Gaze of Jesus," Pope Francis portrays the historical Jesus as an intimately "earthly" figure, engaged in what he terms a "tangible and loving relationship with the world."[39] Christ, for Francis, existed in "constant touch with nature," lending the contours of creation an "attention full of fondness and wonder."[40] Sensitively, and tenderly, Christ constantly reminded his disciples to attend to each and every creature as irreducibly significant, worthy of our awe.[41]

In imitating this gaze, in taking on the eyes of Christ, we might attune ourselves to both the sacramental nature of creation, as well as its lamentable desecration, with a radically heightened sensitivity. As awakened beholders, in touch with the revelatory textures of our world, receptive and responsive to the sacramental beings with whom we share this planet, not only will we be imitating the gaze of Christ, we will be encountering him. For Francis, "The very flowers of the field and the birds which his human eyes contemplated and admired are now imbued with his radiant presence."[42] And, we must add, the pigs, the chickens, the cows, the veal calves howling for mother and meal, they are all imbued with his presence and glory. We are called, then, to take on the gaze of Jesus, to attend to this sacramental world, and to reciprocate the caresses of God, lest we perpetuate desecration, "condemn[ing] ourselves to spiritual and moral loneliness" and "others to want."[43]

Notes

1 Wendell Berry, "The Gift of Good Land," in *The Gift of Good Land: Further Essays Cultural and Agricultural* (Berkeley, CA: Counterpoint, 1981), 281.
2 Berry, "The Gift of Good Land," 281.
3 Berry, "The Gift of Good Land," 281.
4 Peter Singer, *Animal Liberation* (New York, NY: HarperCollins, 2002), 95.

5 Patriarch Bartholomew, "Global Responsibility and Ecological Sustainability," Closing Remarks, Halki Summit I, Istanbul, June 20, 2012. Cited in Pope Francis, *Laudato Si': On Care for Our Common Home* (Vatican City: Libreria Editrice Vaticana, 2015),§ 8. In an earlier address, delivered during the presentation ceremony for the Sophie Prize in 2002, Bartholomew asserts that the "*original sin of humanity*" in relationship to the natural world is our "*refusal to accept the world as a sacrament of communion with God and neighbor* [emphasis his]." See: Patriarch Bartholomew, "The Sophie Prize," in *Cosmic Grace, Humble Prayer: The Ecological Vision of the Green Patriarch Bartholomew I*, edited by John Chryssavgis (Grand Rapids, MI: Wm. B. Eerdmans Publishing Co., 2009), 284.
6 Pope Francis, *Laudato Si'*, §84.
7 Pope Francis, *Laudato Si'*, §84.
8 Pope Francis, *Laudato Si'*, §2.
9 Pope Francis, *Laudato Si'*, §2.
10 Pope Francis, *Laudato Si'*, §2.
11 Pope Francis, *Laudato Si'*, §2; §5; §217; §219; §220. In advocating an "ecological conversion," Pope Francis is echoing Pope John Paul II, who calls for this in his general audience of January 17, 2001. John Paul II declares that "we must … encourage and support the 'ecological conversion' which in recent decades has made humanity more sensitive to the catastrophe to which it has been heading" (§4). In her *Ask the Beasts: Darwin and the God of Love* (London, England: Bloomsbury, 2014), Elizabeth Johnson stresses that a genuine conversion to the earth must include three "discrete turnings" simultaneously: an intellectual turning, an emotional turning, and an ethical turning (258). Intellectually, conversion to the earth requires a shift from an anthropocentric worldview to a theocentric one, which would be capacious enough to include the non-human in the sphere of what is perceived to be "religiously meaningful and valued" (258). Emotionally, conversion to the earth involves a shift from the "delusion of the separated human self" and the "isolated human species" to a "felt affiliation" with all beings (258). Ethically, conversion to the earth consists of an expanded vision of the moral sphere, so that "vigorous moral consideration" can be extended to the entirety of creation (259).
12 Leonardo Boff, *Sacraments of Life, Life of the Sacraments*, trans. John Drury (Washington, DC: The Pastoral Press, 1975), 92.
13 Aristotle Papanikolaou, *The Mystical as Political: Democracy and Non-Radical Orthodoxy* (Notre Dame, IN: University of Notre Dame Press, 2012), 2–3.
14 Pope Francis, *Laudato Si'*, §233.
15 Pope Francis, *Laudato Si'*, §233.
16 Pope Francis, *Laudato Si'*, §235.
17 Pope Francis, *Laudato Si'*, §236.
18 Pope Francis, *Laudato Si'*, §236.
19 Pope Francis, *Laudato Si'*, §236.
20 Pope Francis, *Laudato Si'*, §236.
21 Pope Francis, *Laudato Si'*, §236.
22 Michael J. Himes, "'Finding God in All Things': A Sacramental Worldview and Its Effects," in *As Leaven in the World: Catholic Perspectives on Faith, Vocation, and the Intellectual Life*, ed. Thomas M. Landy (Franklin, WI: Sheed and Ward, 2001), 100.
23 Himes, "Finding God in All Things," 100.
24 Himes, "Finding God in All Things," 99.
25 Himes, "Finding God in All Things," 99.
26 Himes, "Finding God in All Things," 99.
27 Leonardo Boff, *Sacraments of Life*, 8.
28 Thomas Berry, "The Earth Community," in *The Dream of the Earth* (San Francisco, CA: Sierra Club Books, 1988), 11.
29 John Hart, *Sacramental Commons: Christian Ecological Ethics* (Lanham, MD: Rowman & Littlefield Publishers, 2006), 1.
30 Hart, *Sacramental Commons*, 61.

31 Hart, *Sacramental Commons*, 12.
32 Hart, *Sacramental Commons*, 3.
33 Wendell Berry, "How to Be a Poet," *Poetry* 177, no. 3 (January 2001): 269–70. Cited in: Douglas Christie, *The Blue Sapphire of the Mind: Notes for a Contemplative Ecology* (New York, NY: Oxford University Press, 2013), 28.
34 Matthew Scully, *Dominion: The Power of Man, the Suffering of Animals and the Call to Mercy* (New York, NY: St. Martin's Griffin, 2002), 29.
35 Scully, *Dominion*, 29.
36 The women and men who do the dirty work of desecration are often amongst the most marginalized and vulnerable members of society, and are subjected to immense health and safety hazards. In *For the Love of Animals: Christian Ethics, Consistent Action* (Cincinnati, OH: Franciscan Media, 2013), Charles Camosy notes that most factory farm workers are poor, generally do not belong to unions, and are often undocumented immigrants or temporary foreign workers (95). Their vulnerability renders them less likely, and less capable, to "complain or quit in light of the horrific and damaging tasks they are required to complete" (95). These workers, he notes, are at risk for bacterial infections, as well as other extreme health problems, including upper respiratory issues due to the inhalation of toxic gas (95). Francis, echoing Leonardo Boff, declares that we must be careful to "hear *both the cry of the earth and the cry of the poor*" (§49). These two entangled cries are all too often muffled in unison.
37 Scully, *Dominion*, 282.
38 Scully, *Dominion*, 44.
39 Pope Francis, *Laudato Si'*, §100.
40 Pope Francis, *Laudato Si'*, §97.
41 Pope Francis, *Laudato Si'*, §96.
42 Pope Francis, *Laudato Si'*, §100.
43 Berry, "The Gift of Good Land," 281.

Bibliography

Berry, Thomas. "The Earth Community." In *The Dream of the Earth*, 11. San Francisco, CA: Sierra Club Books, 1988.

Berry, Wendell. "How to Be a Poet." *Poetry* 177, 3(January2001): 269–270.

Berry, Wendell. "The Gift of Good Land." In *The Gift of Good Land: Further Essays Cultural and Agricultural*, 281. Berkeley, CA: Counterpoint, 1981.

Boff, Leonardo. *Sacraments of Life, Life of the Sacraments*. Translated by John Drury. Washington, DC: The Pastoral Press, 1975.

Camosy, Charles. *For Love of Animals: Christian Ethics, Consistent Action*. Cincinnati, OH: Franciscan Media, 2013.

Christie, Douglas. *The Blue Sapphire of the Mind: Notes for a Contemplative Ecology*. New York, NY: Oxford University Press, 2013.

Hart, John. *Sacramental Commons: Christian Ecological Ethics*. Lanham, MD: Rowman & Littlefield Publishers, 2006.

Himes, Michael J. "'Finding God in All Things': A Sacramental Worldview and Its Effects." In *As Leaven in the World: Catholic Perspectives on Faith, Vocation, and the Intellectual Life*, edited by Thomas M. Landy Franklin, 91–103. Franklin, WI: Sheed and Ward, 2001.

Johnson, Elizabeth. *Ask the Beasts: Darwin and the God of Love*. London, England: Bloomsbury, 2014.

Papanikolaou, Aristotle. *The Mystical as Political: Democracy and Non-Radical Orthodoxy*. Notre Dame, IN: University of Notre Dame Press, 2012.

Patriarch Bartholomew. "Global Responsibility and Ecological Sustainability." Closing Remarks. Halki Summit I. Istanbul. June 20, 2012.

Patriarch Bartholomew. The Sophie Prize." In *Cosmic Grace, Humble Prayer: The Ecological Vision of the Green Patriarch Bartholomew I*, edited by John Chryssavgis, 312–313. Grand Rapids, MI: Wm. B. Eerdmans Publishing Co., 2009.

Pope Francis. *Laudato Si': On Care for Our Common Home*. Vatican City: Libreria Editrice Vaticana, 2015.

Pope John Paul II, "General Audience: Wednesday 17 January 2001." Vatican.va. Accessed April 8, 2016. http://w2.vatican.va/content/john-paul-ii/en/audiences/2001/documents/hf_jp-ii_aud_20010117.html.

Scully, Matthew. *Dominion: The Power of Man, the Suffering of Animals and the Call to Mercy*. New York, NY: St. Martin's Griffin, 2002.

Singer, Peter. *Animal Liberation*. New York, NY: HarperCollins, 2002.

2.6

"A LAMB AS IT HAD BEEN SLAIN"

Mortal (animal) bodies in the Abrahamic traditions

Marjorie Corbman

One of the companions of the Prophet Muhammad, Jabir ibn Abd Allah, narrated a story about an instance in which the Prophet suddenly stopped walking through the crowded marketplace of Medina. He was staring at a lamb's body, hanging in front of them, waiting to be sold. This slaughtered animal was not a perfect specimen; as the hadith tells us, "it had very short ears." The Prophet lifted up one of its ears and turned to his friends, asking who among them would buy the carcass for a single silver coin. They replied: They wouldn't buy the animal even for less. The Prophet asked again: Would they take the animal for free? Their second response was more assertive: Even if the animal were alive, they wouldn't want him or her due to the flawed ears. The animal was of no use to them; what's more, the animal was dead. It is only after this complete rejection on their part of this unusable, worthless, lifeless body that the Prophet delivers the punchline: "By Allah, this world is more insignificant in the eye of Allah than [this dead lamb] is in your eye."[1]

The power of the story rests in its enforced double-take, the swift movement from the companions' total differentiation of the dead lamb from themselves to the total identification the Prophet urges his friends to accept between the mortality and imperfections present in the slaughtered animal and in themselves. Moreover, the Prophet in the hadith only successfully makes its case because of a context of faith in the magnanimity and compassion of God. Without Muhammad's companions already-existing assumption that God, as the Qur'an repeatedly declares, is "is compassionate and merciful to humanity" despite human weakness and limitations (e.g., 2:143), the story would end in despair, in an assertion that humans and the world in which we live are ultimately as valueless as that hanging corpse—in other words, that from a God's-eye-view, we are trash.

On the contrary: God's primary names, the same tradition tells us, are Mercy and Compassion. The Prophet thus grants his friends new eyes with which to see themselves, the world, and the dead animal in front of them—all are equally dependent on, and wholly undeserving of, the mercy, care, and kindness of God. The dead lamb's life is just like theirs, from its embarrassingly obvious inadequacies to its final impotence in the face of death. No matter how worthless the lamb may appear to us, the animal is a life from God, one that like ours is mysteriously and, though inexplicably, persistently sustained by God's care and protection.

Before we can derive contemporary meaning from this story, however, we have to recognize the radical distance between us and the participants of Jabir ibn Abd Allah's narrative. It is not likely, had the Prophet Muhammad appeared not in seventh-century Arabia but rather in the contemporary United States, that the events of this story would have transpired, for one simple reason: We no longer see dead bodies. The regular contact of people with animal bodies during the times in which the sacred texts of the Abrahamic traditions were written prompted spiritual and moral questions that we, as beneficiaries of an agricultural system that deliberately operates invisibly from the public, can easily ignore.

Thus, I will argue here that in order for advocates of animal rights to mobilize widespread support for more just food systems, there must first be a shift in perception of the world. We have to change how we *see* food; we have to use imagination to restore the lines of vision that have purposefully been occluded from our view. We have to see dead bodies. Even those among us who refrain from eating animal bodies or animal products have to consciously work against our culturally enforced illusion that death can be banished from our lives or from our plates. Confronting the reality of animal death would fundamentally alter how our society relates to animals as food, but it also would change how we understand ourselves as mortal, relational beings.

In this chapter, I will begin with a discussion of how the absence of animal bodies from our perception affects us, using especially the thought of Marxist theorist Guy Debord. Following this, I will move to an overview of how encounter with animal bodies in the traditions of Judaism, Christianity, and Islam influenced attitudes about the treatment of non-human animals. Those of us who are members of these religious communities hold texts sacred that reflect a more direct encounter with animals than what our food systems currently allow us. Thus, I will argue that imaginative readings of these texts can be a vital tool in reshaping our limited perception of ourselves as embodied beings in relationship with others. Before we can help non-human animals, we have to see them in their lives and deaths, their play and their pain, their total and undeniable reality, their bodies.

Mortal bodies and the spectacular society

In order for the meat industry to maintain its crucial role in the world economy, more than hiding the cruelty of the industry's practices is necessary. The public's desire to eat meat at grossly unsustainable levels is manufactured, requiring not only massive amounts of capital invested in marketing and lobbying for meat production but also, in the case of the United States, government subsidies totaling nearly $40 billion per year.[2] In order to reverse this trend, effective and creative organizing with the aim of fracturing the domination of the meat industry over the public's food consumption is necessary. Yet this organizing cannot only target multinational corporations and government representatives, as the power of the meat industry rests upon a pervasive coercion of public consciousness, grounded in the replacement of reality as actually experienced in human and animal bodies with an illusory version of the same.

In order to illustrate this far-reaching divorce from reality, take the extreme example of an advertisement produced by Burger King in the year 2000, as part of a promotional tie-in for the animated motion picture, *Chicken Run*. The children's film charmingly and powerfully portrays an escape plan launched and executed by a community of chickens living on a farm in rural England. Meanwhile, of course, Burger King makes its profits selling the products of an industry that forces billions of chicks every year into unbearable conditions. Each Burger King TenderGrill chicken sandwich or carton of chicken nuggets represents the mutilation and torture of countless members of the species of which Ginger, Babs, Rocky, or any of the lovable Claymation characters in *Chicken Run*, represent.

Not only is this reliance on animal suffering not visible in the Burger King advertisement, but an alternative reality is substituted for it. The voiceover invites its audience to "be part of the greatest escape ever hatched," as children playing with plastic chickens and the tools the birds used to escape from their farm appear on the screen. "You get one with every tasty Burger King kids' meal you buy."[3] In the world depicted by this advertisement, chickens as *animals* simply do not exist as *food*. The advertisement only makes sense within a version of reality, as Carol J. Adams has definitively shown, in which the concept of a *chicken* has no relationship whatsoever to the concept of a chicken sandwich.[4] The images of golden-brown, crispy morsels of deep-fried chicken displayed on a child's plate earlier in the advertisement are meaningless; they are bodiless. They have no broken wings, no severed beaks, no osteoporosis caused by lack of movement, no infections, no slit throats, no blood, no life or death. They have no reality except as objects of consumption.

In order to understand how the acceptance of this inaccurate depiction of reality is constructed and disseminated, it is instructive to consult Guy Debord's description of the "falsification of life" endemic to what he calls the "spectacular" society.[5] Debord uses the term "spectacle" to describe the process in which images, in the form of "news or propaganda, advertising or the actual consumption of entertainment" maintain and reflect the "unreality" of our society and its core devotion to the processes of commodification.[6] He argues that capitalist society has advanced to a point in which we continue to believe that constant production is necessary despite the fact that what is being produced far outpaces our actual needs and desires. This is the reason that false perception is necessary. In order to keep the system functioning, we have to perceive ourselves as always wanting more. At the same time, we have to perceive everything and everyone else as a potential commodity.

Debord insists that this manufacturing of false consciousness is omnipresent and virtually unassailable. We cannot undo our commitment to unreality without addressing the fervent devotion to commodification at the heart of our social worldview. In this, intriguingly, Debord's critique of contemporary society is largely theological. The commodity, he argues, is the false god of capitalist society, a deity that demands loyalty and worship over and against reason and our own individual agency. It is the illusion that determines the world, a "transcendent spirit" that plunges its followers into ever-escalating levels of absurdity.[7] Ultimately, Debord charges, this unreality severs us from our mortal bodies, from actual life and actual death, replacing the fact that we are going to die with only the question of "how much the appearance of life can be maintained in the individual's encounter with death."[8]

Debord in this draws heavily from Marx's argument that capitalism can only perpetuate itself through what he called a "mystical" attachment[9] to something that does not, in the physical world, actually exist, that is, exchange value. What Marx saw as the dangerous fiction at the heart of our economic system is that somewhere in the process of buying, selling, comparing and competing, the commodity stops existing as a real object, and the labor that produced it stops existing as real labor.[10] Marx's primary concern with this societal acquiescence to economic fiction is the way in which it is used to exploit labor, but his critique is wider-reaching than this. It is an indictment of a worldview that holds most sacred what does not exist.[11]

Writing a century after Marx, Debord describes the expansion of this worldview, in the wake of increasing industrialization and globalization, to ever-rising levels of irrationality, reflecting the "complete colonization of social life" by the commodity.[12] This is the economic and social context that prevents us from seeing animals *as* animals, that manufactures desire for meat far beyond what the planet can sustain, and that excuses the treatment of animals as receptacles of lifeless material—neither really alive nor really dead—valuable only in their potential to be bought and sold.

"The life is in the blood": slaughter and sacrifice

Given the ways in which physical bodies in a non-commodified form are made invisible in our society, the Hebrew Scriptures can at first glance shock us with what seems to be a virtually uninterrupted narrative of slaughter, body parts, and death. The Book of Leviticus reads as especially inaccessible to many Jews and Christians in the modern world due to its thorough cataloguing of practices of animal sacrifice as well as its frequent descriptions of disease or bodily discharges. Most scholarly volumes today on Leviticus begin with some form of apology for the subject matter, as in the first sentence of J. Edward Owens' introduction to his commentary on the book: "Leviticus often seems a strange and antiquated text to the modern reader."[13] Nor is this aversion limited to contemporary Christian readers: For instance, Rabbi Maurice D. Harris' introduction to his commentary on the text is aptly titled, "Cozying Up to the Most Avoided Book in the Bible."[14]

Leviticus' unpopularity, however, is a relatively recent phenomenon, at least in Jewish communities. Historically, the text has been used as the first to introduce children to the study of Torah, and this is still the practice in many observant communities of Orthodox Jews. The centrality of this text to the tradition, particularly with its emphasis on animal sacrifice, clashes with our modern expectations about what constitutes religion. Animal sacrifice has not been practiced by Jewish communities since the destruction of the Second Temple in 70 CE, the only location in which these sacrifices could lawfully be conducted, but ongoing study of the Temple and the sacrifices commanded to be performed in it remains integral to observant Jewish practice.

To many, the concept of ritual animal sacrifice can appear cruel or unnecessary, an abandoned vestige of allegedly "primitive" religion. This attitude, however, largely misunderstands the historical role of animal sacrifice and at the same time reflects the refusal to acknowledge death and bodies present in many contemporary societies. Rabbi Harris illustrates the hypocritical aspect of our cultural repugnance towards animal sacrifice by quoting a young student in his Hebrew school classroom:

> Well, which do you think is more moral? Doing a sacred ritual and dealing with God every single time you kill an animal for its meat, or anonymously shoving millions of animals into crowded pens and cages... [and] then cutting up their body parts, shrink-wrapping them in plastic and lining the walls of grocery store refrigerator cases with a horror show of dead animal body parts?[15]

Unlike us, Harris notes, the ancient Israelites regularly saw animals' blood and body parts in public. What's more, "the people who witnessed the priests going through these procedures knew that they were made up of essentially the same parts."[16] The requirement for the process of taking another life was confronting the reality of one's own too-similar death.

The nature of sacrifice as explained in the Hebrew Scriptures and particularly Leviticus conveys a sense of mild terror at the prospect of killing other living beings. This is only to be expected within a society in which animals were raised and cared for by families, as can be seen, for instance, in the prophet Nathan's moving description in the second book of Samuel of a fond relationship shared between a poor man and his only significant possession, a small ewe lamb:

> The poor man had nothing but one little ewe lamb, which he had bought. He brought it up, and it grew up with him and with his children; it used to eat of his meager fare, and drink from his cup, and lie in his bosom, and it was like a daughter to him.
>
> *(2 Samuel 12:3)*

In order to mitigate the fear of trespassing sacred boundaries in slaughtering an animal, strict limitations upon the action were adopted. Anyone of the Israelite camp, so we are told in Chapter 17 of Leviticus, who slaughters an animal without bringing the animal before the presence of God as an offering, is "guilty of bloodshed" and must be cut off from the people (Lev. 17:4).

The connection between shedding human blood and shedding animal blood recalls the words of the covenant God made with Noah in the ninth chapter of Genesis, in which God explains that animal blood cannot be eaten because life—whether human or animal—is sacrosanct:

> You shall not eat flesh with its life, that is, its blood. For your own lifeblood I will surely require a reckoning: From every animal I will require it and from human beings, each one for the blood of another.
>
> *(Gen. 9:4–5)*

Indeed, the chapter of Leviticus cited above repeats the same ban against consuming blood: "The life of every creature" it reminds us "its blood is its life" (Lev. 17:14). It is commanded instead that the blood of every animal killed for food be poured down onto the earth and covered with soil, that is, buried and returned to God (v. 13). It is this deep concern for the holiness of life that the biblical scholar Jacob Milgrom argued constitutes the core meaning of the "purity" that was sought through the Israelites' sacrificial system.[17]

The Book of Deuteronomy relaxes the strictures against killing without offering the animal bodies to God, thereby centralizing all sacrificial worship at the Temple in Jerusalem (Deut. 12:15–27).[18] Yet mindfulness of the potential of transgressing God's ownership of all life through killing other beings pervades the Hebrew Scriptures and was preserved in the detailed laws of *shechita*, or animal slaughter, which until the modern period were as visible to observant Jews as the sacrificial rites in the Temple had been.[19] While contemporary practice of *shechita* at times utilizes the same destructive and inhumane practices as the rest of the meat industry, the origin and history of the laws convey attitudes towards animal life greatly at odds with these harmful systems.[20] In the ideal form of the practice, animal death was brought ritually into the presence of God in full consciousness that the fragile, slaughtered bodies on display reflected humans' identical mortality.

The slaughtered lamb on the throne

The close affinity in scriptural texts between the bodies of human worshippers and the bodies of the animals brought for slaughter perhaps gives us a clue how to interpret the mysterious fact that the final image of God given by the Christian Scriptures is of a lamb with its throat slit, reigning from God's glorious throne. In the fifth chapter of the Book of Revelation, the author of the text describes a vision in which he saw the multitudes surrounding the divine throne joyfully exclaim to the wounded animal reigning from it: "Worthy is the Lamb that was slaughtered to receive power and wealth and wisdom and might and honor and glory and blessing!" (Rev. 5:11). Artists that have attempted to portray this scene have had to work carefully to reflect its seemingly contradictory aspects—conveying the majesty and power of the figure being worshipped while the description given by the text seems to connote only the helplessness, blood, and pain of an innocent animal. The lamb, of course, is Jesus, but we should not move too quickly from this knowledge away from the image of the lamb without contemplating what significance this equivalency has for actual animal bodies. Of course, the author of the Book of Revelation is clearly drawing an analogy: Just like the lambs slaughtered on Passover whose blood was shed in place of the Israelite firstborns, the blood shed by Jesus on his way to resurrection grants power to those whom he loves over death and over all the world's cruelties.[21]

The metaphor only works, however, when the sacrificial worldview as described above is taken for granted, so that human blood and human life are understood as interchangeable with animal blood and animal life. Revelation's depiction of Jesus as a lamb only retains its original power insofar as we can make the imaginative connection the author of the text wanted his hearers to draw between three separate examples of mutilated and blood-drained bodies: 1) those murdered unjustly by the Roman Empire, 2) animals that have been sacrificed, and 3) Jesus, who is "the firstborn of the dead, and the ruler of the kings of the earth" (Rev. 1:5). Once again, this parallel was much more easily understood by those who regularly witnessed human and animal death and therefore understood the relationship between our bodies and those of non-human animals. In displaying the slaughtered lamb on the throne, the author connects Jesus' exaltation to the future exaltation of his murdered followers. Yet it is also precisely a *lamb's* body that is exalted, that is given new life. Life belongs to God, not us; God can glorify any body, no matter how weak.

Slaughter and prayer

Ritualized animal sacrifice ended for the early Christian community at the same time that it ended for Jews, following the destruction of the Temple.[22] Thus, it is hard for Christians or Jews to comprehend except through the imagination how rituals like those described in their religious texts can transform the ways in which animal bodies are seen or conceptualized.

For Muslims, however, representatives of the other major Abrahamic tradition, the connection remains in place. On Eid al-Adha, the feast that ends the hajj—the yearly pilgrimage to Makkah—many Muslim households all over the world slaughter an animal as they recall how God provided an animal for the Prophet Ibrahim to sacrifice in place of his son Ismail. While a household can retain one third of the meat gained from this sacrifice to eat at home, the remainder must be shared, preferably among the poor, those who would not be able to purchase meat on their own.[23] This follows the example of the Prophet Muhammad, who asked his companions not to save more than three days' worth of meat for themselves during a year in which many were hungry.[24]

This annual contact with animal death is profoundly counter-cultural when practiced in countries in which meat production is systematically hidden from the consumers of meat. The Muslim author and environmentalist Ibrahim Abdul-Matin beautifully describes the relationship between witnessing an Eid al-Adha sacrifice and advocating for the ethical treatment of animals in his book, *Green Deen* (2010). He tells a story of visiting a family farm with a friend whose business provides ethically raised halal meat, in order to help him slaughter turkeys before the American holiday of Thanksgiving. After reciting the blessing over the first turkey they slaughtered together—*bismillah, allahuakbar*; in the name of God, God is great—Abdul-Matin recalled seeing an animal sacrificed for the first time at the age of eight. He remembered watching the lamb after he or she was killed, looking as if he or she had fallen asleep.

This memory helped him understand what he was doing as he killed the turkeys, which is to say, something gravely serious. "It's like prayer," his friend tells him. "You don't want it reduced to a mechanical action … The idea is to understand the value of it. You are not supposed to let the process become rote." Abdul-Matin writes of his deep consciousness as he recited the blessing over the animals of his need for God's permission, of the necessity of respecting the other life in his hands. He describes his gratitude that the turkeys lived lives that reflected their dignity, ones that were happy and free. Echoing the words of his friend, he recalls, "I felt like I was in a state of prayer."[25]

Conclusion

In relating this experience as described by Abdul-Matin, I am not arguing that individual participation in the slaughter of ethically-raised farmed animals by those who consume meat is the solution for reforming the deeply unethical practices of our food industry. Industrialized agricultural and meat production require massive overhauls in standards and creative rethinking of structures; systemic change, rather than an emphasis on a change of behavior of individual consumers, is needed. In order for systemic change to happen, however, as I have argued above, we need a change in perception. The feeling of utter *seriousness* Abdul-Matin recognized in himself as he mindfully approached an animal with a knife reflects that he was learning to see the animal not as a commodity, as lifeless material, but as a body, which also means as a being that will die. Without recognizing our own fragility reflected in animal bodies, the limits of our power over other mortal beings, we will not understand or effectively be able to challenge the unethical systems in which we are embedded.

Most of us no longer regularly interact with the animals constantly marketed to us as commodities, and for many people, doing so is not accessible. Yet it is possible for texts and storytelling to keep practices that have fallen into disuse alive. There is no better example of this than how Jewish tradition, after the destruction of the Temple, has sustained the memory of the former sacrificial rites through imaginative and repetitive study of texts. Each year on Yom Kippur, as only one example of this, the section of the liturgy entitled the Avodah comprehensively recounts the rites of atonement formerly conducted in the Temple. In light of how our current societal and economic structures alienate us from animal bodies, a similar practice for us might be returning to the texts of our faith traditions that describe animals and animal death in all their seemingly distasteful detail. Doing so will remind ourselves of the reality of their bodies and their mortality, so similar to our own. As we return to these passages continually and mindfully, we might begin to fracture the unifying hold of a commodified mentality on our consciousness, and be brought closer to the One who gave us life and to whom our souls, our bodies, and our blood must ultimately return.

Notes

1 "Sahih Muslim 42:7059," Sunnah.com, accessed July 12, 2016, http://sunnah.com/muslim/55.
2 See, e.g., David Robinson Simon, *Meatonomics: How the Rigged Economics of Meat and Dairy Make You Consume Too Much* (San Francisco, CA: Conari Press, 2013).
3 Burger King, "Burger King Ad—Chicken Run 1 (2000)," YouTube, posted by that90sguy, August 28, 2014, accessed July 13, 2016, www.youtube.com/watch?v=LMc9z_7SEns.
4 See, e.g. Carol J. Adams, *The Sexual Politics of Meat* (New York, NY: Maiden Lane, 2010), 66–69. Adams' underscoring of the ways in which marginalized groups (i.e., animals and women) are particularly absented from their bodies and realities helpfully complicates and challenges the Debordian emphasis on the totality of a commodified worldview: Some bodies and realities are far more commodified and exploited than others. For the purposes of this chapter, however, Debord's theories are used to highlight the need for complete conversion of a commodifying perception in which both bodies and death do not exist.
5 Guy Debord, *The Society of the Spectacle*, trans. Donald Nicholson-Smith (New York, NY: Zone Books, 1994), 45.
6 Debord, *The Society of the Spectacle*, 13.
7 Debord, *The Society of the Spectacle*, 44.
8 Debord, *The Society of the Spectacle*, 114–115.
9 Karl Marx, *Capital: Volume 1: A Critique of Political Economy*, trans. Ben Fowkes (London, England: Penguin Books, 1990), 164.
10 Marx, *Capital*, 128.
11 It should be noted that how Marx described the "religious" heart of capitalist devotion to the commodity as "commodity fetishism" carried with it clear colonialist overtones. However, the tendency of capitalist ideology to replace physical objects and beings with abstract ideas remains instructive. See Sylvester A. Johnson, *African American Religions, 1500–2000: Colonialism, Democracy, and Freedom* (Cambridge, England: Cambridge University Press, 2015), 105–106.
12 Debord, *The Society of the Spectacle*, 29.
13 J. Edward Owens, *Leviticus*, The New Collegeville Bible Commentary, Vol. 4 (Collegeville, MN: Liturgical Press, 2011), 5.
14 Maurice D. Harris, *Leviticus: You Have No Idea* (Eugene, OR: Cascade Books, 2013), xvii.
15 Harris, *Leviticus*, 36.
16 Harris, *Leviticus*, 37.
17 Jacob Milgrom, *Leviticus: A Book of Ritual and Ethics* (Minneapolis, MN: Fortress Press, 2004), 12–13.
18 See Milgrom, *Leviticus*, 15.
19 See Rabbi Zalman Schachter, "Foreword," in *Vegetarianism and the Jewish Tradition*, ed. Louis A. Berman (New York, NY: Ktav Publishing House, 1982), xiv.
20 Many Jewish communities have argued for the importance of acknowledging animal and human dignity in abiding by *halachic* standards, as can be seen for instance in the call for

hekhsher tzedek/magen tzedek (ethical certification for kosher food) among Jewish communities in the US. See "The Magen Tzedek Standard," Magen Tzedek, accessed September 30, 2016, www.magentzedek.org/.

21 The "lamb of God" analogy (in Revelation and in John 1:29) is not exhausted by the paschal example but also could be understood in reference to other sacrifices described in the Torah that could be performed with lambs. However, for the sake of space and in recognition of the dominance of the paschal theme in early Christian literature, I have not elaborated on this here. For an example of the importance of this theme in ancient Christianity, see Melito of Sardis' *Peri Pascha*.

22 See Daniel Ullucci, *The Christian Rejection of Animal Sacrifice* (New York, NY: Oxford University Press, 2012) on anti-sacrificial attitudes among Christians only becoming prevalent in the century following the destruction of the Temple.

23 "Eid al-Adha, Festival of Sacrifice," in *Asian American Religious Cultures*, Volume 1, eds. Jonathan H. X. Lee, Fumitaka Matsuoka, Edmond Yee, and Ronald Nakasone (Santa Barbara, CA: ABC-CLIO, 2015), 357–259.

24 "Sahih Bukhari 7:68:476," Sunnah.com, accessed July 15, 2016, http://sunnah.com/bukhari/73.

25 Ibrahim Abdul-Matin, *Green Deen: What Islam Teaches about Protecting the Planet* (San Francisco, CA: Berrett Koehler Publishers, 2010), 174–177.

Bibliography

Abdul-Matin, Ibrahim. *Green Deen: What Islam Teaches about Protecting the Planet*. San Francisco, CA: Berrett Koehler Publishers, 2010.

Adams, Carol J. *The Sexual Politics of Meat*. New York, NY: Maiden Lane, 2010.

Burger King. "Burger King Ad—Chicken Run 1 (2000)." YouTube. Posted by that90sguy. August 28, 2014. Accessed on July 13, 2016. www.youtube.com/watch?v=LMc9z_7SEns.

Debord, Guy. *The Society of the Spectacle*. Translated by Donald Nicholson-Smith. New York, NY: Zone Books, 1994.

"Eid al-Adha, Festival of Sacrifice." In *Asian American Religious Cultures*, Volume 1. Edited by Jonathan H. X. Lee, Fumitaka Matsuoka, Edmond Yee, and Ronald Nakasone, 357–359. Santa Barbara, CA: ABC-CLIO, 2015.

Harris, Maurice D. *Leviticus: You Have No Idea*. Eugene, OR: Cascade Books, 2013.

Johnson, Sylvester A. *African American Religions, 1500–2000: Colonialism, Democracy, and Freedom*. Cambridge, England: Cambridge University Press, 2015.

Magen Tzedek. "The Magen Tzedek Standard." Magen Tzedek. www.magentzedek.org.

Marx, Karl. *Capital: Volume 1: A Critique of Political Economy*. Translated by Ben Fowkes. London, England: Penguin Books, 1990.

Milgrom, Jacob. *Leviticus: A Book of Ritual and Ethics*. Minneapolis, MN: Fortress Press, 2004.

Owens, Edward J. *Leviticus*. The New Collegeville Bible Commentary. Vol. 4. Collegeville, MN: Liturgical Press, 2011.

Schachter, Rabbi Zalman. "Foreword." In *Vegetarianism and the Jewish Tradition*. Edited by Louis A. Berman. New York, NY: Ktav Publishing House, 1982.

Simon, David Robinson. *Meatonomics: How the Rigged Economics of Meat and Dairy Make You Consume Too Much*. San Francisco, CA: Conari Press, 2013.

Sunnah.com. "Sahih Bukhari 7:68:476." Accessed July 15, 2016. http://sunnah.com/bukhari/73.

Sunnah.com. "Sahih Muslim 42:7059." Accessed July 12, 2016. http://sunnah.com/muslim/55.

Ullucci, Daniel. *The Christian Rejection of Animal Sacrifice*. New York, NY: Oxford University Press, 2012.

2.7

CATTLE HUSBANDRY WITHOUT SLAUGHTERING

A lifetime of care is fair

Patrick Meyer-Glitza

Introduction

The abolitionist position within the animal rights movement has recently been challenged by Sue Donaldson and Will Kymlicka,[1] as also by Alasdair Cochrane.[2] These authors allow for the keeping of agricultural animals under specific circumstances[3] or "without Liberation."[4] Lacto-vegetarians, on the other hand, are implicated in the slaughtering of cows that are not milked anymore and of fattening male cattle, although they often do not know about it. In Germany, milked cows have a live span of 5.3 years[5] – or about 4.7 years if the calf losses and heifer deaths are taken into account. In organic farming the live span is about 6.8[6] or 6.2 years if the calf losses and heifer deaths are taken into account. Also, the dairy bull calves of organic farms are usually sold to and fattened on conventional farms.

While the abolitionist position is supported by many vegans, for vegetarians the question arises, if it is possible to secure a right of life for cattle and also to consume a reduced amount of milk products. This non-slaughtering milk production would also allow organic farmers the benefits of cow manure within their farm system and to integrate the benefits of fodder legumes into crop rotation.

Method

In line with the small population of farms without slaughter that are milking, qualitative-interpretative methods were chosen. These methods are suited for emerging topics and phenomena.[7] Interviews within five case studies – four in Europe and one in India – were conducted, using narrative and semi-structured interview techniques. The interviewed pioneers have at least fifteen years' experience in developing their farm system. During the ongoing research the cases were selected by theoretical sampling[8] while drawing on emerging results and opportunities, looking for minimal and maximal contrasts among the farms. In addition, the research was initially focused on farms that produce milk, and also farm sanctuaries with cattle came into focus. The farms work as certified organic farms or rely very much on organic farming principles. Notably, in the last two to four years the number of sanctuaries keeping cattle has increased in Germany and Western countries. Most of these sanctuaries also keep other farmed animals.

The data of each case study was coded and categorized separately,[9] using the coding system of grounded theory and the software Atlas.ti. Additionally, text-segments were analyzed by sequential micro analysis,[10] leading to the development of categories and subcategories along their respective properties and dimensions. All the while, memos were generated and refined. The analyzed single cases were anonymized and compared across-case and along the categories within the three given interests of research: The biography of the farmers, the ethics of action and the management of the agri-system.[11] The development of the farms has been followed up since the first interviews in 2008/2009.

The two emerging systems of farming are:

1. The care-system, which is summarized by its five basic rules and its sanctuary function.
2. The agri-system of husbandry and animal products.

Combined, these two worlds make up the agri-care-system.

Basic rules of the care-system

The care-system draws on animal rights,[12] especially right of life, and on the ethics of care.[13] The ethics of care initially looks at significant individual relations among people, including those who cannot sufficiently care for themselves, like children, or patients in the healthcare sector.[14] But there also emerged an ethics of care focussing at significant relations of people to animals like farmed animals.[15] In regard to caring professions Beauchamp and Childress[16] say that "The *ethics of care* emphasizes traits valued in intimate relationships such as sympathy, compassion, fidelity, and love. Caring, in particular, refers to care for, emotional commitment to, and deep willingness to act on behalf of persons with whom one has a significant relationship." The five basic rules of the care-system are:

1 Universal

All cattle, that is, both sexes, all ages, in different health statuses and handicapped cattle, are kept on the farm and are cared for. They are part of the moral community. This basic rule is mainly expressed by the sanctuary function of rescue–protection–care: Of rescuing the animals from slaughter, neglect and abuse; of protecting the cattle from ever being slaughtered or sold; and of constant care (see Lifetime of care).

2 Unconditional

Productivity is no precondition for the animals' right of life and all cattle are entitled to the same benefits as the "productive" animals. Although most of the case studies welcome or expect at least some cattle products, this is no precondition. The life of the cattle, their being alive, is the main product. As with the cattle, all other farmed animals that come on the farm have the (same) right of life and care.

3 Lifetime of care

The animals are cared for during their "complete life"[17] which includes old age, illness and dying. The cattle will not be (re-)commodified[18] but cared for, resembling institutional roles of old age homes,[19] asylum[20] homes, hospitals, workshops for handicapped, kindergartens, hospices and homes for psychological support. Uncommon forms of illness like arthritis, cancer, etc. have to be dealt with e.g. in geriatric and palliative care, while there is not much experience by many vets regarding these illnesses in cattle.

The lifetime of care is a specification of the moral status including palliative care towards the end of life and as a lifetime–progression–integrity (see below). While the Hare Krishna farms do not practice euthanasia, the other farms practice a "preference-respecting"[21] euthanasia.

4 Familizing

The (time) intensity of care, especially for dying animals often resembles that for human family members. The animals are looked at as distinct individualities and treated as part of the enlarged family. This is a metaphor as well as a functional role of the animals and also can be described – in line with attachment theory[22] – as a bond between man and the individual animal resembling the bonds among the core family members. In regard to animals, attachment is usually looked at as bonds between companion animals and humans.[23] Regarding cattle, the bond depends on the time of the cattle staying on the farm, the number of cattle, the use, function and medical history of the cattle, the daily physical contact with the cattle[24] and the amount of work done by hand. In other words: Within a familized "local human/animal culture"[25] on the farm, humans and animals move along a continuum towards (or also away from) each other in their individual pace and timeframe.

5 Prevention

The farmers not only rescue animals from slaughtering but also work to become a model of how to live with farmed animals and prevent slaughtering. They spread the word by selling products (e.g. dung products, calendars, T-shirts), welcoming visitors, the media and the press as well as networking. The development of the farm as a working model and as a public relation initiative includes vocational training and political campaigning like one project, which initiated a trial at the Indian Supreme Court to ban plastic bags in public.

The agri-system

Adopting non-slaughtering policies on a cattle farm leads to the development of a new farm system. This includes adapting to an average lifetime of the cows that more than doubles from 5.3 years[26] to 12.5 years (see Table 2.7.1), ranging from 11.4 years for the milked cows to 16.4 years for the sanctuaries. That the cows of the sanctuaries get older than the milked cows is in part due to the fact that few calves are born at the sanctuaries and that on the other hand calve mortality (10%) is included in the figures of the two case studies milking. With an average lifetime of 10 years for oxen, the lifetime for the male cattle is raised to roughly about 7:1 compared to the usual fate of fattened bulls.

A unique feature of this husbandry system is the length of lactation, which averages 2.83 years for case study 3 (Table 2.7.2). On another farm, some of the cows even have spontaneous seasonal lactations, without being in calf again and after e.g. 2–3 years of being dried up. The long lactations go along with a replacement rate in case study 3 for milked cows as low as 10.2%, which is only about a third of conventional farms.

Assuming that the milked cows have two calves (and lactations) in their lifetime to replace themselves and an additional ox, they would be milked for 5.7 years and retire for 3.8 years in case study 3. Thus, for every year of the 5.7 years being milked, these cows also carry the cost for 0.63 years of retirement and of 1.48 male

TABLE 2.7.1 Age of cattle dead and alive in years[27]

Cattle	*Case Study 1*	*Case Study 3*	*Case Study 4*	*Case Study 5*	*Mean all case studies*	*Mean, the two milking farms*	*Mean sanctuaries*
Died female cattle	14.5 (n = 4)	12.1 (n =19.5)	19.0 (n = 3)	8.6 (n = 5)	**12.5 (n = 32)**	**11.4 (n = 25)**	**16.4 (n = 7)**
Died male cattle	5.7 (n = 3)	9.9 (n = 13.5)	17 (n = 2)	0	**10.0 (n = 19)**	**9.9 (n = 14)**	**10.2 (n = 5)**
Died cattle	10.7 (n = 7)	11.2 (n = 33.0)	18.0 (n = 5)	10.8 (n = 5)	**11.8 (n = 50)**	**11.1 (n = 38)**	**13.7 (n = 12)**

TABLE 2.7.2 Milk yield per lactation and year

Farm	*Milk yield/year*	*Milk yield/lactation in l*	*Length of lactation in years*	*Daily milk yield in l*
Case study 3	3,200	9,055	2.83	8.8

cattle (oxen, male calves and breeding bulls). With increased age of the cattle, the veterinary and medical costs also rise, while the bodies of the dead cattle (carcasses) are not used in any way.[28]

For case study 3, the lactations average 9,055 l in 2.83 years with about 3,018 l as an average per year (see Table 2.7.2).

Compared to the treatment of cats and dogs,[29] veterinarians have little experience in geriatric and palliative care for cattle. The non-slaughtering farms do pioneering work in geriatric homeopathy, caring for the handicapped, old and dying cattle, e.g. turning over downers with a stationary crane or a "vest" attached to the front loader, and in dealing with unfamiliar forms of illness.

The herds usually have a steady and often mixed composition. The cattle usually graze during day and night and are kept in deep litter in the stable during winter. The feeding is up to 1–2 kg concentrate per day for the milked cows, which could be even less in regard to the relatively low milk production.

Adapting to the new farm system and producing the same amount of milk for consumption – excluding the substantial amount of milk for rearing the calves (see below) – would require in Germany about 8.1 times[30] the number of cattle as in conventional dairies and about 5.8 times the number of the conventional cattle including beef cattle.

Because Germany has a trade surplus in dairy products you would need less than 5.3 times the amount of dairy cattle for 100% milk self-sufficiency from slaughtering-free dairies.[31] With a solely non-slaughtering dairy production you could produce an estimated 39% of milk and 38% of cheese products and also 30% of butter of the national consumption in 2012. These percentages presume the substitution of all beef cattle by a slaughter-free dairy.

As there is also much less concentrate feed used in the cattle husbandry without slaughtering, less arable land will be needed to feed the same amount of cattle. This additional arable land could partly be used for grass-clover as part of the crop rotation, feeding the cows as well as the soil in an ecological way of farming.

A dairy husbandry without slaughtering entails a radical reduction of milk production while milk becomes an expensive by-product of the animal life. The life of the animal, or its being alive is the main product of the farms, especially at the sanctuaries. Dung is a valuable by-products for four of the five case studies, milk for two of them. New dung-products of the farming sanctuaries are fermented fertilizers, massage oil, incense sticks, soap and distilled urine. Most of the farms grow and sell produce.

The bull calves are castrated, kept on the farm and partly used for draft.[32] The calves of the sanctuaries have, and the calves of dairies should have[33] free access to the mother-cow to suckle for about 10 months until natural weaning[34] takes place. The daily milk for the calves can be calculated during the first 4 months as 12 kg of milk and from the 5th month on until the end of the 10th month as 9 kg of milk daily. This ends up with a total amount of suckled milk of 3,111 kg. In contrast to the usual practice even in organic dairy farming, the rearing of the calves with the dam includes the rearing of the bull calves in all case studies.

Some of the non-slaughtering farms that do not milk their cattle, use their manure for growing horticulture or other cultures. This is the case with two of the case studies, the farm of Christina Menicocci[35] in Italy, the Canadian farm of Mike Lanigan[36] and was also the case with the German horticulturalist Willmann. On the Menicocci farm, cattle and other animals roam and graze between the olive-trees and vines. The cows of the horticulturalist Willmann were sponsored by people who are placed by the association Lebenlassen e.V.[37]

There still remain dilemmas like castration, euthanasia (or for the Hare Krishna farms how to care for dying animals without euthanasia), the ecological footprint and the feeding of the cats and dogs for the case studies.

Outlook

The combination of the five basics of the care system and the tenets of the agri-system constitute the agri-care-system, which includes the right of life and allows for manure and a radically reduced amount of milk production.

Cattle husbandry without slaughtering is suitable for farms with permanent pasture, as in Germany 27.7% of agricultural land is permanent pasture,[38] or when no or a little amount of concentrate is fed, as also for organic farming where grass-clover is a most common part of crop rotation.

Minimizing the killing of mammals, birds, insects and amphibians while working the land to grow fodder remains a challenge. This can partly be achieved by using techniques like cutter bars and by adapting the time, frequency, height and way of cutting, as also by ways of scaring off the mammals.

Some of the case studies should minimize their dependency on buying fodder. The farms gain sustainability working together with(in) communities, e.g. as CSAs (Community Supported Agriculture) like the Gita Nagari Farm.[39] These communities support the development of new products and help securing a financial basis. Finding people to work with the oxen in order to integrate and finance the oxen as part of the farm system remains a central challenge. Although the cattle in a non-slaughtering system – especially during winter – are still confined to a certain degree, the case studies have the strong tendency to dissolve the ambivalence of humans towards companion animals and farmed animals.

Notes

1 Sue Donaldson and Will Kymlicka. *Zoopolis, A Political Theory of Animal Rights* (Oxford, England: Oxford University Press, 2011), 81ff.
2 Alasdair Cochrane, *Animal Rights without Liberation* (New York, NY: Columbia University Press, 2012).
3 Donaldson and Kymlicka, *Zoopolis*, 122ff.
4 Cochrane, *Rights without Liberation*.
5 Vereinigte Informationssysteme Tierhaltung w. V. (VIT), *Trends Fakten Zahlen 2014* (Verden, 2015), 18.

6 KTBL, *Faustzahlen für den Ökologischen Landbau.* Darmstadt (Kuratorium für Technik und Bauwesen in der Landwirtschaft e.V., 2015), 488.
7 Gabriele Rosenthal, *Interpretative Sozialforschung. Eine Einführung* (Weinheim und München, Germany: Juventa, 2005), 18.
8 Anselm L. Strauss and Juliet M. Corbin, *Grounded Theory: Grundlagen Qualitativer Sozialforschung* (Weinheim, Germany: Beltz, Psychologie Verlags Union, 1996), 148–165.
9 Kathy Charmaz, *Constructing Grounded Theory: A Practical Guide through Qualitative Analysis* (Thousand Oaks, CA: Sage Publication), 2006.
10 Rosenthal, *Interpretative Sozialforschung*, 213.
11 Patrick Meyer-Glitza and Ton Baars "Non-killing Cattle Husbandry," in *Tackling the Future Challenges of Organic Animal Husbandry: 2nd Organic Animal Husbandry Conference, Hamburg, Trenthorst, 12–14 September, 2012*, ed. Gerold Rahmann and Denise Godinho (Braunschweig, Germany: vTI Landbauforschung, Sonderheft 362, 2012), 184–187.
12 Tom Regan, *The Case for Animal Rights* (Berkeley, CA: University of California Press, 2004).
13 Josephine Donovan and Carol J. Adams, "Introduction," in *The Feminist Care Tradition in Animal Ethics*, eds. Josephine Donovan and Carol J. Adams (New York, NY: Columbia University Press, 2007), 1–15.
14 Tom L. Beauchamp and James F. Childress, *Principles of Biomedical Ethics*, 6th ed. (Oxford, England: Oxford University Press), 2008, chapter 2.
15 Donovan and Adams, *Introduction*, 1–15.
16 Beauchamp and Childress, *Principles of Biomedical Ethics*, 36.
17 Hilal Sezgin, *Artgerecht ist nur die Freiheit: Eine Ethik für Tiere oder Warum wir umdenken müssen* (München, Germany: C.H. Beck, 2014), 178.
18 cf. Rhoda M. Wilkie, "Sentient Commodities and Productive Paradoxes: The ambiguous Nature of Human-livestock Relations in Northeast Scotland," *Journal of Rural Studies* 21 (2) (2005).
19 cf. Donaldson and Kymlicka, *Zooplois*, 104–108.
20 Lee Hall, *On Their Own Terms. Bringing Animal-Rights Philosophy Down to Earth* (Darien, CT: Nectar Bat Press, 2010), 49.
21 Regan, *Animal Rights*, 119.
22 Signal Zilcha-Mano, Mario Mikulincer and Phillip R. Shaver, "An Attachment Perspective on Human–pet Relationships: Conceptualization and Assessment of Pet Attachment Orientations," *Journal of Research in Personality* 45 (4) (2011).
23 Zilcha-Mano Mikulincer and Shaver, *Attachment Perspective.*
24 Bettina B. Bock et al., "Farmers Relationship with Different Animals: The Importance of Getting Close to the Animals," *International Journal of Sociology of Agriculture and Food* 15 (3) (2007): 117ff.
25 Pär Segerdahl, "Can Natural Behavior be Cultivated? The Farm as Local Human Animal Culture," *Journal of Agricultural and Environmental Ethics* 20 (2007): 185.
26 Vereinigte Informationssysteme Tierhaltung, *Trends Fakten Zahlen*, 18.
27 Case study 2 is not included in the table as there is no such data available for the Indian case study
28 Although fighting for and practicing cow protection, Mohandas (Mahatma) Gandhi propagated to use a tannery to make cow protection economically feasible for the farmers. Gandhi himself had one. See: Florence Burgat, "Non-Violence Towards Animals in the Thinking of Gandhi: The Problem of Animal Husbandry," *Journal of Agricultural and Environmental Ethics* 14 (2004): 227, 237, 242.
29 Wilfried Kraft, ed., *Geriatrie bei Hunden und Katzen*, 2nd ed. (Stuttgart, Germany: Parey Verlag in MVS Medizinverlage Stuttgart, 2003).
30 The oxen of the non-slaughtering farms are included since most of the oxen on these farms are not working animals.

31 Bauernverband, ed., *Situationsbericht 2014/15 Trends und Fakten zur Landwirtschaft* (Berlin, Germany: Bauernverband, 2015), 22ff.
32 The Indian case study did run an ox-adoption-scheme for several years, which it has now dropped. They gave rescued draft oxen to local farmers in exchange for a deposit and a legal agreement not to slaughter or sell the oxen. Thus the farmers fed the oxen, could work them and give the oxen back to the sanctuary when they are too old or ill.
33 While one of the milking farm weans the calves after birth, they now want to introduce the rearing of the calves with the dam and natural weaning. The other milking farm weans the calves after six to seven months and the cow-calves have a restricted suckling contact around milking time.
34 Victor Reinhardt, *Untersuchungen zum Sozialverhalten des Rindes. Eine zweijährige Beobachtung an einer halbwilden Rinderherde (Bos indicus).* (Stuttgart, Germany: Birkhäuser, Reihe Tierhaltung 10, 1980), 16. Waiblinger, Susanne, Johannes Baumgartner, Marthe Kiley-Worthington, and Knut Niebuhr, "Applied Ethology: The Basis for Improved Animal Welfare in Organic Farming," in *Animal Health and Animal Welfare in Organic Agriculture*, eds. Mette Vaarst, Stephen Roderick, Vonne Lund, and William Lockeretz (Wallingford, England: CABI Publishing, 2004), 117–161.
35 Menicocci, accessed February 3, 2014, www.veganitaly.com/wine_producer/vegan_agriculture.html.
36 Farmhouse Garden Animal Home, accessed October 22, 2017, www.facebook.com/pg/farmhousegardenanimalhome/about/?ref=page_internal.
37 "Lebenlassen" means "let live." lebenlassen e.V., accessed January 29, 2017, www.lebenlassen.de.
38 Bundesministerium für Ernährung und Landwirtschaft (BMEL), *Statistisches Jahrbuch über Ernährung, Landwirtschaft und Forsten der Bundesrepublick Deutschland 2014.* 58. ed. (Münster-Hiltrup, Germany: Landwirtschaftsverlag, 2014), 92.
39 Gita Nagari Yoga Farm, Accessed February 8, 2017, www.theyogafarm.com/p/csa.html.

Bibliography

Bauernverband, ed. *Situationsbericht 2014/15 Trends und Fakten zur Landwirtschaft.* Berlin, Germany: Bauernverband, 2015.

Beauchamp, Tom L., and James F. Childress. *Principles of Biomedical Ethics.* Sixth edition. Oxford, England: Oxford University Press, 2008.

Bock, Bettina B., Marjolein M. van Huik, Madeleine Prutzer, Florence Kling-Eveillard, and Anne Dockes. "Farmers Relationship with Different Animals: The Importance of Getting Close to the Animals." *International Journal of Sociology of Agriculture and Food* 15(3) (2007): 109–125.

Boltanski, Luc, and Laurent Thévenot. *Über die Rechtfertigung. Eine Soziologie der kritischen Urteilskraft.* New edition. Hamburg Germany: Hamburger Edition, HIS, 2014.

Bundesministerium für Ernährung und Landwirtschaft (BMEL). *Statistisches Jahrbuch über Ernährung, Landwirtschaft und Forsten der Bundesrepublick Deutschland 2014.* 58. ed. Münster-Hiltrup, Germany: Landwirtschaftsverlag, 2014.

Burgat, Florence. "Non-Violence Towards Animals in the Thinking of Gandhi: The Problem of Animal Husbandry." *Journal of Agricultural and Environmental Ethics* 14(2004), 223–248.

Charmaz, Kathy. *Constructing Grounded Theory: A Practical Guide through Qualitative Analysis.* Thousand Oaks, CA: Sage Publication, 2006.

Cochrane, Alasdair. *Animal Rights without Liberation.* New York, NY: Columbia University Press, 2012.

Donaldson, Sue, and Will Kymlicka. *Zoopolis. A Political Theory of Animal Rights*. Oxford, England: Oxford University Press, 2011.

Donovan, Josephine, and Carol J. Adams. "Introduction." In *The Feminist Care Tradition in Animal Ethics*, edited by Josephine Donovan, and Carol J. Adams, 1–15. New York, NY: Columbia University Press, 2007.

Hall, Lee. *On Their Own Terms. Bringing Animal-Rights Philosophy Down to Earth*. Darien, CT: Nectar Bat Press, 2010.

Kraft, Wilfried, ed. *Geriatrie bei Hunden und Katzen*. Second edition. Stuttgart, Germany: Parey Verlag in MVS Medizinverlage, 2003.

KTBL. *Faustzahlen für den Ökologischen Landbau*. Darmstadt, Germany: Kuratorium für Technik und Bauwesen in der Landwirtschaft e.V., 2015.

Meyer-Glitza, Patrick, and Ton Baars. "Non-killing Cattle Husbandry." In *Tackling the Future Challenges of Organic Animal Husbandry: 2nd Organic Animal Husbandry Conference, Hamburg, Trenthorst, 12–14 September, 2012*, edited by Gerold Rahmann and Denise Godinho, 184–187. Braunschweig, Germany: vTI Landbauforschung, Sonderheft 362, 2012.

Regan, Tom. "Sentience and Rights." In *Animals, Ethics and Trade. The Challenge of Animal Science*, edited by Jacky Turner and Joyce D' Silva, 79–86. London, England: Earthscan, 2003.

Regan, Tom. *The Case for Animal Rights*. Berkeley, CA: University of California Press, 2004.

Reinhardt, Victor. *Untersuchungen zum Sozialverhalten des Rindes. Eine zweijährige Beobachtung an einer halbwilden Rinderherde (Bos indicus)*. Stuttgart, Germany: Birkhäuser, Reihe Tierhaltung10, 1980.

Rosen, Aiyana, and Sven Wirth. "Tier_Ökonomien? Über die Rolle der Kategorie „Arbeit" in den Grenzziehungspraxen des Mensch-Tier-Dualismus." In *Tiere Bilder Ökonomien. Aktuelle Forschungsfragen der Human Animal Studies*, edited by Chimaira – Arbeitskreis für Human-Animal Studies, 17–42. Bielefeld, Germany: Transcript Verlag, 2013.

Rosenthal, Gabriele. *Interpretative Sozialforschung. Eine Einführung*. Weinheim und München, Germany: Juventa, 2005.

Segerdahl, Pär. "Can Natural Behavior be Cultivated? The Farm as Local Human Animal Culture." *Journal of Agricultural and Environmental Ethics* 20(2007): 167–193.

Sezgin, Hilal. *Artgerecht ist nur die Freiheit: Eine Ethik für Tiere oder Warum wir umdenken müssen*. München, Germany: C.H. Beck, 2014.

Strauss, Anselm L., and Juliet M. Corbin. *Grounded Theory: Grundlagen Qualitativer Sozialforschung*. Weinheim, Germany: Beltz, Psychologie Verlags Union, 1996.

Statistisches Bundesamt. *Land- und Forstwirtschaft, Fischerei. Betriebe mit ökologischem Landbau. Landwirtschaftszählung/Agrarstrukturerhebung 2010*. Wiesbaden: Statistisches Bundesamt, Fachserie 3 Reihe, 2.2.1, 2011.

Vereinigte Informationssysteme Tierhaltung w. V. (VIT). *Trends Fakten Zahlen 2014*. Verden, 2015. Accessed August 9, 2015. www.vit.de/fileadmin/user_upload/wirsindvit/jahresberichte/vit-JB2014-gesamt.pdf.

Waiblinger, Susanne, Johannes Baumgartner, Marthe Kiley-Worthington, and Knut Niebuhr. "Applied Ethology: The Basis for Improved Anim Welf in Organic Farming." In *Animal Health and Animal Welfare in Organic Agriculture*, edited by Mette Vaarst, Stephen Roderick, Vonne Lund, and William Lockeretz, 117–161. Wallingford, England: CABI Publishing, 2004.

Wilkie, Rhoda M. "Sentient Commodities and Productive Paradoxes: The Ambiguous Nature of Human-livestock Relations in Northeast Scotland." *Journal of Rural Studies* 21 (2) (2005): 213–230.

2.8

ARE INSECTS ANIMALS?

The ethical position of insects in Dutch vegetarian diets

Jonas House

Introduction

Climate change and a rising world population are contributing to ever-increasing pressure on global food security. In this context, considerable effort is being expended within the arenas of policy, academia and business to develop solutions to the challenge of feeding the world's population in a more "sustainable" fashion than the global agri-food network does at present. One such proposed solution is the use of insects as a source of animal feed and human food in Europe and the US (henceforth "the West"), the defining statement of which was a report published by the FAO in 2013.[1] This report outlined the state of the art of research into the consumption of insects – or "entomophagy" – and has provoked a great deal of interest in the subject.

The central argument of the report, and of much (although not all) subsequent academic research and commercial discourse, is that the Western replacement of conventional sources of animal protein with insect protein would yield significant environmental benefits. Insect species consumed by humans around the globe are generally high in protein and nutrients (indeed, many species are comparable to beef in these respects), low in fat, and require considerably less land and water during rearing than the animals that are currently raised for food in the West.[2] Further, the four insect species developed for human consumption in the Netherlands since around 2007 – crickets, grasshoppers, mealworms and buffalo worms – have a high feed conversion ratio relative to farmed animals,[3] and, as well as requiring less feed than ruminant species, are able to subsist on a wider range of feed types.[4] As such, insects are argued to represent a healthy, nutritious, and environmentally friendly source of protein,[5] and it is these points that tend to be emphasized in the discourse surrounding edible insects. Insects are also often held to represent a more *ethical* choice than some existing foods such as meat from conventional farmed animals.[6]

So far, there has been a relative lack of attention to the ethical dimensions of insect consumption.[7] As this chapter attempts to demonstrate, ethical considerations appear to play a relatively significant role in the consumption of currently available insect products, and are related to an individual's ethical dietary orientation, both for those wishing to reduce their meat consumption and for self-defined vegetarians. The latter group are the primary focus here, but both groups are discussed.

The chapter is based on interviews with 33 consumers of the *Insecta* range of insect-based convenience foods. Containing around 14% ground-up buffalo worms,[8] these foods were available in branches of a Dutch national supermarket chain during 2015.[9] The participants are organized into "meat eaters" (5 participants), who made no special effort to reduce their meat consumption; "meat reducers" (17), who deliberately refrained from eating meat for one or more days a week (typically 1–3 times); and self-defined "vegetarians" (11). Of the vegetarians, some ate no animals other than insects (4); the others also ate fish (7).

This chapter argues that the introduction of the buffalo worm into the European agri-food network gives us two important insights into vegetarianism in the Netherlands. These are:

1. Vegetarianism is not a fixed or static concept, and both its motivations and form may change over time. For some people, vegetarianism is motivated by environmental rather than animal welfare concerns. This form of vegetarianism does not preclude the consumption of animals, and thus animals whose rearing, slaughter and consumption is perceived as being "good" for the environment (or "better" than alternatives) are deemed an ethically permissible source of food.
2. For some vegetarians who *are* motivated to reduce their meat consumption by animal welfare concerns, insects are still positioned as an acceptable food source, chiefly because of their perceived lack of sentience and/or incapacity to suffer, but also for other reasons. This type of assessment often appears to be based on a kind of folk taxonomy of species, akin to the Aristotelian "great chain of being."

Empirical material is presented to explain each point in turn. However, my primary intention is to demonstrate how the introduction of a novel animal species to the human food system in an industrialized Western country "problematizes" ethically oriented diets, by raising a number of hitherto largely unexplored questions about animal ethics, and illuminating a number of taken-for-granted assumptions about the nature of animal life on which ethical diets frequently appear to be based. As such, the latter portion of the chapter is dedicated to highlighting two important further questions regarding the ethical treatment of animals that are raised by the appearance of the buffalo worm in Dutch supermarkets.

Vegetarianism: A fluid concept

Vegetarianism is not a unified adherence to a fixed set of culinary, ethical, and other principles, but rather a fairly loose term indicating a diet with a broadly ethical inflection. Such diets can differ in motivation and form over time, between individuals, and often *within* individual accounts at different points in one's "vegetarian career."[10] Research with self-defined vegetarians in the UK has identified a sort of continuum of diets, with total abstention from meat at one end to occasional consumption of meat at the other.[11] In a nationally representative US sample, only a third of those who self-defined as vegetarian did not occasionally eat poultry.[12] Clearly, the term is relatively flexible in usage, and can accommodate different social and practical considerations[13] as well as the ethical variances that are the focus of the present chapter. Indeed, such flexibility has led to suggestions that vegetarianism might be better thought of as an "orientation" rather than a completely consistent practice,[14] which I would argue is a persuasive conceptual move. Among the present study participants, their particular form of vegetarianism permitted the consumption of certain animal species.

Within my group of participants, both meat reduction and vegetarianism was reportedly mostly environmentally motivated. Environmentally motivated vegetarianism did not completely preclude the consumption of animals. For those vegetarians who were prepared to eat both fish and insects, this was usually because those species were perceived as less environmentally damaging than farmed animals. Willem (vegetarian, eats fish), said:

> It felt quite natural for us to stop [eating meat]. Not because we thought, oh those poor animals are gonna get killed, but, it's like, we don't need it. And then there's the other side of it that, how much food and water is needed to produce one kilo of beef, compared to soya, and compared to insects. [...] [O] ur generation's probably gonna have sufficient food, but our next, and the next after that, they won't have, if they continue like this.

Self-reported vegetarians who did not eat any animals other than insects often did so because they had concerns about overfishing or depriving local people of food to eat – in addition to similar environmental concerns about farmed animals – but did not have these concerns about insects. Angela, a vegetarian who did not eat fish, explained "I don't eat fish at all" due to concerns about "the welfare of the rest of the people in the world," particularly in areas where industrial fishing has reduced the availability of fish for local people.

Whether vegetarians ate fish or not, those who were primarily environmentally motivated often did not have a problem with eating animals per se, but rather with the environmental impact of their production. Notably the ethical justification for not eating particular species in these cases appeared to be predominantly anthropocentric, given that ethical concern was directed at the wellbeing of future human generations. As Gijs (vegetarian, eats fish) said, "[I do] all the things I can help to make the planet, for my children and grandchildren, a nice place." I return to this point in the discussion section of this chapter.

Are insects animals? The ontology and ethics of the buffalo worm

The identification of an environmentally motivated form of vegetarianism that permits the consumption of some animal species is interesting, because it implies that animal welfare motivations – a traditionally large facet of vegetarian diets – are not shared by everyone who aims to reduce or eliminate meat from their diets, even if they self-define as vegetarian.

However, among the vegetarians I spoke to, there *were* still some who were motivated primarily or substantially by animal welfare. For these people, the positioning of insects as edible was justified ethically on the grounds that insects were deemed to be of low moral standing, for a number of reasons. Predominant among these was insects' perceived lack of sentience or capacity to suffer, although for some people, insects were simply not "proper" animals. In many ethical accounts of insects, it appeared that their "low" ontological position relative to conventional farmed animals was the grounds for their diminished ethical standing. This is not an original discovery: Brock Bastian et al., for example, observed that people find it easier to deny mind and thus moral standing to animals that are not cows, sheep or pigs.[15] However, the data presented here provide further insight into how this process works in the context of ethically-oriented food consumption.

Participants were questioned directly about whether they perceived insects as animals, as well as a number of more indirect questions about why they were prepared to eat certain species and not others. Although the primary focus of this chapter is vegetarians' ethical assessments of insects, some data from those who eat meat is presented to contextualize the findings.

The question of whether or not insects are animals was answered slightly differently by people who regularly ate meat and by those who label themselves as vegetarian. For meat eaters, the answer was generally a straightforward yes. For example, Dorieke (meat eater) said:

> I'm not a vegetarian so for me it doesn't really matter if it's a fish or it's a pig or it's an insect. It's all animals.

One speculates as to whether the relative straightforwardness of this type of classification is to do with there being nothing "at stake" in the categorization of insects as animals, if one's diet already involves the frequent consumption of other species.

By contrast, for vegetarians, as well as many of those who ate meat but attempted to reduce their consumption, there was often a disjuncture between their rationalized classification of insects as animals and the way in which, practically speaking, insect foods were categorized and integrated into food practices in a broadly similar way to vegetarian products. Pieter (meat reducer), for example, ate insects on his "no meat" days, although he was quite emphatic that these were distinct from his "vegetarian Mondays." For Pieter, insects were not "meat," but were still animals of some sort. Angela (vegetarian, no fish) ate them in the same way as the vegetarian convenience foods she otherwise ate five or six nights a week. Jelmer (meat reducer) had the following exchange with my research assistant, which sums up the ambiguity fairly well:

INTERVIEWER: OK. And does the insect burger belong to the three days a week [that you eat] no meat, or does it belong to the meat days?

JELMER: Oh… yes that's a good one. That one belongs to the no-meat days.

[…]

INTERVIEWER: Does not belong to meat. All right. So you do not see insects as a meat product, then?

JELMER: Yes. Actually I do! So that is … how to put it … not a very consistent thought of mine. Haha. Uhm … Most people discussed how rationally, *of course* insects are animals; but it seems in terms of the practical intelligibilities that help to structure eating practices, for many people they are closer to plants. When pushed to classify insects, people often describe them as animals, but also as somehow crucially different. For example, Els (meat reducer) thought that:

> They are animals, but not animals like the real animals.

Similarly, Margreet (vegetarian, eats fish) said:

> Well of course they're animals. But I don't think, well I think that they don't have so very much brain. So in that respect I think they're more like plants or something […] Insects are … well. I don't, I don't really consider them being animals.

The grounds provided for distinguishing between animals of moral standing and insects were varied, and included factors such as the difficulty in identifying or connecting with insects; that one could not cuddle insects; that insects were perceived to have low intelligence or intellectual capacity, as well as a less developed nervous system, and thus had an inability to feel pain; that insects were perceived to have less emotions than other animals; that they had no demonstrable social behavior; that in general their needs are low; that they are not "real" animals; and that they are abundant ("they are everywhere"). Thomas (meat eater) suggested that insects' small size made it easier not to care about eating them, stating that

> Yeah they probably also have feelings, but since they're so small and insignificant actually I think it's easier to eat them than a cow or a pig or anything.

Willemijn (vegetarian, no fish) explained that as insects were not available to eat when she first became vegetarian, they are difficult to classify ethically. She felt that "there is some difference" between insects and other animals, but found this hard to explain. She added:

> Right now I don't have the feeling that ... that I am violating my being-vegetarian by eating insects. But I do not have a logical explanation for that.

Willemijn's comments here succinctly capture the ambiguity surrounding the ethical position of insects in vegetarian diets that was evident across the group's responses. In addition to such ambiguity, people's ethical accounts of animals also indicated the pervasiveness of an implicit hierarchy of sentience, with humans at the top, descending through mammals, birds, fish, and insects. This also seemed to go beyond sentience, however, and appeared to relate also to some indefinable quality of being. Buffalo worms evidently occupy an ambiguous ontological position relative to other animals, which is not entirely reducible to assessments of their capacity for subjective experience. People drew the line regarding where to stop eating things at different points along this implied hierarchy, but many made reference to the idea that moral concern was less of an issue the further away animals are perceived as being from humans.

Such an implicit hierarchy bears a striking resemblance to the long-established notion of the "great chain of being," which seemed to be the structuring principle of many ethically oriented diets. This idea, which is Aristotelian in origin, is essentially the notion that life is hierarchically ordered. Classically, this hierarchy was held to have God at the highest point, descending down through Man – usually Man, of course, rather than Woman – and then through animals, plants, and inanimate objects. The idea of a hierarchy *within* the category of "human" has been discussed by Joanna Bourke in her excellent historical assessment of the Western concept of humanity.[16] The present data suggest that internal hierarchies may obtain within the category of "animal" as well. In general, participants were quite happy to confirm that their conceptions of animal life were based on a notional hierarchy. Femke (meat eater) even referred to the idea explicitly, saying, for her, there is a hierarchy of animals that runs down from cows, through sheep, chickens, fish and so on.

These kind of assumptions were also evident in relation to fish. Indeed, many people also seemed to have difficulty classifying fish and accounting for them ethically. While a fuller treatment of this point is beyond the scope of the present chapter, there is some indication that fish are also positioned as creatures of low moral standing because of their fundamental difference from humans, in a similar way to insects – a sort of low ontological standing which provides the grounds for reduced moral concern.

For a particular group of self-reported vegetarians, then, eating particular animals is acceptable. For some who are relatively unconcerned by animal welfare, this is on environmental grounds. For others, for whom animal welfare is still a prevailing concern, this is on the grounds that some species are below the threshold of moral concern, and accounts of this ethical positioning seem to indicate a kind of folk hierarchy that at least partially structures ethical diets. Such findings raise a number of interesting questions regarding the ethical treatment of animals, and the way in which ethically oriented diets are formed and justified. Here, I indicate two pertinent areas for further discussion and investigation.

Insects and the precautionary principle

One important ethical question raised by the prospect of introducing insects as human food is whether or not we should extend the precautionary principle to them. That is, do we owe animals a duty of care if we do not know for sure that they're incapable of experiencing pain or suffering? Writing on animal ethics more broadly, Andrew Linzey has argued that in cases where we do not fully know, we should generally give animals the benefit of the doubt.[17]

There is a relative paucity of research into insects' capacity for subjective experience or ability to feel pain. An early review concluded that available evidence did not suggest that insects were capable of feeling pain,[18] although this view has been questioned by subsequent research.[19] In a recent article, Andrew Barron and Colin Klein have argued that insects are capable of subjective experience.[20] Although this work has been criticized,[21] clearly the debates on insect pain and subjectivity are far from settled. Evidently, more work needs to be done to establish clearer grounds for arguing for or against insect sentience, a point that for many may have implications for ethical decisions made regarding the treatment and consumption of insects. For example, as Gjerris et al. have argued in an article about the ethics of insect production for food and feed,

> As long as there is only little knowledge about the capacity of insects to experience better or worse welfare, the informal logical fallacy of *argumentum ad ignorantiam* should be avoided, i.e. absence of proof should not be misunderstood as proof of absence.[22]

By contrast, Bob Fischer argues that vegans should *not* extend the precautionary principle to insects, contending that given the uncertainty regarding insect sentience, ethical priority should be given to species that we *know* (or have sufficient evidence to suspect) are capable of experiencing pain or suffering: Namely, those mammalian and bird species we know to be harmed by plant agriculture.[23] He argues further that in the light of the number of insect deaths resulting from plant agriculture, less harm may actually result from the rearing and consumption of insects that have been specifically bred for the purpose,[24] although this particular argument may be dependent on insects being reared on waste streams rather than plant-based animal feed.[25] Conventional animal feed is still currently used by a number of European breeders,[26] and the use of post-consumer waste or manure as a substrate for rearing insects is currently banned in the EU.[27] As such, Fischer's stimulating account is, nevertheless, unlikely to spell the end of ethical discussions around the consumption of insects. Clearly these issues warrant further investigation and debate.

Here, it is worth emphasizing the need to disaggregate the category of "insects" within ethical treatments of insects as a potential food source. This need has elsewhere been identified in the literature[28] in the light of a tendency to treat "insects" as a discrete and homogeneous category (which does, of course, also include the present chapter). Gjerris et al. make the point that ethical considerations are likely to be species-specific:[29] Perhaps a stronger ethical case can be made for eating beetle larvae than adult crickets or grasshoppers, which are killed at a later stage of development. Such a decision would no doubt involve the work of both biologists and ethicists. Clearly, this is one area that requires greater attention as particular insect species are proposed as a potentially widespread source of food.

Anthropocentric vegetarianism

A further question to arise from the foregoing data concerns the ethical orientation of vegetarian diets: Are both the environmental and animal welfare motivations for vegetarianism that I have discussed anthropocentric?

Environmentally motivated vegetarianism seems to be anthropocentric in the sense that the ethical concern is primarily directed at humans, and the survival of the human species. A number of people, for example, made reference to leaving the planet a good place for their children, or future generations, to grow up in. Although of course the idea of maintaining the health of the earth is certainly not an ethical position that totally disregards the welfare of animals, ethical priority does appear to be afforded to humans, rather than animals per se.

By contrast, vegetarianism that is primarily motivated by animal welfare concerns seems, on the face of it, to be less easy to conceive of as anthropocentric. Nevertheless, I would suggest that this may be a useful way of conceptualizing it, because ethical care still seems to radiate out or trickle down from humans, and evidently stops at a certain point. Joanna Bourke's concept of a "limited economy of sympathy" is relevant here.[30] Bourke's term refers to the tendency to extend ethical care to those who are perceived as ontologically proximate to humans, but also points out how crucially, this care is in limited supply. The idea can perhaps be usefully extended to the analysis of animal ethics as well. For many, it seems that ethical concern is based on the identification of certain species with humans, but not others. In a sense, humans are still prioritized, as are creatures perceived to be ontologically "close" to them. After a certain point, animals are exempted from moral concern: The economy of sympathy is exhausted. This ethical approach could be contrasted with a more inclusive ethics in which humans are considered to be just one facet of the broader sphere of animal and plant life.

In bringing my discussion to a close, I wish to emphasize that it is not my intention to try and poke holes in people's efforts to adopt a more ethical diet. All of the people I spoke to were making commendable efforts to eat more ethically, which was manifestly not always particularly easy. I wish simply to point out what appear to be prominent structuring principles within the establishment and maintenance of a particular set of Western vegetarian diets, and to suggest how these ultimately appear to rely on established philosophical notions of the prominence of humans (and species that are perceived to be "close" to humans). Future efforts towards the development of a more inclusive animal ethics must acknowledge the prevalence of this hierarchical apprehension of animal life. The case of the buffalo worm offers us a useful way of understanding the fundamental malleability of vegetarianism, its different dimensions, and perhaps some of its limitations.

Notes

1 Arnold van Huis et al., *Edible Insects: Future Prospects for Food and Feed Security* (Rome, Italy: FAO, 2013).
2 van Huis et al., *Edible Insects*, 64.
3 van Huis et al., *Edible Insects*, 60.
4 van Huis et al., *Edible Insects*, 60–61.
5 Agreement as to the benefits of insect consumption is not unanimous, and questions have been raised as to just how "sustainable" insects are. See, for example, Mark E. Lundy and Michael P. Parrella, "Crickets Are Not a Free Lunch: Protein Capture from Scalable Organic Side-Streams via High-Density Populations of *Acheta Domesticus*," *PLoS One* 10, no. 4.
6 Examples include Giles Dexter, "Insects on the Menu for Ethical Eaters," *Stuff* (blog), June 17, 2016, www.stuff.co.nz/life-style/food-wine/81116872/insects-on-the-menu-for-ethical-eaters; Carol Duncan, "Eating Insects. Ethical, Sustainable and Crunchy," *ABC* (blog), May 22, 2013, www.abc.net.au/local/photos/2013/05/22/3764772.htm; Sean Moloughney, "Edible Insects: Nutritional Value and Ethical Appeal," *Nutraceuticals World* (blog), 29 August, 2014, www.nutraceuticalsworld.com/contents/view_online-exclusives/2014-08-29/edible-insects-nutritional-value-ethical-appeal.
7 Notable exceptions include Bob Fischer, "Bugging the Strict Vegan," *Journal of Agricultural and Environmental Ethics* 29 (2016), and Mickey Gjerris, Christian Gamborg, and Helena Röcklinsberg, "Ethical Aspects of Insect Production for Food and Feed," *Journal of Insects as Food and Feed* 2, no. 2 (2016).
8 Buffalo worms, or "lesser mealworms," are not actually worms. They are the larvae of the darkling beetle *Alphitobius diaperinus*.
9 The range of foods can be viewed at http://web.archive.org/web/20170109124949/http://www.damhert.be/en/shop/producten?categorie=insecta. As of 2018, only the burger product is still in production. For more details about the research, including the process of ethical review, see Jonas House, "Consumer Acceptance of Insect-Based Foods in the Netherlands: Academic and Commercial Implications." *Appetite* 107.
10 Alan Beardsworth and Teresa Keil, "The Vegetarian Option: Varieties, Conversions, Motives and Careers," *The Sociological Review* 40, no. 2 (1992): 271. See also Matthew Ruby, "Vegetarianism. A Blossoming Field of Study," *Appetite* 58 (2012).
11 Beardsworth and Keil, "The Vegetarian Option"; A. D. Beardsworth and E. T. Keil, "Vegetarianism, Veganism, and Meat Avoidance: Recent Trends and Findings," *British Food Journal* 93, no. 4 (1991).
12 Thomas Dietz et al., "Values and Vegetarianism: An Exploratory Analysis," *Rural Sociology* 60, no. 3 (1995): 539.
13 For example, if a meal cooked by friends contains meat but a vegetarian individual does not wish to be rude, they may eat it, while still considering themselves to be vegetarian. Some elaboration of this point can be found in Beardsworth and Keil, "The Vegetarian Option," 1992; 263–265, and Swinder Janda and Philip J. Trocchia, "Vegetarianism: Toward a Greater Understanding," *Psychology and Marketing* 18, no. 12 (2001): 1216–1220.
14 Janda and Trocchia, "Vegetarianism," 1206.

15 Brock Bastian et al., "Don't Mind Meat? The Denial of Mind to Animals Used for Human Consumption," *Personality and Social Psychology Bulletin* 38, no. 2 (2012).
16 Joanna Bourke, *What it Means to be Human: Reflections from 1791 to the Present* (London, England: Virago Press, 2011).
17 Andrew Linzey, *Why Animal Suffering Matters: Philosophy, Theology, and Practical Ethics* (New York, NY: Oxford University Press, 2013).
18 C. H. Eisemann et al., "Do Insects Feel Pain? A Biological View," *Experientia* 40 (1984).
19 For a review of literature in this area see Gjerris et al., "Ethical Issues," 104–105.
20 Andrew B. Barron and Colin Klein, "What Insects Can Tell Us About the Origins of Consciousness," *Proceedings of the National Academy of Sciences* 113, no. 18 (2016).
21 For an overview of the criticisms of Barron and Klein's article, and their response, see Colin Klein and Andrew B. Barron, "Reply to Adamo, Key et al., and Schilling and Cruse: Crawling Around the Hard Problem of Consciousness," *Proceedings of the National Academy of Sciences* 113, no. 27 (2016).
22 Gjerris et al., "Ethical Issues," 105.
23 Fischer, "Bugging the Strict Vegan," 258.
24 Fischer, "Bugging the Strict Vegan," 260–261.
25 Fischer, "Bugging the Strict Vegan," 261.
26 A. Hubert and T. Arsiwalla, "Implementation of EU Food and Feed Safety Standards by the Insect Sector: Overview of Production Practices and IPIFF Guiding Principles." Presentation at International Platform of Insects for Food and Feed (IPIFF) Workshop, Brussels, April 26, 2016.
27 M. D. Finke et al., "The European Food Safety Authority Scientific Opinion on a Risk Profile Related to Production and Consumption of Insects as Food and Feed," *Journal of Insects as Food and Feed* 1, no. 4 (2015).
28 J. Evans et al., "'Entomophagy': An Evolving Terminology in Need of Review," *Journal of Insects as Food and Feed* 1, no. 4 (2015).
29 Gjerris et al., "Ethical Issues," 105.
30 Bourke, *What it Means*, 100–114.

Bibliography

Barron, Andrew B., and Colin Klein. "What Insects Can Tell Us About the Origins of Consciousness." *Proceedings of the National Academy of Sciences* 113, 18(2016): 4900–4908.

Bastian, Brock, Steve Loughnan, Nick Haslam, and Helena R. M. Radke. "Don't Mind Meat? The Denial of Mind to Animals Used for Human Consumption." *Personality and Social Psychology Bulletin* 38, 2(2012): 247–256.

Beardsworth, A. D., and E. T. Keil. "Vegetarianism, Veganism, and Meat Avoidance: Recent Trends and Findings." *British Food Journal* 93, 4(1991): 19–24.

Beardsworth, Alan, and Teresa Keil. "The Vegetarian Option: Varieties, Conversions, Motives and Careers." *The Sociological Review* 40, 2(1992): 253–293.

Bourke, Joanna. *What it Means to be Human: Reflections from 1791 to the Present*. London, England: Virago Press, 2011.

Dietz, Thomas, Ann Stirling Frisch, Linda Kalof, Paul C. Stern, and Gregory A. Guagnano. "Values and Vegetarianism: An Exploratory Analysis." *Rural Sociology* 60, 3(1995): 533–542.

Eisemann, C. H., W. K. Jorgensen, D. J. Merritt, M. J. Rice, B. W. Cribb, P. D. Webb, and M. P. Zalucki. "Do Insects Feel Pain?—A Biological View." *Cellular and Molecular Life Sciences* 40, 2(1984): 164–167.

Evans, J., M. H. Alemu, R. Flore, M. B. Frøst, A. Halloran, A. B. Jensen, G. Maciel-Vergara et al. "'Entomophagy': An Evolving Terminology in Need of Review." *Journal of Insects as Food and Feed* 1, 4(2015): 293–305.

Finke, M. D., S. Rojo, N. Roos, A. van Huis, and A. Yen, "The European Food Safety Authority Scientific Opinion on a Risk Profile Related to Production and Consumption of Insects as Food and Feed." *Journal of Insects as Food and Feed* 1, 4(2015): 245–247.

Fischer, Bob. "Bugging the Strict Vegan." *Journal of Agricultural and Environmental Ethics* 29, 2 (2016): 255–263.

Gjerris, Mickey, Christian Gamborg, and Helena Röcklinsberg. "Ethical Aspects of Insect Production for Food and Feed." *Journal of Insects as Food and Feed* 2, 2(2016): 101–110.

House, Jonas. "Consumer Acceptance of Insect-Based Foods in the Netherlands: Academic and Commercial Implications." *Appetite* 107(2016): 47–58.

Hubert, Antoine and Tarique Arsiwalla, "Implementation of EU Food and Feed Safety Standards by the Insect Sector: Overview of Production Practices and IPIFF Guiding Principles." Presentation at International Platform of Insects for Food and Feed (IPIFF) Workshop. Brussels. April 26, 2016.

Janda, Swinder, and Philip J. Trocchia. "Vegetarianism: Toward a Greater Understanding." *Psychology and Marketing* 18, 12(2001): 1205–1240.

Klein, Colin, and Andrew B. Barron. "Reply to Adamo, Key et al., and Schilling and Cruse: Crawling Around the Hard Problem of Consciousness." *Proceedings of the National Academy of Sciences* 113, 27(2016): E3814–E3815, doi: 10.1073/pnas.1607409113.

Linzey, Andrew. *Why Animal Suffering Matters: Philosophy, Theology, and Practical Ethics*. New York, NY: Oxford University Press, 2013.

Lundy, Mark E., and Michael P. Parrella. "Crickets Are Not a Free Lunch: Protein Capture from Scalable Organic Side-Streams via High-Density Populations of Acheta Domesticus." *PLoS one* 10, 4(2015):e0118785, doi: 10.1371/journal.pone.0118785.

Ruby, Matthew B. "Vegetarianism. A Blossoming Field of Study." *Appetite* 58, 1(2012): 141–150.

van Huis, Arnold, Joost van Itterbeeck, Harmke Klunder, Esther Mertens, Afton Halloran, Giulia Muir, and Paul Vantomme. *Edible Insects: Future Prospects for Food and Feed Security*. Rome, Italy: FAO, 2013.

PART III

The human and environmental costs of institutionalized killing

3.1

OUR AMBIVALENT RELATIONS WITH ANIMALS

Jeanette Thelander

There are several reasons for not eating meat, including environmental reasons, individual and public health reasons, ethical reasons, and more. Yet, on a global scale, people eat more meat than ever. According to the UN, this is a major problem. Already in 2006, the report *Livestock's Long Shadow* pointed out that meat consumption was a bigger problem from an environmental point of view, than global transports, including air-traffic.[1] At the same time, Western societies are becoming both more animal friendly (when it comes to companion animals) and more animal abusive (when it comes to farmed animals). There seems to be a lot of empathy for animals, yet people choose to hurt them, kill them and eat them. This ambivalence is likely to create conflicts of psychological as well as cultural nature, which calls for both societal and individual strategies in order to create meaning and logic to life.

From an ethnological point of view, everyday life is the most exciting area of research. This is because it is the everyday life that reflects our culture the most. It is the obvious and natural norms that are the most intriguing and often worthwhile to study – because they reveal our culture's secret codes and messages. Or, to put it in another way, they reveal the way we are thinking – without thinking of it.

Two extremes are worthwhile to keep in mind when trying to reveal the cultural norm of meat eating in Western societies: Vegans and cannibals. Only cannibals and vegans are consistent in their choosing of foods. To vegans, no meat of any kind – or other animal products for that matter – is acceptable to eat. Cannibals, on the other hand, are just as consistent, but to them all meat – even human flesh – is accepted as food. However, most people are not vegan and even fewer are cannibals. Most people eat meat, or at least some meat, sometimes. So what social and cultural factors influence our choice of foods? Is it obvious where to draw the line?

In order to find out how meat eating is reflected as a norm in the Western culture and how this norm is maintained, theory from the feminist and gender research field could be of interest. Like women, animals are not the social norm, and are often referred to as being of less value (than white male, which is considered to be the norm). Carol Adams' concept of "The absent referent," which describes a process of making *someone* into *something*, is a useful tool.[2]

Theoretical inspiration can also be drawn from the pedagogical field. The Swedish pedagog and researcher, Helena Pedersen, deals with the attitudes against animals in schools, with a specific interest in the process where individual animals cease to exist as individuals and instead are seen as representatives for their species. According to Pedersen, this makes sense of the absurd quote from the South Park television series: "We have to kill the animals so that they won't die!" found on the cover page of some Swedish learning material.[3]

Drawing the line

One way of getting to know how people think and feel about animals and meat, is by asking them. This was done in a Swedish study addressing questions about eating or not eating both human and non-human meat.[4]

In the study, seven semi-structured deep interviews were made. Two of the interviews were made with public animal welfare inspectors, three interviews with high school students and two interviews with people chosen at random. Of particular interest is that one of those interviewees turned out to be a trained chef. Four out of the six interviewees had companion animals of their own, being dogs, cats, a horse and a rabbit.

A method often referred to as "participant observation"[5] was also used in the study. To a great extent, the study also consists of a literary analysis and study of fiction and nonfiction literature, both contemporary and dating back as far as to the 1960s when a Swedish ethnologist published her, at the time, ground-breaking research on horse meat and slaughter in Sweden since the 900s.[6] It has been more than half a century since it was first published, but it is still valid. Although meat consumption up until recently has risen steadily in Sweden, the amount of horse meat on a yearly basis is astonishingly small, only 0.2 out of total 87 kg/capita (< 0.2%). Cultural factors behind this phenomenon, like religious beliefs, are similar in some other countries where the horse has had a special position in the historical context, e.g. Hungary and Ireland. Other sources in the study include a wide range of texts: academic literature, art and film, newspaper articles, commercial advertisements and websites.

The empirical material described above is relatively small and generalizations should be made with great precaution. The findings, however, support the hypothesis that people in the Western culture have an ambivalent relation to animals, which could be described by the psychological term "cognitive dissonance," a state of mind that occurs when an individual's attitudes and behavior diverge.[7]

In order to bridge the diverging gap between attitudes and behavior towards animals, two types of strategies are widely used: 1) creating distance, and 2) drawing lines. This is, for example, expressed in the material when a Swedish teenager describes a scene from a visit to a European country where she has family. She describes how she and some other children play with a chicken out in the street all day, being kind to him/her and having fun. But in the late afternoon, the chicken is slaughtered and served for the evening's dinner. The informant cannot force herself to actually eat the meat from the chicken. To her, the chicken has become an individual, a sentient being with feelings and desires of its own, with a right to live its life to full length, just like any human being. The meat has been transformed from being food from an absent referent to dead body parts of a formerly living being. Meat consumption is, in other words – at least to her – only possible when there is a distance between the one eating and the one being eaten. Meeting eye to eye eliminates this distance and makes it impossible to go on maintaining the cultural norm of meat eating.

Quality of the relationship human–animal

Some of the informants also expressed their love for animals. Most of them could not imagine themselves eating their own companion animals. Notable is that one of them could; "if the situation was that I would die otherwise, but I do not know if I could live with myself afterwards if I did." As a contrast, all of the informants could imagine themselves eating another, unknown, human being "if necessary."

When analyzing the material, it became clear that drawing the line between humans and animals is not at all as easy as perhaps believed at first. Humans can be seen as mammals among other mammals. Categorizing someone as "human" or "animal" does not automatically label him or her as "food" or not. On the contrary, what defines someone as eatable or not, depends on the quality of the relationship and the context. It seems easier to consume an unknown human being than a beloved companion animal. It is also easier to eat an unknown, anonymous cow than a cherished riding horse.

A study carried out at the Swedish University of Agricultural Sciences[8] indicated that both dog- and cat carers have a close, emotional relationship to their companion animals. Six out of ten dog carers and 75% of the cat carers stated that their companion animal was "very close" to them. Today, the majority (figures from different studies vary between 57–70%) of the Swedish dog carers state that they let their dog sleep in their bed. In the study above, 95% of the cat carers let their cat sleep in their bed. In other words, it seems as if both the dogs and the cats have come as close as they can get to humans, both physically and emotionally. According to Adams, this strongly influences how the animal is perceived, as someone rather than something.

Context matters

The context is also of importance when it comes to decide who can be eaten or not. In the novel *The Road* [9] is a horrific passage of cannibalism. The main character, a father trying to survive in a post-apocalyptic world, all for his son, steps down into the basement of an abandoned house. He discovers several naked persons, some have already had their limbs cut off, in waiting. It is a world where people eat people, in order to survive and the people in the basement have been captured and labelled as "food" for others, stronger and better armed. It is fiction, but could it possibly occur in real life? It is not unlikely.

One of the most well-known cases of modern cannibalism occurs in the context of the plane crash in the Andes in 1972 where members of a rugby team from Uruguay had to consume body parts from other passengers in order to survive.[10] This seems to be the kind of context the interviewees have in mind when answering that they could imagine themselves eating other humans "if necessary."

Many people – and certainly all of the informants in the Swedish study – consider humans to be animals among other animals. Biologically, there is little difference between mammals. All mammals have a heart, blood circulating, a central nervous system, the same type of tissues (bone, muscle etc.) and so forth. The line between humans and animals is a cultural construction and there are many signs that the line is being blurred.

This is due to several reasons, one of them being that we know a lot more about animals today than some decades ago. Today, we know that animals – at least mammals and birds, which are the most commonly studied – have feelings and emotions just like humans do. They have the ability to think, and even plan for the future. They have needs and desires, they play and learn from one another, not to mention their ability to communicate. The prairie dog (genus *Cynomys*) is a species that is considered to have the most sophisticated animal language. Not only have they got different alarm calls for different species, like humans and coyotes, they can also describe the size, shape and color of individual predators. When presented with a new kind of threat, the prairie dogs create new alarm calls, including new descriptions.[11]

In the past few decades a lot of research has been made on dogs, showing that they are extremely good at reading people, even better than humans themselves, they have got empathy, a strong sense of justice and even a sense of the passing of time.[12] This is also valid for Caledonian crows, who are also fully capable of deceit and practice this when hiding food from others. Many animals (at least the ones studied) have self-consciousness and theory of mind, including pigs, horses, sheep and magpies.

There is a growing awareness that all animals, including humans, are very similar, both biologically and cognitively. The differences that exist are exactly this: Differences, without specific value. Perceiving the world in different ways does not mean that one way is better or more valuable than the other. Logically, this makes a lot of people treat their companion animals as members of the family, caring and looking after them in much the same way as one would do with one's children. This is quite the opposite of the strategy of creating distance.

Keeping the distance

The great paradox is that this loving and caring seems to be reserved only for the family and the closest relationships. This is demonstrated in the global factory farming where animals are treated in ways that would be considered appalling and an absolute abomination, had companion animals been treated in the same way. As an example, pigs have proven to be as intelligent as or even more intelligent than dogs, yet they are treated in horrific ways and finally eaten, not unusually by people considering themselves as being animal friendly.

Professor Keith Kendrick, formerly head of the Laboratory of Cognitive and Behavioural Neuroscience at the Babraham Institute in Cambridge, has studied cognition in sheep for more than 20 years. In an interview during a visit to Sweden in 2008, he admitted that some of his research was controversial. Since sheep are intelligent and presumably have both the ability to suffer and to feel joy, the use of sheep raises ethical and moral questions. These questions are not easily addressed politically and of course the findings also pinpoints the issue of using sheep as research animals.[13]

During the last decades, the world has been flooded with new scientific findings in the animal cognition field. The ones that have had the greatest impact are the ones carried out on traditional companion animals, such as dogs. Findings about farmed animals, such as cattle or pigs have not reached the public to the same extent. This could be due to the fact that the public does not really want to know.

By distancing oneself from "the others" (production animals), refusing to think about them as sentient beings, it is possible to continue functioning in a culture where meat eating is the norm. The psychological mechanism at work is the same as the one that often starts functioning in war-time. By dividing people in terms of nationality, borders are put up between "us" and "them." We do not hurt one another, but it is acceptable, at times even rewarding, to hurt "the others." Here, as well, it is the context and the quality of relationships that guide most people's behavior. There will always be people taking stands because of moral or ethical principles, but most people want to identify themselves as a part of the majority and live their lives in adjustment to the overall norm.[14]

Knowledge about the intelligence and about the social and emotional abilities of farmed animals would probably add to the cognitive dissonance that already exists. If one of the main strategies to cope with the psychological incongruence, the making of distance, fails, then there would be two alternatives: Either the anxiety that follows the cognitive dissonance would increase or another strategy would be necessary. One such strategy could be to adjust behavior according to the attitudes, e.g. eat less meat or no meat at all. This would call for quite a change, both on an individual level and on the societal level, and one of the hardest things to do for a human being is to change.

According to Lumbert, conformity is a predominant human trait that influences most of our everyday culture. But the norm is not static, and the norm of meat eating is challenged in a lot of ways. Public health issues, like cardiac disease, cancer and even antibiotic resistance calls for changes both in consumer habits and production conditions.[15] New findings in the field of animal cognition paves the way for a new kind of awareness. From an environmental perspective, the meat eating, or at least the level of it, is far from sustainable.[16]

It is possible that the future norm will look different from today's. There are signs that change is already on its way. In Sweden, it seems like meat consumption is no longer on the rise. During 2017, the total meat consumption was 85.5 kg/capita annually, compared to the years 2010–2015, where it stabilized around 87.5 kg/capita, according to the Swedish Ministry of Food and Agriculture.

The second largest hamburger chain in Sweden, Max Burgers, recently opened up a totally meat-free restaurant during a big music festival in Gothenburg and has both vegan and vegetarian burgers on the menu on a daily basis. In a television interview, the CEO of the company, Christoffer Bergfors, has stated that the shift to vegetarian and vegan dishes is the greatest transformation during the company's history, dating back to the 1960s. He also said that the change is "absolutely necessary."

Is there hope for the billions of farmed animals that suffer in today's farm factories? That depends on how people in general will go about the changes that the UN and other worldwide institutions call for. The Western world suffers from increasing anxiety, showing in growing numbers of psychiatric diagnoses. This could be the result of the harm done, not only to other living beings, but to the environment as a whole.[17] It seems as if both of the strategies to cope with the cognitive dissonance deriving from the consumption of animals are failing. The line between humans and animals is blurred and the distance at different levels between humans and animals seems to decline. There is a call for a new strategy, and that strategy could be facing up to reality, meeting eye to eye with the absent referent and stop the eating of meat.

Notes

1 Henning Steinfield et al., *Livestock's Long Shadow: Environmental Issues and Options* (Rome, Italy: Food Agriculture Organization of the United Nations [FAO], 2006), 3–6.
2 Carol J. Adams, *The Sexual Politics of Meat: A Feminist-Vegetarian Critical Theory* (New York, NY: Continuum, 2010), 13.
3 Helena Pedersen, *Animals in Schools: Processes and Strategies in Human-Animal Education* (West Lafayette, IN: Purdue University Press 2010), 110.
4 Jeanette Thelander, "Cannibals and Vegans: Distances and Borders in the World of Meat" (MA diss., University of Uppsala/Gotland University College, 2011), 1–63.
5 Barbara Kawulich, "Participant Observation as a Data Collection Method," *Forum Qualitative Social Research* 6, no 2 (May 2005): 1–2.
6 Brita Egardt, *Hästslakt och Rackarskam* (Stockholm, Sweden: Nordiska Museet, 1962).
7 Leon Festinger, *A Theory of Cognitive Dissonance* (Stanford, CA: Stanford University Press, 1962), 3.
8 Emilia Bolin, "Hundägares och Kattägares Relation till Sitt Djur" (Veterinary diss., University of Agricultural Sciences, 2003), 19.
9 Cormac McCarthy, *The Road* (New York, NY: Knopf), 197.
10 Paul Piers Read, *Alive: The Story of the Andes Survivors* (New York, NY: Avon Books).
11 Con Slobodchikoff et al., *Prairie Dogs: Communication and Community in an Animal Society* (Cambridge, MA: Harvard University Press, 2009).
12 Therese Rehn, "The Role of the Emotional Relationship with Humans on Dog Welfare" (Ph. Lic. diss., Swedish University of Agricultural Sciences, 2011).
13 Keith Kendrick, personal interview with author, July 2008.

14 Samantha Lumbert, "Conformity and Group Mentality: Why We Comply," *Rochester Institute of Technology*, last modified November 2005, www.personalityresearch.org/papers/lumbert.removed.
15 World Health Organization, "Diet, Nutrition and the Prevention of Chronic Diseases," *WHO Technical Report Series 916* (Geneva, Switzerland: WHO, 2003), 21.
16 Steinfield et al., *Livestock's Long Shadow*, 23.
17 Theodore Roszak, *The Voice of the Earth: An Exploration of Ecopsychology* (New York, NY: Simon and Schuster 1992), 324–327.

Bibliography

Adams, Carol J. *The Sexual Politics of Meat: A Feminist-Vegetarian Critical Theory*. New York, NY: Continuum, 2010.

Bolin, Emilia. "Hundägares och Kattägares Relation till Sitt Djur." Veterinary. diss. Swedish University of Agricultural Sciences, 2003.

Egardt, Brita. *Hästslakt och Rackarskam*. Stockholm, Sweden: Nordiska Museet, 1962.

Festinger, Leon. *A Theory of Cognitive Dissonance*. Stanford, CA: Stanford University Press, 1962.

Kawulich, Barbara. "Participant Observation as a Data Collection Method." *Forum: Qualitative Social Research* 6, 2(May 2005): 1–19.

Lumbert, Samantha. 2005. "Conformity and Group Mentality: Why We Comply." *Rochester Institute of Technology*. Last modified November2005. www.personalityresearch.org/papers/lumbert.removed.

McCarthy, Cormac. *The Road*. New York, NY: Knopf, 2006.

Pedersen, Helena. *Animals in Schools: Processes and Strategies in Human-Animal Education*. West Lafayette, IN: Purdue University Press, 2010.

Read, Piers Paul. *Alive: The Story of the Andes Survivors*. New York, NY: Avon Books, 1975.

Rehn, Therese. "The Role of the Emotional Relationship with Humans on Dog Welfare." Ph.Lic. diss. Swedish University of Agricultural Sciences, 2011.

Roszak, Theodore. *The Voice of the Earth: An Exploration of Ecopsychology*. New York, NY: Simon and Schuster, 1992.

Slobodchikoff, Con, Bianca S. Perla and Jennifer Verdolin. *Prairie Dogs: Communication and Community in an Animal Society*. Cambridge, MA: Harvard University Press, 2009.

Steinfield, Henning, Pierre Gerber, Tom Wassenaar, Vincent Castel, Mauricio Rosales and Cees de Haan. *Livestock's Long Shadow: Environmental Issues and Options*. Rome, Italy: Food and Agriculture Organization of the United Nations, 2006.

Thelander, Jeanette. "Cannibals and Vegans: Distances and Borders in the World of Meat." MA diss. Uppsala University/Gotland University College, 2011.

World Health Organization. "Diet, Nutrition and the Prevention of Chronic Diseases." *WHO Technical Report Series 916*. Geneva, Switzerland: WHO, 2003.

3.2

FROM DEVOURING TO HONORING

A Vaishnava-Hindu therapeutic perspective on human culinary choice

Kenneth Valpey (Krishna Kshetra Swami)

Introduction

In this chapter I propose to regard the current state of most human dietary practice as both cause and symptom of a collective condition of *disease*—what might be called a *civilizational* disease or, indeed, a pandemic. Taken in this light, we may be led to conclude that there is warrant for urgently seeking an effective therapy leading to an eventual cure for this dangerous condition. Here, I suggest Indic dharma practices and yoga practices as a significant potential contributors to a viable program of therapy that may be promising as a way toward such a cure.[1] Of course, whenever we encounter disease, we search for causes. I would suggest, quite simply, that this diseased condition is brought about, aggravated, and sustained by the rampant habitual inclusion of animal flesh in the daily, or near-daily, diet of vast human populations. The direct consequence of this quite unprecedented culinary preference for animal flesh is the infliction of violence on a scale that is unimaginable to us, all the more made unimaginable by the meat industry's thorough success in making its violent practices largely invisible to the public. To address this violence requires that we address culinary habits; and to alter culinary habits consequentially, beyond mere cosmetic adjustments, the human community must surely draw on all available cultural and religious resources that may be of any help. Further, to consider such help in therapeutic terms calls attention to the application of a programmatic practice, rather than a simple reliance on one-time reasoned changes of mind.

Two key therapeutic principles from Indic traditions to be considered are (1) an important component of the ancient Vedic and post-Vedic Dharma tradition, namely, the practice of Fivefold Sacrifice (*pañca-yajña*) by which a moral community is established and sustained; and (2) aspects of the classical Yoga tradition, by which identification of the most basic cause of human suffering and the Yoga practices for its removal are articulated. Yet, as we will see, these two principles by themselves cannot be expected to appeal to moderns. They may, however, be drawn together, enriched, and transformed by a third therapeutic tradition rooted in India, namely, Bhakti. Particularly by applying principles of bhakti (devotion), a viable program of civilizational therapy can be conceived.

Here I will initially sketch the practice of the Fivefold Sacrifice and its rationale, suggesting how this may be regarded in therapeutic terms, as facilitating the engagement of persons in a broadened sense of moral community. Next, I will consider relevant aspects of classical Yoga as it serves both to identify the essential cause of human suffering—the individual and collective diseased condition—and the process of its cure. Finally, I will draw these strands together in the context of the Bhakti—devotional yoga—tradition, as represented in the foundational and paradigmatic Hindu text, the Bhagavad-gītā. In my concluding reflections, I suggest how the devotional yoga tradition of Bhakti might be brought into conversation with the important recent stream of ethical thought, the ethics of care, which is characterized by an emphasis on practice over ethical deliberation.

Establishing moral community: the fivefold sacrifice (*pañca-yajña*)

Early Indic Dharma tradition saw the need for easily performed, remedial measures to help human beings reform their habits of poorly relating to the environment to imbibe a healthy, inclusive, non-harming worldview.[2] The key term here is *yajña*, usually translated as "sacrifice." Yajña takes numerous forms, an important one being the *pañca-yajña*, which is conceived as a fivefold process of acknowledging existential debt and creating moral community.[3]

By "existential debt" I mean the human being's innate condition of being dependent, in a deep sense, on other beings. Humans, as recipients of goods and benefits from their environment, tend to become avaricious, amassing more than required and failing to show gratitude in acknowledgement of dependence. The Dharma tradition is a means of curbing this tendency, first by identifying five types of beings on whom we humans are inherently dependent and to whom we are therefore indebted. Katrin Stamm summarizes these debts (*rinam*) as follows:[4]

> [A] person is indebted to the *deva-s*, the managers of the forces of nature, for supplying the means to sustain his or her body (*deva-rinam*); to the seers of yore, the *rishi-s*, and the teachers who received and then passed on the knowledge about the ultimate meaning of life and the means to attain it … (*rishi-rinam*); to the *pitri-s*, or former generations who helped him or her to be what and where s/he is now (*pitri rinam*); to the goodwill and support of his and her fellow humans (*nri-rinam*) and to all living beings who help that person to sustain him- or herself (*bhu-rinam*).

The same sacred texts prescribe daily means for acknowledging and repaying these debts, namely, by performing the fivefold sacrifice. This includes a fire ritual with oblations (typically, of uncooked grains and clarified butter) to *deva-s*; daily study, recitation, and teaching of sacred texts, as sacrifice to the *rishi-s* (in particular, the seers of the sacred hymns collectively referred to as Veda); making offerings to the ancestors, honoring one's elders, and cherishing progeny, as sacrifice to the *pitri-s*; hosting a guest who is not a family member but, rather, is a stranger, and to give land, clothing, and food to those in need, as sacrifice to humankind (*nri*); and finally, feeding other living beings, both domesticated and free living, by making feed available to them, as a sacrifice to nonhuman living beings (*bhuta*).

As a remedial practice that addresses human beings' avaricious tendencies, the fivefold sacrifice also addresses the notion of *karma*—the idea that human actions invariably bring about consequences for the actor, whether immediate or eventual. With this understanding, linked to the matter of existential debts and their repayment, humans practicing the fivefold sacrifice may be led into thinking in terms of *quid pro quo* exchanges. On a surface level of understanding, this may be so; and indeed, with regard to the first of the five sacrifices, offered to nature's invisible managers (*devas*), the Bhagavad-gītā (3.12) admonishes human beings sternly that to not offer the gifts of the *devas* back to the *devas* in sacrifice makes one a thief. Conversely, if one makes appropriate offerings, one is regarded as freed from sin (from transgressive karma)—specifically, according to commentators, the sin of committing violent actions in the course of maintaining one's household.[5]

But on a deeper level, to acknowledge existential debt by performance of the fivefold sacrifice is to demonstrate gratitude for having received someone's goodwill and trust. As Stamm argues, goodwill and trust place one in a moral obligation to be oneself reliable, trustworthy, and benevolent in return. Particularly in relation to unseen agents, she notes, "Humans and nature's agents, the *deva-s*, thereby form a *moral community* of mutual trust and responsibility, mutual obligations, reciprocal acknowledgement of rights and duties and the general demand for inter-personal regard."[6] Crucial to the notion of moral community embodied in the fivefold sacrifice is that it extends beyond the sphere of human beings to include, ultimately, the whole range of living beings, seen and unseen; such inclusiveness prompts practitioners to allow the sense of gratitude to flow freely beyond its common expressions limited to persons one experiences direct benefit from.

Clearly, the inclusion of unseen beings in moral community points to the Dharma tradition's implied metaphysics, whereby behind all existence is the principle of consciousness. In this light we must recognize that the moral community thus described reaches quite beyond that conceived by Immanuel Kant, for whom duties and obligations are limited to persons directly experienced.[7] This is both the challenge and the opportunity of the Dharma tradition, promising to facilitate a powerful therapy toward curing the civilizational disease at the deepest level, namely, ignorance of the relationality implied in consciousness that underlies all of creation.

Rehabituation through Yoga

I have identified the practice of fivefold sacrifice with the Dharma tradition of India, but it can also be related to our second therapeutic principle, namely, the

Yoga tradition. Specifically, as suggested in the Bhagavad-gītā, it embodies the practice of *karma-yoga*, the spiritual discipline of right action. As such, its transformative character is highlighted, whereas its identification with the Dharma tradition highlights its remedial character.

To appreciate the transformative potential of the Yoga tradition, we do well to begin with rudiments of its metaphysics. As with Dharma traditions, for Yoga, the metaphysical point of departure is the centrality of consciousness as the basis of all life. Yet in the case of Yoga an important distinction is made. Living beings (*puruṣa*) have in common their root existence in consciousness, whereas the manifest world (*prakṛti*) is essentially non-conscious. Due to ignorance of their nonmaterial identity, atemporal living, conscious beings are entrapped in non-conscious matter—each conscious being residing in a temporal body (including the "mind"—faculties of thought, memory, etc.); and this condition perpetuates ignorance and misery. Among all living beings, however, by virtue of their self-awareness, humans have the capacity to attain existential freedom through yoga principles and yoga-related practices.

As the proactive complement to Sāṅkhya's metaphysical analysis, Yoga proposes a praxis by which humans may fully realize, or actualize, themselves in relation to—and yet separate from—the world of matter. The soteriological trajectory of Yoga is to attain—or better, to recover—absolute freedom (*kaivalya*) from the manifold and perpetually binding conditions of the world.[8] To become free from bondage in classical Yoga is to undo habits of thought and action that are binding as a consequence of action (*karma*) and hence crippling, which is to say, diseased. Yet the attainment of freedom from bondage is best seen as a welcome byproduct of Yoga's goal, which is the attainment of perfect self-awareness, technically referred to as *samādhi*. In the context of our discussion in terms of therapy, one could speak of *samādhi* as the condition of perfect health.

In Patañjali's Yogasūtra (YS), a key text of classical Yoga, the first of four chapters outlines the causes of the diseased, or dis-integrated human condition; it then outlines the features of this state of healthy being that it calls *samādhi*. Gregory Fields summarizes:[9]

> Yoga seeks to counteract dis-integration, whose forms include physical illness and mental distress, to help bring about higher states of knowing and being. *Samādhi* is the quintessential form of reintegration—the recovery of Unity. The polarity of integrated higher knowledge and states of disability opens perspectives on meanings of well-being, both psychophysical and spiritual. *Vyādhi* [disease] implies dis-integration or fragmentation. Yoga counteracts dis-integration, which manifests as physical and mental dysfunction and distress.

To reach the state of *samādhi* is, doubtless, an arduous process, as described in the Yogasūtra. But the substance of this process is the stilling of the "fluctuations of the mind" (*citta-vṛtti-nirodhaḥ*—YS 1.2), to be accomplished by the well-known eight-limbed progression of practices (*aṣṭāṅga-yoga*, initially listed at YS 1.29). And within these practices, it is the very first aspect of the first limb ("restraint"—*yama*) that identifies the therapeutic character of Yoga in terms of our present concern. This first and foremost principle of self-restraint is called *ahiṁsā*, non-harming.[10] It is this ethical principle that is regarded as the gateway to both individual clarity of mind and, I would argue, the recovery of social functionality, peace, and deep sustainability of the environment.[11]

This centrality of non-harming in Yoga therapy is underscored when it is considered in relation to the fifth of the eight yoga limbs, namely, "withdrawal" (*pratyāhāra*). Sense engagement with sense objects is seen as the sphere of habit through which the mind remains always unsettled. From the perspective of Yoga, contemporary consumer culture could be characterized as a crippling mass-habituation to the opposite of the practice of sense withdrawal. Indeed, consumer culture could well be defined in terms of "over-consumption" or over-indulgence, quite direct translations of the Sanskrit term *atyāhāra*. But *atyāhāra* can also be construed as "indiscriminate consumption," particularly in the matter of eating.

Thus the combined principles of nonharming and sense withdrawal act as a rejection of and, indeed, a protest against consumer culture, based on the understanding that the satisfaction of living beings in a deep sense cannot come from consumption of temporal matter. At the same time, these two principles serve the development of moral community that is nurtured by the fivefold sacrifice, by pursuing the aim of shrinking and ultimately erasing the gap between the reality of biological violence and the ideal of yogic cessation of violence.[12] Thus the yoga practitioner is determined to minimize all forms of harm to other beings as far as possible and, in this spirit, to abstain entirely from eating animal flesh.[13] Again, the state of perfection, or perfect health—*samādhi*—would be regarded as the state in which one is entirely freed from violent actions and their implications.

Curative devotion

Fundamental to both the Dharma tradition of *pañca-yajña* and the classical Yoga tradition is an acknowledgement that a conscious principle underlies all creation. By this understanding, the notion that all living beings would best be included in moral community becomes a compelling proposition, grounded in a metaphysical reality that rejects a reductionist ontology constituted solely of matter and energy. Likewise, by this understanding, a process of "rehabituation," based on principles of restraint enshrined in the Yoga tradition, becomes conceivable, as it renders plausible the conscious self's separateness from—and hence ultimate non-dependence upon—the bodies of other living beings.

And yet, however compelling the moral vision enshrined in these Dharma and Yoga traditions might be, those who might take up such practices and self-disciplines as I have here described are surely few indeed. What hope is there for the vast majority of people who will not adopt such practices?

Neither the ritual practices of *pañca-yajña* nor protracted personal austerities of *ahiṁsā* and *pratyāhāra* are, in themselves, likely to be attractive forms of collective therapy in the modern world. Rather, I would suggest, it is these principles and practices placed in support of a third Indic tradition, namely Bhakti (devotion), that a viable civilizational therapy may be conceived and implemented.

The Bhakti tradition, as set out particularly in the Bhagavad-gītā and developed significantly in other Bhakti texts (especially the Bhāgavata Purāṇa), incorporates the therapeutic benefits of the Dharma and Yoga traditions. Further, however, the Bhakti tradition invites the trans-temporal conscious self (*puruṣa*, also referred to as *ātmā*) to experience relishable aesthetic relationship (*rasa*) with its trans-temporal counterpart, the supreme self or over-soul (*paramātmā*). Such a relationship is not based on ritual or austerity, but on reciprocal affection. To be sure, this affection is nurtured and fructified with the aid of ritual and austerity, but the practice of these in the spirit of bhakti springs from inherent willingness and eagerness of the essentially unfettered *ātmā* to be consequentially reformed to enable transcendent relationship.

In the Bhagavad-gītā, an extended dialogue between the warrior Arjuna and his friend and charioteer Krishna, Krishna identifies himself as *paramātmā*, the supreme self. As such, as a basic practice of bhakti he urges the devoted offering of vegetarian food to him (9.26). When such acknowledgement of the food's source is made in a spirit of devotion, Krishna says, he "eats" it. And the "remainder" of such simple sacrifice becomes sanctified food that can be relished without fear of negative karmic consequences.

Similarly (and immediately following the previous stanza), Krishna urges that *whatever* one does, whatever one eats, whatever one sacrifices or gives in charity, and whatever austerity one might practice, are all best done as offerings to him, the transcendent self (9.27). These practices may be regarded as deeply therapeutic, surely fostering moral community and freedom from self-destructive worldly attachment; but they are understood to do so by way of divine participation that leads to aesthetic pleasure in ever-deepening loving exchange. Indeed, Krishna indicates this dynamic two stanzas later: "I am equal to all beings, neither averse nor affectionate; yet those who resort to me with devotion are in me, and I am in them" (9.29). It is this interconnection of self and supreme self that awakens the full comprehension of personhood that, in turn, enables one to experience the common personhood of living beings in all their many forms. Thus moral community transforms into devotional community that opens out to expanding spheres of compassion and care.

Again, as with the Dharma and Yoga traditions, one may well doubt that the way of Bhakti would attract the masses of people required to effect a civilizational change of habit. Yet in this case, Krishna seems to anticipate this doubt, calling attention to the need for enlightened leadership (Bg. 3.21). "Whatever a prominent person does, the people will follow; what example such person exhibits, all the world pursues." The Bhāgavata Purāṇa, another important bhakti text (far more extensive than the Bhagavad-gītā) elaborates on this idea, placing responsibility squarely on heads of state to practice bhakti themselves as the basis for educating their citizenry and extending protection to animals. According to the

Bhāgavata Purāṇa, leaders serve both these purposes best when they themselves comprehend and practice the life-affirming principles of bhakti, having themselves undergone the therapeutic treatment by which their innate aesthetic relish of divine reciprocity is awakened. As educators by example as much as precept, leaders are regarded as qualifying themselves as true leaders of human society, elevating all citizens to the pursuit of human beings' highest potential, far beyond the degrading and crippling functions of mindless consumption that causes equally mindless violence.

Concluding reflections

I have suggested that the Bhakti tradition—rooted in India but translatable to any culture of the world—may be most helpful in contributing toward a therapeutic approach to the clearly diseased condition of contemporary global civilization. The salient feature of this diseased condition is the vastly predominant, yet unnecessary, habitual dietary practice of consuming animal flesh, for which billions of animals are slaughtered annually and by which, in large measure, the eco-system of the entire world is being devastated.

Yet one may still wonder how the Bhakti tradition might relate to Western traditions, particularly regarding animal ethics in relation to the eating of animals. In this connection, I wish to suggest an avenue in contemporary ethical thought to help us in considering the application of the Bhakti tradition as a useful approach to civilizational therapy. This is the ethics of care, which has only in recent decades developed through various streams of feminist thought. As the name suggests, the basic idea is that ethics is best rooted in our natural experience either as persons cared for or as caregivers. Although the ethics of care is generally focused on human-human relationality, it has also been shown to hold promise in extending the circle of care to animals.[14]

To extend the ethics of care to animals could be accomplished with the aid of the Bhakti tradition. But to make a connection between the ethics of care and the Bhakti tradition, it is first necessary to consider the gendered character of the three traditions discussed here. As Virginia Held, a writer on ethics of care, notes, "Virtually all [traditional communities] are patriarchal,"[15] and are therefore not conducive to the practice of an ethics of caring. Both Dharma and Yoga traditions can be seen as largely patriarchal in ethos. However, the Bhakti tradition is arguably rooted (especially in its later developments) in a metaphysics of gender equality (ultimate divinity being both masculine and feminine, as in the case of Rādhā and Krishna being eternal counterparts of ultimate being). Further, it is predicated on what might be called an inversion of divine–human relationality, in which the devotee becomes the carer for the divinity; or even further, where the (male) divinity depends on, even pines for, the favor of the (female) devotee.

Space does not permit a proper elaboration on just how principles of the Bhakti thought tradition might relate with and expand the scope of the ethics of care.[16] Suffice to say here that the "devotional turn" that constitutes the Bhakti tradition promises to place an ethics of care on a solid metaphysical foundation of relationality from which it becomes not only possible, but straightforwardly and rationally compelling, to extend caring to the environment in general and to animals specifically. Importantly, the Bhakti tradition may provide, along with the metaphysical grounding for an ethics of care, a comprehensive practice by which individuals and societies can effectively, freely, and permanently let go of the civilizational disease-perpetuating habit of eating meat. And an important principle that facilitates precisely the letting go of the inclination for tasting meat is also articulated in the Bhagavad-gītā: "Sense objects recede for the embodied soul who abstains from them, but taste remains; yet even taste ceases on seeing something better" (Gītā 2.59, trans. Goswami). The "something better" to which Krishna refers is the "taste" (*rasa*) of experiencing transcendent relationality by the practice of Bhakti. It is this practice that promises to transform the human habit of devouring animals to the "something better," which brings about the appropriate honoring of animals by not eating them, and also the honoring (regardful eating) of vegetarian food that has been sanctified by preparing it for and offering it to the transcendent person who is understood to be behind the principle of consciousness, which is, in turn, understood to be foundational to all of existence. Such a change of attitude from devouring to honoring in culinary practice points to a deeper quality of culinary choice that can inspire human beings to become truly and fully human as we learn to relate respectfully and joyfully, rather than exploitatively and hatefully, to all of creation.

Notes

1 For a detailed discussion of the long history of the therapeutic paradigm in Indian philosophy, see Wilhelm Halbfass, "The Therapeutic Paradigm and the Search for Identity in Indian Philosophy," in *Tradition and Reflection: Explorations in Indian Thought* (Albany, NY: SUNY, 1991), 243–257.
2 The term *dharma* finds its Sanskrit verbal root in **dhṛ*—to carry, maintain, preserve—leading to a rich constellation of meanings across the various Indic religious traditions. In the present context, dharma points to the set of routine religious obligations, especially for householders, and especially as delineated in Sanskrit texts called Dharmaśāstras. As Katrin Stamm ("On the Moral and Spiritual Implications of the Fivefold Sacrifice: Metaphysical Obligations to the Divine in Nature According to Vaishnava Texts and in Contrast to Immanuel Kant." *Journal of Vaishnava Studies,* 24/1 (Fall) 2015, 91–120) notes (94), the fivefold sacrifice referred to here is described in *Manu-Samhita* III.67 and *Brihadaranyaka-Upanishat* 1.4.16, as well as being referred to in *Rigveda* X.124.1, X.52.4, *Bhagavadgita* III.13 and *Manu-Samhita* III.118.
3 Typically translated in English as "sacrifice," the semantic field of the term *yajña* only partially overlaps that of the English term. However, the overlap is relevant to our discussion, to the extent that both carry the sense of giving up something or oneself to a higher entity or purpose.
4 Stamm, "Fivefold Sacrifice."
5 Bhānu Swami, trans., *Sārārth-varṣiṇī-ṭīkā: The Bhagavad-gītā Commentary of Śrīla Viśvanātha Cakravartī Ṭhākura* (Chennai, India: Sri Vaikunta Enterprises, 2003), 116.
6 Stamm, "Fivefold Sacrifice," 95.
7 Stamm, "Fivefold Sacrifice," 100, citing Immanuel Kant *The Metaphysics of Ethics*, trans. J. W. Semple (Edinburgh, Scotland: T. and T. Clark, 1886).

8 As Julian Woods explains in the context of discussion on the epic Sanskrit text, the *Mahābhārata*, "[T]he 'freedom' involved is not of any function or faculty of the ego (such as a 'will'), but of the human spirit—a very different matter. Epic freedom (or *mokṣa*) points beyond what we might recognize as the human 'person' to a freeing of the bonds that bind that person to the things and beings of the world itself." Julian F. Woods, *Destiny and Human Initiative in the Mahabharata* (Albany, NY: SUNY, 2001), 4.
9 Gregory P. Fields, *Religious Therapeutics: Body and Health in Yoga, Āyurveda, and Tantra* (Albany, NY: SUNY, 2001), 123.
10 As is well known, Mahatma Gandhi made the term *ahiṁsā* into a household word by his application of the concept to his movement for Indian independence. His inspiration for incorporating this principle in political activism is generally credited to his early contact with Jainism, in which *ahiṁsā* is a central tenet of righteous living.
11 *Non-harming* is the first of five "restraints" (*yama*), which is the first "limb" of the well-known eight-limbed system of yoga practice, *aṣṭāṅga*-yoga. It is worth noting that in his celebrated commentary on the Patañjali Yogasūtra, Vyāsa considers harmlessness to be rooted (*tan mūla*) in the four remaining restraints and five observances (*niyama*) that constitute the second limb of *aṣṭāṅga-yoga*. The four remaining restraints are truthfulness (*satya*), abstinence from theft (*asteya*), continence (*brahmacārya*), and non-acceptance of gifts (*aparigraha*). The five observances are cleanliness/purity (*śauca*), contentment (*santoṣa*), penance/austerity (*tapa*), study (*svādhyāya*), and full-aspiration-after-Īśvara (the supreme Lord; *īśvara-praṇidhāna*) (Baba 2010, 55–57). By connecting all the restraints and practices through harmlessness, Vyāsa seems to suggest that the Yogasūtra shows moral coherence.
12 As do Indic thought traditions generally, the Yoga tradition recognizes that all physical bodies subsist on other bodies, and therefore some violence is bound to involved in the act of preparing and eating food. The essential step in reducing violence in eating is to reject the flesh of animals. Jainism's ultimate answer to the problem of violence is *sallekhana*—a ritual process of fasting to death (only with the approval of elders of one's religious community). This practice is not formally known in other Indic traditions, although it may not be uncommon in informal practices, usually involving terminal disease.
13 Significantly, in the important commentary to Patañjali's Yogasūtras by Vyāsa, *ahiṁsā* is defined as "the absence of oppression towards all living beings by all means and for all times" (Bangali Baba, *The Yogasūtra of Patañjali with the Commentary of Vyāsa*, Delhi, India: MLDB, 2010, p. 55, commentary to YS 1.30). The comprehensiveness of this definition clearly includes the slaughter of animals; hence, the notion that "happy animals" can be slaughtered for human use would be rejected.
14 The present confinement of the ethics of care to human–human relations is indicated, for example, in Held's observation, "It is the relatedness of human beings, built and rebuilt, that the ethics of care is being built to try to understand, evaluate, and guide." Virginia Held, *The Ethics of Care: Personal, Political, and Global* (Oxford, England: Oxford University Press, 2006), 30. However, animal ethics has also become a subject of attention for some ethics of care feminists. See Josephine Donovan and Carol J. Adams, eds., *Feminist Care Tradition in Animal Ethics: A Reader* (New York: Columbia University Press, 2007).
15 Held, *The Ethics of Care*, 19.
16 For detailed expositions on one particular Bhakti tradition, that of Caitanya Vaiṣṇavism (also known as Gauḍīya Vaiṣṇavism), see Ravi M. Gupta, ed., *Caitanya Vaiṣṇava Philosophy: Tradition, Reason and Devotion* (Aldershot, England: Ashgate, 2014).

Bibliography

Baba, Bangali. *The Yogasūtra of Patañjali with the Commentary of Vyāsa*. Delhi, India: MLDB, 2010.

Donovan, Josephine, and Carol J. Adams, eds. *Feminist Care Tradition in Animal Ethics: A reader*. New York: Columbia University Press, 2007.

Fields, Gregory P. *Religious Therapeutics: Body and Health in Yoga, Āyurveda, and Tantra.* Albany, NY: SUNY, 2001.

Goswami, H. D. *A Comprehensive Guide to Bhagavad-gītā with Literal Translation.* Gainesville, FL: Krishna West, 2015.

Gupta, Ravi M., ed. *Caitanya Vaiṣṇava Philosophy: Tradition, Reason and Devotion.* Aldershot, England: Ashgate, 2014.

Halbfass, Wilhelm. *Tradition and Reflection: Explorations in Indian Thought.* Albany, NY: SUNY, 1991.

Held, Virginia. *The Ethics of Care: Personal, Political, and Global.* Oxford, England: Oxford University Press, 2006.

Stamm, Katrin. "On the Moral and Spiritual Implications of the Fivefold Sacrifice: Metaphysical Obligations to the Divine in Nature According to Vaishnava Texts and in Contrast to Immanuel Kant." *Journal of Vaishnava Studies*, 24, 1 (Fall) 2015, 91–120.

Swami, Bhānu, trans. *Sārārtha-varṣiṇī-ṭīkā: The Bhagavad-gītā Commentary of Śrīla Viśvanātha Cakravartī Ṭhākura.* Chennai, India: Sri Vaikunta Enterprises, 2003.

Woods, Julian F. *Destiny and Human Initiative in the Mahabharata.* Albany, NY: SUNY, 2001.

3.3

THE OTHER GHOSTS IN OUR MACHINE

Meat processing and slaughterhouse workers in the United States of America

Rebecca Jenkins

Introduction

More and more people are choosing to avoid animal products based on a myriad of ethical issues that surround the animal agriculture industry. In the U. S. alone, it is estimated that there are more than seven million vegetarians and vegans.[1] One issue coming to the forefront of these ethical discussions in recent years is the impact that eating animals has on *people*. These more anthropocentric discussions have predominantly concerned the environmental degradation that animal agriculture causes,[2] and the impact that consuming animal products has on human health.[3]

In addition to these environmental and health based conversations, a new issue is beginning to emerge in mainstream vegan outreach and advocacy. As a growing number of vegans are becoming more intersectional in their advocacy,[4] and as the global meat industry continues to grow,[5] the welfare of those employed by the animal agriculture industry[6] is coming to light.[7, 8] This worker-oriented conversation may evoke memories of Upton Sinclair's classic American novel: *The Jungle* [9] for some people. This American journalist and writer hoped to expose the harsh conditions and exploitation endured by immigrants in Chicago and other industrialized American cities at the turn of the century. Many readers were shocked by the unsanitary practices described in this book. The novel certainly was impactful, and contributed to outcry that led to reforms including the U.S. Meat Inspection Act. However, Sinclair had hoped that the novel would illicit more sympathy for the unseen suffering of the workers. He famously said of the public reaction "I aimed at the public's heart, and by accident I hit in the stomach." A century has passed, but the plight of meat industry workers is finally becoming more of mainstream conversation within animal advocacy.[10] It appears that the time may now be ripe for the public's heart, or at least the animal advocacy movement's heart, to open to the plight of these workers.

Outline

In this chapter, I contend that the concealed nature of the facilities in which we kill and butcher the animals eaten in the United States not only has a severely negative impact on the welfare of the animals, but also the people who make their living by killing them. This secrecy raises not just ethical concerns but also legal ones. This chapter discusses the multiple human rights violations that face many slaughterhouse and meat processing plant laborers in the United States. I discuss some of the major human rights issues concerning slaughterhouses and meat processing workers across the United States. These major concerns are workplace safety, workers compensation, and limitations on freedom of association. Upon conclusion of the human rights discussion, this chapter will move to discuss two other areas of concern for meat industry workers. It first touches on some of the Environmental Justice concerns regarding the communities in which these facilities are placed. Finally, this chapter will move to some of the psychological and social impacts of these facilities on both individual workers and their wider communities.

Underlying racism

Before moving forward in this discussion, an acknowledgement of the role that race plays in these problems is imperative. Many of the case studies we will look at demonstrate how an extremely large proportion of those who work in slaughterhouses and meat processing plants are people of color, and/or immigrants, both documented and undocumented. In addition to this fact, the communities that bear the environmental and social burdens of housing these meatpacking facilities and slaughterhouses, are often low-income communities and communities of color.[11] It is important to be aware that this is symptomatic of a much broader societal issue regarding racial injustice and the treatment of immigrants in the United States.

The role of immigration status in worker's welfare

Immigration issues permeate through all of the problems discussed in this chapter. Despite the fact that immigrant workers theoretically enjoy the same national and international legal protections, and that many have legal permission to work in the U.S., immigrants are often extremely vulnerable to employer coercion.[12] A significant amount of documented workers do not speak English fluently and are hesitant for this and other reasons to navigate the complicated procedures to vindicate their rights.[13]

Exacerbating these problems is the reality that many legal immigrants have family and friends who are undocumented and who they do not seek to draw attention to. Therefore, while vulnerability is most acute for undocumented workers, oftentimes those working lawfully are equally reluctant to vindicate their rights. Research suggests that this results in employers lacking incentive to protect such rights, knowing that complaints to labor authorities are unlikely.[14]

Immigration status is also directly linked with workplace health and safety, more specifically. A 2004 Investigate report showed that Mexican workers in the U.S are 80% more likely to die in the U.S. than their U.S. born counterparts. The report also showed that this number had increased dramatically in the preceding decade.[15] These immigrant workers are some of the most vulnerable members of American society.

Human rights

International Human Rights Law

According to a submission by Human Rights Watch to the Office of the United Nations High Commissioner for Human Rights Committee on Migrant workers,[16] immigrant workers make up the majority of the labor force in the U.S. meat and poultry industry. Human Rights Watch posits that the widespread human rights abuses are directly linked to the vulnerable immigration status of most workers.

In spite of the development of a comprehensive body of international law affirming a range of rights to *all* workers,[17] Human Rights Watch's *Blood Sweat and Fear* [18] report revealed an epidemic of cases in which employers' treatment of workers violated international human rights law as well as U.S. law reflecting those standards. This report highlights how ineffective enforcement by U.S. authorities compounds this by failing to remedy such abuses. In other cases, U.S. law itself fails to meet these international norms.

I Worker health and safety

Many aspects of modern day meat and poultry production are rife with hazards to life, limb and health.[19] Hard numbers support anecdotal evidence of the extreme dangers in this industry. One Nebraska expert described how the industry has the highest rate of injury and illness in the entire manufacturing sector. In 2001, the reported injury and illness rate was two and a half times greater than the average manufacturing rate and almost four times high than the overall rate for private industry.[20] An understanding of the impact that immigration status can have on reporting begs the question of how accurate this number is. It is likely a conservative number, as underreporting of injuries is certainly extremely common in this industry.[21] Line speed and long working hours are two features of this work, which make it exceptionally dangerous.

Line speed

Meatpackers maximize the volume of animals that they process by increasing the speed at which they go through the processing line. This raises obvious concerns for the welfare of the animals we put through this process.[22] For the purposes of this worker-oriented discussion, line speed is also important. Workers often cite the speed of the lines as the most dangerous feature of their work.[23] A Southern Poverty Law Center survey found that 78% of workers they asked said the line speed make them feel unsafe and 99% of workers said they had no opportunity to voice their concerns about the speed.[24]

These numbers and statistics can be shocking. They can also seem somewhat abstract. The power of the anecdote has been instrumental in bringing more awareness to the issues these workers face. Human stories can help us remember the deep individual human suffering involved in this industry. The story of Jesus Soto Cabajal is one tragic example of this. In July 2000, Jesus was cutting rounds of beef from hindquarters coming down the line at him every six seconds towards the end of his shift at Excel Corp.'s meatpacking plant in Schuyler, Nebraska, when tragedy struck. An investigative report explained that although no one witnessed the exact moment of Jesus' fatal injury, a knife punctured his chest above the left collarbone where the jugular vein returns blood from the head to the heart.[25] Excel was not fined for this death because no federal agency standards covered the circumstances that killed Jesus, according to the Occupational Safety and Health Administration. This story illustrates that it is not just poor conditions, long hours and more generalized forms of suffering that these workers endure, but sometimes even gruesome fatalities.

Long hours

In spite of the rapid modernization of the meat industry, many of these jobs involve lifting and manipulating very heavy animals, animal parts and equipment. The volume of animals being processed is hard to imagine for those of us who have never been inside a slaughterhouse. A legal case involving a Smithfield hog processing plant in Tar Heel, North Carolina, revealed that workers kill and cut up 25,000 hogs per day in this facility alone.[26]

This work is not just physically demanding, but also often involves exceptionally long workdays. United States labor law permits employers to require mandatory overtime with no limit with risk of dismissal if workers refuse. Most countries limit how much overtime can be required without consent of employees on a daily or weekly basis.[27] Time demands are particularly worrying considering the dangerous and physical nature of this sort of work. Research shows that risk increases at a worrying rate with risk being more than double at the end of a 12-hour shift when compared to an eight-hour shift.[28] When accidents inevitably occur, one would hope that an injured worker would be entitled to fair compensation. However, this is another area where human rights violations are rampant.

II Workers' compensation

The issue of workers' compensation for those in the meat and poultry processing plants is multifaceted and complex. While compensation for work related injuries and illnesses is a crucial part of the international human rights standards for workers, HRW's *Blood, Sweat and Fear* found that there were systematic failings of the industry in this regard. Some areas of concern were failings to recognize and report claims, delaying claims, wrongfully denying claims and threatening and taking reprisals against workers who sought compensation.[29] Workers' compensation is a states-based system. This report found that state government authorities were often guilty of not sufficiently informing workers of their rights, not adequately enforcing their rights under compensation statutes and not adequately enforcing anti-retaliation provisions designed to protected workers against dismissal for exercising their rights under workers' compensation laws.[30] Even when the rights are exercised, compensation is often inadequate. At the time of the publication of the Human Rights Watch report, workers' compensation weekly benefits leave for injured workers fell below the poverty line in 16 states.[31]

The subnational nature of workers compensation law, which varies by state, might have something to answer for in this context. It has come under scrutiny from analysts who believe that it has led to a "race to the bottom" between states in terms of cutting benefits and making eligibility rules stricter to retain and attract business.[32]

III Freedom of association

The Human Rights Watch investigation that examined working conditions and compensation problems also looked into organizing rights for workers in this industry. The same report found systematic interference with workers' freedom of association and right to organize trade unions. Some of the conduct reported is legal under U.S. labor law, which allows aggressive campaigning against employee's self-organization in violation of international standards. One example of this is the legality of mandatory "captive audience" meetings in which employers can inveigh against workers self-organizing and speak about potential workplace closure so long as the discussion does not constitute a "threat" of closure. U.S. labor law also allows permanent replacement of workers who exercise the right to strike, in violation of international protections on the right to strike.[33]

There have also been cases where some of the biggest meat processing companies have crossed the line regarding U.S. law on organizing rights.[34] One such case concerned a meat processing plant in North Carolina. In this case, a union representation election campaign was in motion. Smithfield organized the deployment of local police forces and the company employed security officials on their facility on the Election Day. Two workers were assaulted by Smithfield personnel on the Election Day and brought a legal case against the company.[35] Smithfield later established its own special police force under state law that created an atmosphere that many believe to have further chilled workers organizing efforts here. Today, this police involvement is particularly concerning in light of recent events of police brutality towards people of color in the U.S. Again, the underlying issue of systematic racism is hard to ignore in this context.[36]

Many commentators argue that the continuing failure of the U.S. administration and Congress to enact legislation to correct these problems puts the U.S. squarely in violation of its international obligations.[37] This lack of legislative action has led to a situation where some of the most vulnerable workers in the country, those for whom the right to freedom of association is most crucial, are afraid to exercise that right.

Environmental justice

Speaking more broadly about the environmental impacts of modern animal agriculture and meat processing, it is undisputed that this industry is having a very significant and negative impact on our environment. Its contribution to climate change is significant and many experts in this field agree that the effects of climate change are felt most acutely by those in some of the poorest nations in the world who have contributed to climate change the least.[38] Those who study environmental justice are aware that all members of society do not share these environmental costs evenly. Moving the discussion back to the United States, a brief discussion of the dangers of living near a CAFO, slaughterhouse or meatpacking facility (or all three) is necessary.

The issue of race is not only relevant here due to the fact that many workers live in close proximity to these facilities. It is also because these facilities are often situated in communities that have a high population of immigrants or a high percentage of people of color and low-income people.[39] Those who live in the areas surrounding these plants and slaughterhouses bear a disproportionate share of the health and environmental burdens, which these sites cause. For example, according to air quality studies, industrial agriculture causes nearby residents respiratory inflammation, headaches, eye irritation and nausea.[40] A 2015 Environmental Protection Agency publication highlighted further concerns regarding the water quality harms from this industry.[41]

At this point you might be wondering where these slaughterhouses and meat and poultry processing plants actually are. We certainly do not see them every few blocks in major cities like we see fast food outlets, restaurants and grocery stores selling the finished products of this labor. The facilities in question are, unsurprisingly, often located in rural communities and often in low-income communities and communities of color. North Carolina, which was the subject of much discussion in this chapter, is a good example of this. In this state there are 7.2 times as many swine CAFOs located within the areas of highest poverty as compared with the areas of lowest poverty.[42]

The Environmental Justice movement is concerned with the importance of having a safe place to live, work and play. It seems that working inside or living outside a CAFO, slaughterhouse or meat processing facility is in many ways, a very unsafe and unhealthy environment. One of the fundamental principles of Environmental Justice is that workers should enjoy a right to a safe and healthy work environment, without being forced to choose between an unsafe livelihood and unemployment. Unfortunately, this is not the case for many meat processing plant and slaughterhouse workers and their families.[43]

Psychological and social impacts

The impacts of working in a slaughterhouse or meat processing plant go beyond the physical. While research on the psychological effects on slaughterhouse workers is limited, with few studies finished to date, those that exist contain compelling and important information that needs to be recognized and further researched. A recent Oxfam study found that poultry workers commonly develop both depression and anxiety.[44] Another study put the prevalence of depressive symptoms 80% higher among poultry workers than other low-wage workers in the same area.[45]

These psychological impacts are not just experienced on the individual level, but also by the community. A strong correlation between the presence of a large slaughterhouse and high crime rates in U.S. communities has been established by recent research.[46] In a 2007 study, housing an abattoir in a community stood out as the factor most likely to spike crime statistics. According to this study, slaughterhouse workers were often forced to become "desensitized, and their behaviour outside of work sometimes reflected this."[47]

A health care provider serving poultry plant workers at a medical clinic in Northwest Arkansas echoed similar concerns about the communities in which this industry operates. This professional told Human Rights Watch that they saw a lot of psychological problems in workers, along with physical injuries. This individual was of the opinion that the relentless overtime involved caused fatigue and depression. HRW quoted them as saying that "the company gets cheap labor, they maximize production and all the social costs get passed onto the community."[48] Again, as discussed in the context of the disproportionate environmental and public health burdens, these social burdens are often dealt with by some of the most vulnerable communities.[49]

Conclusion

The cement walls of slaughterhouses conceal a great deal of suffering for both the animals that are killed and those who make their living killing. The human rights issues that affect laborers in this industry appear to be rampant in the whole process, from raising the animals to slaughtering and butchering them.

While the physical costs are well documented, it seems that the less tangible but equally severe emotional, psychological and social impacts are becoming increasingly apparent. All of these problems raise a further issue, one more philosophical in nature. If we could not raise, slaughter and butcher the animals, which end up in our plates, is it ethical to ask someone else to do so? The ethical implications of this scenario are further complicated by the sad fact that many of the workers involved often have no other options employment options available. Those paying the price to satiate our appetite for meat are not just the animals but also the often-exploited workers, their families and their wider community.

The name of this chapter was inspired by a recent documentary *The Ghosts in Our Machine*, which shed light on some of the suffering animals endure in modern our industrialized world. The use of the term "machine" was powerful in describing many of the industries predicated on animal exploitation. However, these machines are also the cause of significant human suffering. Most of this suffering is endured by low-income people of color and immigrants, communities and individuals already dealing with great inequity in many aspects of life. While acknowledging animals as silent victims of exploitation in this machine is extremely important, it is also critical in our advocacy to remember that the workers in American meat processing are also often voiceless, vulnerable ghosts in this same machine.

Notes

1 Vegetarian Times Editors, "Vegetarianism in America," *Vegetarian Times*, April 6, 2008, www.vegetariantimes.com/article/vegetarianism-in-america.

2 For examples see, Kip Anderson and Keegan Kuhn. *Cowspiracy: The Sustainability Secret.* Film (Santa Rosa, CA: AUM Films and First Spark Media, 2014); and "Meat and the Environment," PETA, accessed September 13, 2016, www.peta.org/issues/animals-used-for-food/meat-environment/.

3 "Meat Consumption and Cancer Risk," The Physicians Committee, accessed September 13, 2016, www.pcrm.org/health/cancer-resources/diet-cancer/facts/meat-consumption-and-cancer-risk; and "Eating for Your Health," PETA, accessed September 13, 2016, www.peta.org/issues/animals-used-for-food/eating-health/.

4 "Slaughterhouse Workers," Food Empowerment Project, accessed September 13, 2016, www.foodispower.org/slaughterhouse-workers.

5 "Meat and Animal Feed," Global Agriculture, accessed September 13, 2016, www.globalagriculture.org/report-topics/meat-and-animal-feed.html.

6 I have opted to use the term "slaughterhouse" rather than "abattoir" as I believe the language used to describe animals and their bodies is important. I have opted to use the term "meat" rather than "animal flesh" for the most part; this difficult decision was mostly based on convenience. For a more detailed discussion of the significance of language in this context see: Carol J. Adams, *The Sexual Politics of Meat: A Feminist-Vegetarian Critical Theory*, 10th ed. (New York, NY: Continuum, 2000).

7 "More Reasons to Go Vegan," PETA, accessed September 13, 2016, www.peta.org/issues/animals-used-for-food/reasons-go-vegan/.

8 NPR Producers, "Working 'the Chain,' Slaughterhouse Workers Face Lifelong Injuries," *NPR*, August 11, 2016, www.npr.org/sections/thesalt/2016/08/11/489468205/working-the-chain-slaughterhouse-workers-face-lifelong-injuries.

9 Upton Sinclair and Philip M Parker, *The Jungle* (San Diego, CA: Icon Group International, 2005).

10 "Meatpacking in the U.S.: Still a 'Jungle' Out There?" PBS, December 15, 2006, www.pbs.org/now/shows/250/meat-packing.html.

11 David H. Harris Jr., "The Industrialization of Agriculture and Environmental Racism: A Deadly Combination Affecting Neighborhoods and the Dinner Table," Presented to the

Seminar "Environmental Justice: A Challenge for the 21st Century," at the 72nd Annual Convention of the National Bar Association, July 30, 1997, www.iatp.org/files/Industrialization_of_Agriculture_and_Environme.htm.
12 "Immigrant Workers in the United States Meat and Poultry Industry," Human Rights Watch, December 15, 2005, www.hrw.org/legacy/backgrounder/usa/un-sub1005/.
13 Justin Prichard, "Mexican-Worker Deaths are Rising Sharply in the U.S.," *Deseret News*, March 14, 2004, www.deseretnews.com/article/595048917/Mexican-worker-deaths-are-rising-sharply-in-US.html.
14 Maurice Emsellem and Beth Avery, "Racial Profiling in Hiring: A Critique of the 'Ban the Box' Studies," *National Employment Law Project*, August 11, 2016, www.nelp.org/publication/racial-profiling-in-hiring-a-critique-of-new-ban-the-box-studies/; and "Home: National Immigration Law Center," National Immigration Law Centre, accessed August 30, 2016, www.nilc.org.
15 Prichard, "Mexican-Worker Deaths."
16 Human Rights Watch, "Immigrant Workers."
17 The Universal Declaration of Human Rights (UDHR) The International Covenant of Civil and Political Rights (ICCPR) The International Covenant on Economic, Social and Cultural Rights (ICESCR) and the International Convention on the Protection of the Rights of All Migrant Workers and Members of their Families recognize that economic and social rights must exist alongside civil and political rights to ensure full protection of human rights.
18 *Blood, Sweat, and Fear: Workers' Rights in U.S. Meat and Poultry Plants*, Human Rights Watch, January 24, 2005, www.hrw.org/report/2005/01/24/blood-sweat-and-fear/workers-rights-us-meat-and-poultry-plants.
19 Human Rights Watch, *Blood, Sweat, and Fear.*
20 Human Rights Watch, *Blood, Sweat, and Fear.*
21 J. Paul Leigh, James P. Marcin, and Ted R. Miller, "An Estimate of the U.S. Government's Undercount of Nonfatal Occupational Injuries," *Journal of Occupational and Environmental Medicine* 46 no.1 (2004): 10–18. doi: 10.1097/01.jom.0000105909.66435.53.
22 K. Olsson, "The Shame of Meatpacking," *The Nation*, August 29, 2002, www.thenation.com/article/shame-meatpacking/.
23 Deborah Fink, *Cutting into the Meatpacking Line: Workers and Change in the Rural Midwest* (New York, NY: University of North Carolina Press, The, 1998).
24 "Unsafe at These Speeds," Southern Poverty Law Centre, February 28, 2013, www.splcenter.org/20130228/unsafe-these-speeds.
25 Justin Prichard, "Workers Death in a Midwest Town Transformed by Immigrant Laborers," *Associated Press Wire Story*, March 13, 2004.
26 "In the Matter of Smithfield Packing Company Inc. – Tar Heel Division and United Food and Commercial Workers Union Local," (1999) AFL-CIO: 3826–3850.
27 Donald Dowling Jr., "From the Social Charter to the Social Action Program 1995–1997: European Union Employment Law Comes Alive," *Cornell International Law Journal*, no. 43 (1996).
28 Human Rights Watch, *Blood, Sweat, and Fear.*
29 Human Rights Watch, *Blood, Sweat, and Fear.*
30 Human Rights Watch, *Blood, Sweat, and Fear.*
31 Allan H. Hunt, *Adequacy of Earnings Replacement in Workers' Compensation Programs: A Report of the Study Panel on Benefit Adequacy of the Workers' Compensation Steering Committee, National Academy of Social Insurance* (Kalamazoo, MI: W.E. Upjohn Institute for Employment Research, 2005).
32 Martha T. McCluskey, "Efficiency and Social Citizenship: Challenging the Neoliberal Attack on the Welfare State," *Indiana Law Journal* 78 no. 2 (2003): 783–876.
33 The ILO's Committee on Freedom of Association has condemned the "Mackay Doctrine" that was drawn from NLRB v Mackay Radio & Telegraph Co., (1938) 304 U.S 333 as a violation of freedom of association.

34 See for example, "Decision and Direction of the NLRB," *Perdue Farms Inc and UFCW* 328 NLRB 909 (1999); "NLRB Decision and Order," *Tyson Foods Inc and UFCW Local 425*, 311 NLRB 552 (1993); and "NLRB Decision, Order and Direction of Second Election," *Cooking Good Division of Perdue Farms Inc and LIUNA* 323 NLRB 345 (1997).
35 *Ward v Smithfield Packing Company* [2003], (United States Court of Appeals for the Fourth Circuit, 2003).
36 Frederick C. Harris, "The Next Civil Rights Movement?" *Dissent*, Summer 2015, accessed September 13, 2016, www.dissentmagazine.org/article/black-lives-matter-new-civil-rights-movement-fredrick-harris.
37 "ILO Declaration on Fundamental Principles and Rights at Work," International Labour Organization, accessed September 13, 2016, www.ilo.org/declaration/lang–en/index.htm.
38 "Assessing the Environmental Impacts of Consumption and Production: Priority Products and Material," United Nations Environmental Program, 2010, accessed September 13, 2016, www.unep.fr/shared/publications/pdf/dtix1262xpa-priorityproductsandmaterials_report.pdf.
39 Nicole Wendee, "CAFOs and Environmental Justice: The Case of North Carolina," *Environmental Health Perspectives*, 121 (2013): A182–A189, http://ehp.niehs.nih.gov/121-a182/.
40 B. Friedrich and S. Wilson, "Coming Home to Roost: How the Chicken Industry Hurts Chickens, Humans, and the Environment," *Animal Law Review, Lewis & Clark Law School*, 22 (2016).
41 U.S. Environmental Protection Agency, *Environmental Assessment of Proposed Revisions to the National Pollutant Discharge Elimination System Regulation and the Effluent Guidelines for Concentrated Animal Feeding Operations* (Washington DC: Environmental Protection Agency, 2001), www3.epa.gov/npdes/pubs/cafo_proposed_env_assess_ch1-3.pdf.
42 C. J. Hodne, "Concentrating on Clean Water: The Challenge of Concentrated Animal Feeding Operations," *Iowa Policy Project*, April 2005, www.iowapolicyproject.org/2005docs/050406-cafo-fullx.pdf.
43 "Principles of Environmental Justice," Proceedings, The First National People of Color Environmental Leadership Summit xiii, October 1992, Principle 8, www.ejnet.org/ej/principles.html.
44 "Lives on the Line: The High Human Cost of Chicken," Oxfam America, accessed September 27, 2016, www.oxfamamerica.org/livesontheline/.
45 Hester J. Lipscomb et al., "Depressive Symptoms among Working Women in Rural North Carolina," *International Journal of Law and Psychiatry* 30 no. 4–5 (2007): 284–298.
46 One might object that a slaughterhouse town's disproportionate population of poor, working-class males might be the real cause, but Fitzgerald controlled for that possibility by comparing her data to counties with comparable populations employed in factory-like operations.
47 Amy Fitzgerald, Linda Kalof and Thomas Dietz, "Slaughterhouses and Increased Crime Rates: An Empirical Analysis of the Spillover from 'The Jungle' Into the Surrounding Community," *Organization and Environment* 22 no, 2 (2009): 158–184.
48 Human Rights Watch, Interview, Northwest Arkansas, August 15, 2003.
49 Christian E. Weller and Farah Z. Ahmad, "The State of Communities of Color in the U.S. Economy," *Center for American Progress*, October 29, 2013, www.americanprogress.org/issues/economy/report/2013/10/29/78318/the-state-of-communities-of-color-in-the-u-s-economy-2/.

Bibliography

Adams, Carol J. *The Sexual Politics of Meat: A Feminist-Vegetarian Critical Theory*. Tenth edition. New York, NY: Continuum, 2000.

Anderson, Kip and Keegan Kuhn. *Cowspiracy: The Sustainability Secret*. Film. Santa Rosa, CA: AUM Films and First Spark Media, 2014.

"Decision and Direction of the NLRB." *Perdue Farms Inc and UFCW 328 NLRB* 909 (1999).

Dowling Jr., Donald. "From the Social Charter to the Social Action Program 1995–1997: European Union Employment Law Comes Alive." *Cornell International Law Journal* 43(1996).

Emsellem, Maurice and Beth Avery. "Racial Profiling in Hiring: A Critique of the 'Ban the Box' Studies." *National Employment Law Project*. August 11, 2016. www.nelp.org/publication/racial-profiling-in-hiring-a-critique-of-new-ban-the-box-studies/.

Fink, D. *Cutting into the Meatpacking Line: Workers and Change in the Rural Midwest*. New York, NY: University of North Carolina Press, 1998.

Fitzgerald, Amy, Linda Kalof and Thomas Dietz. "Slaughterhouses and Increased Crime Rates: An Empirical Analysis of the Spillover from 'The Jungle' Into the Surrounding Community." *Organization and Environment* 22, 2(2009): 158–184.

Food Empowerment Project. "Slaughterhouse Workers." Accessed September 13, 2016. www.foodispower.org/slaughterhouse-workers.

Friedrich, B. and Wilson, S. "Coming Home to Roost: How the Chicken Industry Hurts Chickens, Humans, and the Environment." *Animal Law Review, Lewis and Clark Law School* 22(2016).

Global Agriculture "Meat and Animal Feed." Accessed September 13, 2016. www.globalagriculture.org/report-topics/meat-and-animal-feed.html.

Harris Jr., David H. "The Industrialization of Agriculture and Environmental Racism: A Deadly Combination Affecting Neighborhoods and the Dinner Table." Presented to the Seminar "Environmental Justice: A Challenge for the 21st Century." At the 72nd Annual Convention of the National Bar Association. July 30, 1997. www.iatp.org/files/Industrialization_of_Agriculture_and_Environme.htm.

Harris, Frederick C. "The Next Civil Rights Movement?" *Dissent*. Summer2015. Accessed September 13, 2016. www.dissentmagazine.org/article/black-lives-matter-new-civil-rights-movement-fredrick-harris.

Hodne, C. J. "Concentrating on Clean Water: The Challenge of Concentrated Animal Feeding Operations." *Iowa Policy Project*. April 2005. www.iowapolicyproject.org/2005docs/050406-cafo-fullx.pdf.

Human Rights Watch, Interview, Northwest Arkansas, August 15, 2003.

Human Rights Watch. *Blood, Sweat, and Fear: Workers' Rights in U.S. Meat and Poultry Plants*. January 24, 2005. www.hrw.org/report/2005/01/24/blood-sweat-and-fear/workers-rights-us-meat-and-poultry-plants.

Human Rights Watch. "Immigrant Workers in the United States Meat and Poultry Industry." December 15, 2005. www.hrw.org/legacy/backgrounder/usa/un-sub1005/.

Hunt, Allan H. *Adequacy of Earnings Replacement in Workers' Compensation Programs: A Report of the Study Panel on Benefit Adequacy of the Workers' Compensation Steering Committee, National Academy of Social Insurance*. Kalamazoo, MI: W. E. Upjohn Institute for Employment Research, 2005.

"In the Matter of Smithfield Packing Company Inc. – Tar Heel Division and United Food and Commercial Workers Union Local." (1999) AFL-CIO: 3826–3850.

International Labour Organization. "ILO Declaration on Fundamental Principles and Rights at Work." Accessed September 13, 2016. www.ilo.org/declaration/lang–en/index.htm.

Leigh, J. P., J. P. Marcin and T. R. Miller. "An Estimate of the U.S. Government's Undercount of Nonfatal Occupational Injuries." *Journal of Occupational and Environmental Medicine* 46, 1(2004): 10–18. doi: 10.1097/01.jom.0000105909.66435.53.

Lipscomb, H. J., J. M. Dement, C. A. Epling, B. N. Gaynes, M. A. McDonald and A. L. Schoenfisch. "Depressive Symptoms Among Working Women in Rural North Carolina:

A Comparison of Women in Poultry Processing and Other Low-wage Jobs." *International Journal of Law and Psychiatry* 30, 4–5(2007): 284–298.
McCluskey, Martha T. "Efficiency and Social Citizenship: Challenging the Neoliberal Attack on the Welfare State." *Indiana Law Journal* 78, 2(2003): 783–876.
National Immigration Law Centre. "Home: National Immigration Law Center." Accessed August 30, 2016. www.nilc.org.
"NLRB Decision and Order." *Tyson Foods Inc and UFCW Local 425, 311 NLRB* 552(1993).
"NLRB Decision, Order and Direction of Second Election." *Cooking Good Division of Perdue Farms Inc and LIUNA 323 NLRB* 345(1997).
NPR Producers. "Working 'the Chain,' Slaughterhouse Workers Face Lifelong Injuries." *NPR*. August 11, 2016. www.npr.org/sections/thesalt/2016/08/11/489468205/working-the-chain-slaughterhouse-workers-face-lifelong-injuries.
Olsson, K. "The Shame of Meatpacking." *The Nation*. August 29, 2002. www.thenation.com/article/shame-meatpacking/.
Oxfam America. "Lives on the Line: The High Human Cost of Chicken." Accessed September 27, 2016. www.oxfamamerica.org/livesontheline/.
PBS. "Meatpacking in the U.S.: Still a "Jungle" Out There?" December 15, 2006. www.pbs.org/now/shows/250/meat-packing.html.
PETA. "Eating for Your Health." Accessed September 13, 2016. www.peta.org/issues/animals-used-for-food/eating-health/.
PETA. "Meat and the Environment." Accessed September 13, 2016. www.peta.org/issues/animals-used-for-food/meat-environment/.
PETA. "More Reasons to Go Vegan." Accessed September 13, 2016. www.peta.org/issues/animals-used-for-food/reasons-go-vegan/.
Prichard, Justin. "Workers Death in a Midwest Town Transformed by Immigrant Laborers." *Associated Press Wire Story*. March 13, 2004.
Prichard, Justin. "Mexican-Worker Deaths are Rising Sharply in the U.S." *Deseret News*. March 14, 2004. www.deseretnews.com/article/595048917/Mexican-worker-deaths-are-rising-sharply-in-US.html.
"Principles of Environmental Justice." Proceedings. The First National People of Color Environmental Leadership Summit xiii. October 1992. www.ejnet.org/ej/principles.html.
Sinclair, Upton and Philip M. Parker. *The Jungle*. San Diego, CA: Icon Group International, 2005.
Southern Poverty Law Centre. "Unsafe at These Speeds." February 28, 2013. www.splcenter.org/20130228/unsafe-these-speeds.
The Physicians Committee. "Meat Consumption and Cancer Risk." Accessed September 13, 2016. www.pcrm.org/health/cancer-resources/diet-cancer/facts/meat-consumption-and-cancer-risk.
United Nations Environmental Program. "Assessing the Environmental Impacts of Consumption and Production: Priority Products and Material." 2010. Accessed September 13, 2016. www.unep.fr/shared/publications/pdf/dtix1262xpa-priorityproductsandmaterials_report.pdf.
U.S. Environmental Protection Agency. *Environmental Assessment of Proposed Revisions to the National Pollutant Discharge Elimination System Regulation and the Effluent Guidelines for Concentrated Animal Feeding Operations*. Washington DC: Environmental Protection Agency, 2001. www3.epa.gov/npdes/pubs/cafo_proposed_env_assess_ch1-3.pdf.
Vegetarian Times Editors. "Vegetarianism in America." *Vegetarian Times*. April 6, 2008. www.vegetariantimes.com/article/vegetarianism-in-america.
Ward v Smithfield Packing Company [2003]. United States Court of Appeals for the Fourth Circuit, 2003.

Weller, Christian E. and Farah Z. Ahmad. "The State of Communities of Color in the U.S. Economy." *Center for American Progress*. October 29, 2013. www.americanprogress.org/issues/economy/report/2013/10/29/78318/he-state-of-communities-of-color-in-the-u-s-economy-2/.

Wendee, Nicole. "CAFOs and Environmental Justice: The Case of North Carolina." *Environmental Health Perspectives* 121(2013): A182–A189. http://ehp.niehs.nih.gov/121-a182/.

3.4

ANIMAL AGRICULTURE AND CLIMATE CHANGE

Tobias Thornes

Global warming brought about by human emissions of greenhouse gasses, combined with the destruction of natural habitats to make space for human activities, is changing living conditions for animals across the world. Provoked by an especially strong "El Niño" event – a natural phenomenon whereby weakened winds in the Pacific allow warm water that is usually confined in the West to spread across the whole ocean, releasing extra heat into the atmosphere – 2015 was easily the warmest year on record,[1] with 2016 set to be even warmer at the time of writing. Many places, especially the Arctic, are getting warmer, on average, year by year. Patterns of Monsoon rain are shifting, tending to make drier regions even drier and wetter regions wetter. Extreme rainfall events are becoming more frequent in Europe and North America and extreme heatwaves are becoming more common across the globe.[2] Such long-term change in temperatures and weather patterns is referred to as "climate change", and at human hands it is taking place more rapidly today than for at least a million years. Centres of biological diversity such as the Amazon Rainforest and the Great Barrier Reef are, as a result, coming under particular stress, and many animals are finding it difficult to cope.[3]

But as well as bearing the brunt of the burden of climate change, non-human animals play a very significant part in the problem, albeit as a result of human exploitation of other species. Thirty percent of anthropogenic greenhouse gas emissions are associated with agriculture, and most of this can be put down to animal agriculture. Flocks of ruminant animals – which include sheep, cows and goats – directly produce methane, a greenhouse gas 20 times more potent than carbon dioxide. Furthermore, domesticated farmed animals require either pasture to graze or food crops to eat instead. This means that under an animal-based human food system, more land needs to be used for agriculture overall, so that all the other emissions associated with farming each acre of land – including habitat destruction through land-use change, nitrogen emissions from fertilisers and pollution from farm machinery – also increase. As food production expands to feed more people and if diets continue to become more meat- and dairy-oriented, agricultural emissions are likely to become even more important in the future. This chapter will explore the magnitude of the contribution made by animal agriculture to climate change, and the steps that could be taken to reduce or eliminate this contribution in an ethical way.

All ruminant animals produce methane gas in the gut when they break down carbohydrates and amino acids. This is released through the mouth and anus in different proportions depending on the species. Inevitably, this puts a heavy environmental cost on all meat and dairy products. For each litre of milk she produces, a typical cow emits 19 grams of methane, which has the equivalent warming effect on the planet as 440 grams of carbon dioxide. Producing one litre of milk is thus about a fifth as damaging to the climate as burning one litre of petrol, without accounting for the additional emissions associated with feeding and milking the cow, processing, packaging and transporting the milk. The methane output of a single cow on a single day would be enough to power a car fuelled by the gas for 5 km. In combination, all the world's cows produce enough methane to fuel 95 600 MW power stations.[4]

In 2011, atmospheric methane concentration stood at 1802.3 parts per billion, around one hundred times less than carbon dioxide. But this is 2.5 times the pre-industrial (that is, prior to 1750, after about which time the industrial revolution meant that human greenhouse gas emissions began to rise) concentration,[5] and levels in excess of 700 parts per billion are unprecedented over at least the past 450,000 years.[6] Although it is shorter-lived – breaking down into carbon dioxide and water in an average of 12 years – than carbon dioxide, each methane molecule is 21 times as potent at trapping heat in the atmosphere.[7] Substantially reducing methane emissions would therefore have a strong impact on global warming in the near-term (this century), even if the longer-term effects are less significant. Ruminants are not the only source of methane: Nearly one third of global emissions are from natural sources such as wetlands, and only about a third of the human contribution is attributable to domesticated animals and their dung; the remainder comes from oil and gas extraction and rice paddies.[8] But animal agriculture is, nevertheless, an important and growing source of this very powerful greenhouse gas.

Various ways of reducing methane emission from ruminants have been proposed, falling broadly into three categories: Trapping the gas released; changing the diet of farmed animals; or improving the animals' well-being to increase the yield per gram of greenhouse gas emitted. If ruminant methane could be trapped and harvested, it would provide a cheap and abundant fuel to replace fossil fuels, hence potentially mitigating global warming in more ways than one. This might be possible for animal dung, but trapping the gas emitted by the animals themselves is probably not feasible at present. It would, of course, be very difficult for animals grazing in the open air, and would probably require the animals to be kept in an airtight artificial environment, bringing with it its own environmental and ethical implications.

An easier alternative is to change animal feed so that the feed produces less methane when digested, for example by using corn- or legume-silage instead of grass-silage. Grain-fed cattle produce less methane than grass-fed ones, but additional environmental costs are associated with growing, processing and transporting grain and switching the cattle from the foods and grazing lifestyle they would naturally have raises welfare concerns. More controversially, there are supplements that can be added to animal feed to reduce methane production, including chloroform, bromochloromethane (banned for its ozone-depleting effects), nitrates and tannins.[9] All these have the potential to do great harm to animals not used to taking them, and the long-term effects on animal well-being remain unclear. Adding vegetable or even animal fats, which do not produce methane when they are digested, may also have an effect, but the environmental costs associated with this type of feed are significant.

These methods treat the animals themselves as little more than reaction vessels, in which the undesirable by-product (methane) can be reduced by changing the input reactants. No thought is given to the animals' feelings and experiences, so long as they continue to yield the desired product (meat, milk or eggs). But the best means of reducing methane output may, in fact, be to tend more carefully to animals' individual needs and improve their well-being. A healthy, happy animal will in many cases produce more milk, eggs or meat than a distressed one, which means that fewer animals are needed – and emissions are lessened – for the same amount of product.[10] However, a focus on animal welfare would require more land and potentially more food to be provided per animal, which may cancel out any benefits for the environment. It appears, therefore, there is no ethical, environmentally friendly way to reduce methane emissions from ruminants at the same time as maintaining or increasing the availability of meat and dairy produce. Herd sizes will need to be dramatically reduced by ceasing the artificial insemination of ruminant farmed animals and preventing them from breeding if this methane source is to be mitigated.

Methane from farmed animals is far from being the only means by which animal agriculture contributes to climate change. Two thirds of agricultural greenhouse gas emissions are indirect emissions resulting from land-use change, fertiliser production and fuel used to grow and transport crops.[11] Farming animals, as opposed to eating crops directly, exacerbates all these emissions sources. In the United States in 2003, it was estimated that around nine billion animals were reared for human consumption per year, consuming seven times as much grain as did the human population.[12] This enormous amount of grain explains why 85 percent of the fresh water used in the USA is swallowed by agriculture, and why 60 percent of the agricultural land there is being overgrazed. A typical cow might require 13 kg grain and 30 kg hay to produce 1 kg beef, all of which in turn needs land and water to produce.

Not all meat production is so inefficient: To produce 1 kg broiler chicken only requires only 2.4 kg grain. Beef is by far the most environmentally damaging component of a meat-based diet. Beef production has been blamed for much of the ongoing destruction of the Amazon rainforest to clear land either for cattle to graze or for soy to be grown to feed to cattle. For cattle reared on newly deforested land, it is estimated that per kilogram of animal carcass 700 kg carbon dioxide equivalent is released. Perhaps as much as 6 percent of global greenhouse gas emissions can be attributed to rainforest destruction to produce soy or beef in South America alone. Brazil is the second-largest producer and largest exporter of beef in the world, and of nearly two million hectares of rainforest lost, on average, per year between 1986 and 2005 in Brazil, up to three quarters was eventually converted into cattle pasture.[13] Brazil is also the world's second-largest producer of soy, which it exports around the world as farmed animal feed. Global production of soybeans more than doubled between 1996 and 2011, with growth especially strong in Brazil and Argentina.[14]

Deforestation is an immensely powerful source of greenhouse gas emissions because it converts a carbon "sink" into a carbon "source". As carbon dioxide concentrations increase in the atmosphere, forests are able to grow more vigorously, drawing down extra carbon dioxide through photosynthesis, and thereby providing a "negative feedback" to lower the concentration again. When forests are chopped down, not only is this feedback mechanism destroyed, but the carbon stored in the forest biomass is released – either by burning or decay – back into the atmosphere. Burning the trees provides an immediate nutritional benefit for the soil, so recently deforested land is often used for crops, including soy animal feed but also grains for human consumption and biofuels. But Amazonian soils are shallow, and within a few years these nutrients are rapidly depleted, driving further deforestation to find replacement land. Abandoned agrarian fields are used for pasture until overgrazing leaves them entirely untenable, so that lush forest ends up as lifeless wasteland.

Land-use pressure from the meat and dairy industries is not unique to the Amazon. On average, to produce 1 kg protein from farmed animals, 6 kg plant protein are required,[15] which means that keeping animals for human consumption inevitably requires much more land than would be required to feed those humans directly. This means that less land is available for free-ranging animals or reforestation that could otherwise remove carbon dioxide humans have previously emitted. Altering local land-use can also exacerbate local climatic changes by, for example, shifting rainfall patterns, removing the shade provided by trees or removing barriers to flooding. As the human population increases over the coming decades, either the extent of cultivated land will increase, contributing to both global and local climatic change, or we shall have to use existing agricultural land more efficiently to be able to provide everyone with food. Currently, two thirds of the world's population live on a mainly plant-based diet. Increasing that proportion would help us to reduce the pressure for more agricultural land.

More animal feed could be produced per acre of land if chemical fertilisers were used more widely or targeted more specifically. However, nitrate fertilisers are themselves a major contributor towards climate change: Nitrous oxide is the third most important anthropogenic greenhouse gas, after carbon dioxide and methane.

Although it is emitted in smaller quantities, nitrous oxide is even more powerful than methane as a greenhouse gas; it is nearly three hundred times as potent at trapping heat as carbon dioxide, and has an average lifetime in the atmosphere of more than a century. The more plant mass we grow with nitrate fertiliser, and the more produce we transport with nitrate-emitting fossil-fuel-powered vehicles, the more nitrous oxide ends up in the atmosphere, irrespective of whether that plant mass is grown over many fields or concentrated into a few. This therefore constitutes another way in which inefficiently producing crops for farmed animals rather than directly consuming them ourselves contributes to climate change.

The final facet of the panoply of climate change contributions made by animal agriculture is the carbon dioxide emissions associated with transport and food processing. To produce and distribute any foodstuff, some amount of energy must be expended, and in the absence of carbon-neutral ways of powering farm machinery, delivery vans and food processing factories, most of which continue to be fed by fossil fuels, this will inevitably contribute to climate change. But just as for land and fertiliser, keeping animals for human consumption magnifies the problem. If animals are fed with grain, this first has to be harvested and transported. Animals destined to be eaten then have to be transported to market, then to the abattoir, where they are slaughtered – often with electricity from fossil-fuel power stations – before their carcasses are processed and the meat is distributed to shops. Likewise, energy must be expended in pasteurising milk, delivering to processing plants, bottling it and shipping it to consumers. One could also count the emissions associated with transporting animals around farms during their lifetimes, and with visits from vets, neither of which occur on an arable farm.

Overall, much less energy would be expended if we were to use the grain fed to the animals to feed humans directly instead. Taking into account all sources of emissions, it is estimated that a plant-based diet requires 2.2 kilocalories (kcal) of energy to be put in to produce 1 kcal of food for humans. By contrast, beef requires 50 kcal per 1 kcal consumed by humans, eggs require 39 kcal, milk requires 14 kcal, and even the meat factories with the lowest associated carbon emissions – "broiler" chickens kept in captivity – require 4 kcal. This is all assuming that the animals are reared according to United States standards.[16] If land has to be cleared to rear the animals, the energy requirements may be even higher.

What remedy can be devised to ease the damage to the climate caused by animal agriculture? One strategy that would allow us to maintain or increase meat and dairy output without increasing emissions would be to try to make the production of meat more land- and energy-efficient. But it is not clear how both these requirements can be satisfied at once. If animals are kept indoors, in a climate-controlled environment on the smallest possible amount of land, their resource demands per kilogram of product could be minimised. But this will not prevent ruminants or their slurry from releasing methane, and would increase the emissions associated with processing animal feed, since the animals would no longer be able to graze at all. It is unclear whether further industrialising animal agriculture would, hence, give rise to much environmental benefit.

Furthermore, animals living in such unhappy and unnatural circumstances might well be less productive, and the welfare issues raised by such "factory farming" must not be underplayed. The motivation behind mitigating climatic change in the first place is to ensure that the world continues to be habitable and to increase the chances that its inhabitants will be able to live joyful and comfortable lives. These inhabitants include humans, but also the sentient non-humans alongside whom we live. To try to prevent climatic change by inflicting suffering upon any sentient population would be to undermine our own motivations for doing so.

It is difficult to see, then, how the greenhouse gas emissions from the agricultural sector can be substantially reduced over the coming decades without a widespread change in human diets. Reducing food waste would help, given that 40 percent of food currently goes to waste in the United Kingdom, as this would reduce demand across the whole sector. But by far the most significant emissions per kilogram of food are associated with meat and dairy produce, especially red meat, which don't constitute the majority of this food waste. Switching unilaterally from red to white meats such as chicken and insects would eradicate the methane emissions associated with ruminants and the pressure for deforestation exerted by the cattle industry, and decrease the amount of land, water and energy required to produce each kilogram of food. But chicken is still twice as energy-intensive to produce per kilogram as is plant-based food,[17] and the most energy-efficient ways of rearing such animals are seldom the most ethical.

To reduce the environmental footprint of a growing human population as much as possible, more radical dietary changes are the most ethical and straightforward option. In 2006, the average human consumed 24.2 kg of meat per year, and this is expected to almost double by 2030. Yet 35 percent of Indians are vegetarian, and meat is by no means essential to a healthy life.[18] In fact, the current levels of meat consumption in the rich world are detrimental to health. A recent Oxford Martin School study found that were everybody simply to stick to the recommended maximum meat intake guidelines, this would prevent five million premature deaths per year, factoring in both health and environmental benefits, a large part of which is the reduction in greenhouse gas emissions. It is estimated that at least one billion US dollars in healthcare costs would be saved. Greater benefits would come from universal vegetarianism, which could prevent seven million premature deaths and cut greenhouse gas emissions by 63 percent.[19] Most of the greenhouse gas emissions associated with animal agriculture come through meat rather than dairy production, and veganism can be less healthy than vegetarianism if not adequately supplemented to prevent nutritional deficiencies,[20] so eliminating milk and eggs from human diets would only increase these benefits slightly, to eight million prevented premature deaths and a 70 percent reduction in agricultural greenhouse gasses.[21]

It is not necessarily the case that entirely eliminating animals from agriculture would be the best option on environmental, cultural, human health or ethical grounds. Keeping a small number of farmed animals on marginal land where food crops for humans cannot be grown, or feeding them parts of crops that humans cannot digest, provides a rich source of manure. This can be used to add nutrients to other fields, spreading nutrients in a way akin to the role that large herbivores would naturally play. Whilst high-intensity arable farms reliant on artificial fertiliser and pesticides are uninhabitable wastelands for most free-roaming animals, low-intensity farming that utilise organic methods can actually be beneficial to animals that have already adapted to centuries or millennia of human influence on the landscape.[22] Domesticated animals help to maintain such environments, the demand for water and grain for such grass-fed ruminants is small and replacing synthetic fertiliser with manure from these animals may partially or wholly offset their greenhouse gas emissions. Small flocks of cows and sheep kept in this way could live happy lives at the same time as providing small amounts of milk and wool to humans. Likewise, chickens or ducks kept free-range in gardens and on farm scrubland can provide small quantities of eggs without the ethical or environmental concerns associated with keeping grain-fed, caged birds.

The mutual benefit of companionship enjoyed by humans and domesticated animals alike across thousands of years of co-existence has been well-documented,[23] and to eliminate domesticated farmed animals entirely from our lives would render humans and other animals bereft of this joy and domesticated species non-existent. Taking milk from cows that have been bred over centuries to require milking by humans, and eggs from birds that have been bred not to incubate them naturally, if done in an appropriately respectful way, causes no harm to these animals. These sources of protein and other nutrients are, furthermore, a beneficial addition to the human diet, helping to prevent the deficiencies that can be associated with total veganism.[24] Although industrialisation has tended to distance "consumers" in richer countries from the farms upon which they depend for their food, many human cultures around the world involve a close relationship between humans and animals that it would cause mutual distress to sever. A lot of these communities – especially in Africa and South Asia – are traditionally vegetarian, and perhaps epitomise a sustainable harmony between mankind and both free and domesticated animals that other parts of the world would do well to replicate.

This chapter has highlighted a number of ways in which animal agriculture contributes to climate change, often through practices that are themselves harmful to animals, to produce higher quantities of animal produce than are healthy for humans to consume. Various means of mitigating these effects have been outlined, but none has been found to be satisfactory on ethical or environmental grounds without substantial reductions of animal products in human diets. Agricultural policy should be directed in a way to bring about a joyful existence both for humans and for other animals, and mitigating climatic change that could harm individuals both near and far is one important component of this. Globally adopting vegan diets would provide the maximum possible reduction in greenhouse gas emissions from agriculture, at the same time as eliminating a vast amount of animal suffering and improving human health. Yet it is important not to mitigate climate change at the expense of the animal well-being that we thereby aim to improve. Although slightly less beneficial from a climate change perspective, an ethical vegetarianism with a limited milk and egg component, where the animals producing these foodstuffs are compassionately treated and individually cared for, may be a way of achieving the greatest well-being for humans and other animals overall.

Notes

1 J. Blunden and D. S. Arndt, eds., "State of the Climate in 2015," *Bulletin of the American Meteorological Society* 97, no. 8 (2016): S1–275.
2 Lisa Alexander, Simon Allen and L. Nathaniel, "Climate Change 2013: The Physical Science Basis Summary for Policymakers," *Intergovermental Panel on Climate Change* 4, no. June (2013): 1–31.
3 Tobias Thornes, "Animals and Climate Change," *Journal of Animal Ethics* 6, no. 1 (2016): 81–88.
4 Keith Lassey, "Livestock Methane Emission: From the Individual Grazing Animal through National Inventories to the Global Methane Cycle," *Agricultural and Forest Meteorology* 142 (2007): 120–132, doi:10.1016/j.agrformet.2006.03.028.
5 Alexander, Allen and Nathaniel, "Climate Change 2013 : The Physical Science Basis Sum."
6 Lassey, "Livestock Methane Emission."
7 Angela R. Moss, Jean-Pierre Juany and John Newbold, "Methane Production by Ruminants: Its Contribution to Global Warming," *Anneles de Zootechnie* 49, no. 3 (2000): 231–253. doi:10.1051/animres:2000119.
8 G. W. Mathison et al., "Reducing Methane Emissions from Ruminant Animals," *Journal of Applied Animal Research* 14 (1998): 1–28.
9 Pierre J. Gerber, Benjamin Henderson and Harinder P. S. Makkar, eds. *Mitigation of Greenhouse Gas Emissions in Livestock Production: A Review of Technical Options for Non-CO2 Emissions* (Rome, Italy, Food and Agriculture Organization of the United Nations [FAO], 2013).
10 Gerber, Henderson and Makkar, *Mitigation of Greenhouse Gas Emissions in Livestock Production*.
11 Tara Garnett, "Where Are the Best Opportunities for Reducing Greenhouse Gas Emissions in the Food System?" *Food Policy* 36 (2011): S23–32, doi:10.1016/j.foodpol.2010.10.010.
12 David Pimentel and Marcia Pimentel, "Sustainability of Meat-Based and Plant-Based Diets and the Environment," *American Journal of Clinical Nutrition* 78, no. 3 (2003): 6605–6635.
13 Christel Cederberg et al., "Including Carbon Emissions from Deforestation in the Carbon Footprint of Brazilian Beef," *Environmental Science and Technology* 45 (2011): 1773–1779, doi:10.1021/es103240z.
14 Erica Geraldes and Fausto Freire, "Greenhouse Gas Assessment of Soybean Production: Implications of Land-Use Change and Different Cultivation Systems," *Journal of Cleaner Production* 54 (2013): 49–60, doi:10.1016/j.jclepro.2013.05.026.
15 Pimentel and Pimentel, "Sustainability of Meat-Based and Plant-Based."
16 Pimentel and Pimentel, "Sustainability of Meat-Based and Plant-Based."
17 Pimentel and Pimentel, "Sustainability of Meat-Based and Plant-Based."
18 Timothy J. Key, Paul N. Appleby and Magdalena S. Rosell, "Health Effects of Vegetarian and Vegan Diets," *Proceedings of the Nutritional Society* 65 (2006): 35–41, doi:10.1079/PNS2005481.
19 Fiona Harvey, "Eat Less Meat to Avoid Dangerous Global Warming," *The Guardian*, March 21, 2016.
20 Winston J. Craig, "Health Effects of Vegan Diets," *American Journal of Clinical Nutrition* 89, no. 5 (2009): 16275–16335, doi:10.3945/ajcn.2009.26736N.
21 Harvey, "Eat Less Meat to Avoid Dangerous Global Warming."

22 Lubos Halada, Doug Evans, Carlos Romao and Jan-Erik Petersen, "Which Habitats of European Importance Depend on Agricultural Practices?" *Biodiversity Conservation* 20 (2011): 2365–2378, doi:10.1007/s10531–011–9989-z.
23 Froma Walsh, "Human-Animal Bonds I: The Relational Significance of Companion Animals," *Family Process* 48, no. 4 (2009): 462–480, doi:10.1111/j.1545–5300.2009.01296.x.
24 Craig, "Health Effects of Vegan Diets."

Bibliography

Alexander, Lisa, Simon Allen and L. Nathaniel. "Climate Change 2013: The Physical Science Basis Summary for Policymakers." *Intergovernmental Panel on Climate Change* 4, June (2013): 1–31.

Blunden, J., and D. S. Arndt, eds. "State of the Climate in 2015." *Bulletin of the American Meteorological Society* 97, 8(2016): S1–275.

Cederberg, Christel, U. Martin Persson, Kristian Neovius, Sverker Molander and Roland Clift. "Including Carbon Emissions from Deforestation in the Carbon Footprint of Brazilian Beef." *Environmental Science and Technology* 45(2011): 1773–1779. doi:10.1021/es103240z.

Craig, Winston J. "Health Effects of Vegan Diets." *American Journal of Clinical Nutrition* 89, 5 (2009): 16275–16335. doi:10.3945/ajcn.2009.26736N.

Garnett, Tara. "Where Are the Best Opportunities for Reducing Greenhouse Gas Emissions in the Food System?" *Food Policy* 36(2011): S23–32. doi:10.1016/j.foodpol.2010.10.010.

Geraldes, Erica, and Fausto Freire. "Greenhouse Gas Assessment of Soybean Production: Implications of Land-Use Change and Different Cultivation Systems." *Journal of Cleaner Production* 54(2013): 49–60. doi:10.1016/j.jclepro.2013.05.026.

Gerber, Pierre J., Benjamin Henderson and Harinder P. S. Makkar, eds. *Mitigation of Greenhouse Gas Emissions in Livestock Production: A Review of Technical Options for Non-CO_2 Emissions.* Rome, Italy: Food and Agriculture Organization of the United Nations [FAO], 2013.

Halada, Lubos, Doug Evans, Carlos Romao and Jan-Erik Petersen. "Which Habitats of European Importance Depend on Agricultural Practices?" *Biodiversity Conservation* 20 (2011): 2365–2378. doi:10.1007/s10531-011-9989-z.

Harvey, Fiona. "Eat Less Meat to Avoid Dangerous Global Warming." *The Guardian*, March 21, 2016.

Key, Timothy J., Paul N. Appleby and Magdalena S. Rosell. "Health Effects of Vegetarian and Vegan Diets." *Proceedings of the Nutritional Society* 65(2006): 35–41. doi:10.1079/PNS2005481.

Lassey, Keith. "Livestock Methane Emission: From the Individual Grazing Animal through National Inventories to the Global Methane Cycle." *Agricultural and Forest Meteorology* 142 (2007): 120–132. doi:10.1016/j.agrformet.2006.03.028.

Mathison, G. W., E. K. Okine, T. K. McAllister, Y. Dong, J. Galbraith and O. I. N. Dmytruk. "Reducing Methane Emissions from Ruminant Animals." *Journal of Applied Animal Research* 14(1998): 1–28.

Moss, Angela R., Jean-Pierre Juany and John Newbold. "Methane Production by Ruminants: Its Contribution to Global Warming." *Anneles de Zootechnie* 49, 3(2000): 231–253. doi:10.1051/animres:2000119.

Pimentel, David, and Marcia Pimentel. "Sustainability of Meat-Based and Plant-Based Diets and the Environment." *American Journal of Clinical Nutrition* 78, 3(2003): 6605–6635.

Thornes, Tobias. "Animals and Climate Change." *Journal of Animal Ethics* 6, 1(2016): 81–88.

Walsh, Froma. "Human-Animal Bonds I: The Relational Significance of Companion Animals." *Family Process* 48, 4(2009): 462–480. doi:10.1111/j.1545-5300.2009.01296.x.

3.5

THE INTENTIONAL KILLING OF FIELD ANIMALS AND ETHICAL VEGANISM

Joe Wills[1]

Introduction

One of the motivating factors for adopting a vegan diet is the belief that it is wrong to inflict unnecessary death on nonhuman animals (hereafter "animals"). Call this the *unnecessary death principle.* Given that one can have an optimally healthy diet consisting exclusively of plants,[2] the practice of killing animals for their flesh and secretions appears to constitute inflicting unnecessary death on them and is therefore morally impermissible according to the unnecessary death principle.[3]

However, as often pointed out, it is not only diets based around animal consumption that involve killing: Plant-based diets also bring about the deaths of animals. When crops are harvested there will inevitably be fatalities. Many animals – such as rabbits, birds, and small rodents – are killed through plant agriculture either because they get crushed under farming machinery, like combine harvesters and wheat threshers, or because they are killed by poisons and traps left by farmers to protect their crops. Some commentators have gone so far as to argue that there are certain variants of omnivorous diets – based around the consumption of some pasture-fed cattle – that actually result in fewer animal deaths than exclusively plant-based ones.[4] If certain omnivorous diets result in fewer animal deaths than exclusively plant-based diets, then it would seem to follow that, all else equal, one should adopt such diets in order to conform to the unnecessary death principle. Call this *the argument from field deaths.*

There are a number of responses that can be made to the argument from field deaths. First, one can dispute its empirical basis. A number of commentators have done just that, arguing that proponents of this thesis have used misleading or inaccurate data or have interpreted it incorrectly.[5] A second line of argument is to point out that all else is not equal: There are many other compelling ethical reasons for adopting a vegan diet other than the avoidance of animal death. These include: The various forms of suffering found in even the most "humane" forms of animal agriculture;[6] the adverse environmental impacts of rearing animals;[7] and the physical and psychological harms to humans associated with working in slaughterhouses.[8]

Whilst these responses have merit they are also somewhat limited by their contingent nature. For it is at least possible to imagine certain variants of omnivorous diets that both avoid many of the adverse impacts of animal agriculture and also result in fewer overall animal deaths than an exclusively plant-based diet. For example, somebody who hunts and catches fish for a certain quantity of their food may conceivably contribute to fewer overall animal deaths than certain exclusively plant-based diets and may also avoid many of the negative outcomes of animal agriculture. At this stage one might appeal to the psychological impact that hunting and consuming meat might have on its participants. It is plausible to suppose that killing animals or consuming their flesh, even in limited contexts, could normalise violence against animals, and this in turn could have negative spill-over effects beyond the confines of limited hunting and fishing. Though intuitively plausible, again, such arguments are hostage to certain empirical realities. It may not always be the case that hunting and fishing lead to negative spill-over effects.

A third strategy to challenge the argument from field deaths is to find a deontological principle that draws a relevant moral distinction between the killing of field animals and the killing of animals for meat that doesn't depend purely on the consequences of each set of practices. One such principle that a number of animal ethicists have appealed to is the Doctrine of Double Effect (DDE).[9] Very roughly stated, the DDE holds that, all else equal, it is worse to intentionally bring about a harm than to bring about a harm as an unintended but foreseeable side-effect of one's actions. The DDE has intuitive appeal. Many people, and nearly all legal systems, hold that intentions, and not merely consequences, are relevant to assessing the wrongness of an action. Thus, in the context of an armed conflict, most people agree that the intentional targeting of civilians is a particularly egregious form of waging warfare, even more so than other military operations that result in the "collateral" deaths of civilians.

The DDE can thus be used to explain the moral difference between plant agriculture and animal agriculture: The former involves growing plant food for consumption, and any deaths that arise are unintended and unwanted; whereas the latter involves intentionally inflicting deaths on animals so we can consume their bodies and secretions. The deaths of farmed animals are, accordingly, not an incidental side effect of animal farming, they are its intended outcome. If this is right, then the unnecessary death principle needs to be refined to *the unnecessary intentional death principle*: It is wrong to inflict unnecessary death on nonhuman animals and it is especially wrong, *pro tanto*, to do so intentionally. Where one must choose between bringing about a death intentionally or foreseeably bringing about death as a side-effect, one should, all else equal, pursue the latter course of action.

The DDE may go a great deal of the way in providing an ethically salient distinction between the deaths of field animals and the deaths of farmed animals. However, there is a subdivision of field animal deaths to which it does not apply, namely those animals (usually small rodents) who are killed *intentionally* through the use of poisons and traps. A particular target of such methods are rats who cause damage to food that we intend to eat, either while the food is growing or after it has been harvested and is being stored. Rats also pose a threat to human health: They are often infected with a bacterial disease that can also infect people, causing a form of jaundice known as Weil's disease. Given that killing rats and other animals exists in both crop farming and food storage practices, it is likely that vegans will, on occasion, pay farmers to intentionally kill animals so that they can consume food, just as omnivores do. At this point, if one wishes to maintain that the ethical difference between a vegan diet and an omnivorous one is qualitative as well as quantitative, a principle distinction will need to be drawn between the intentional killing of farmed animals and the intentional killing of field animals such as rats.

Distinguishing intentional field and farmed animal deaths

Killing farmed animals for food and killing field animals to protect crops both involve forms of *intentional agency*. That is to say, they both involve the killing of animals as the intended goal or purpose of the agent's (i.e. the farmer's) actions. Are there any moral distinctions to be drawn between these types of intentional killings and the consumption habits that support them? I am going to argue that there are at least three grounds for thinking that there are ethical distinctions to be made: The first reason is epistemic in nature, the second appeals to a distinction between contingent and necessary features of a practice, and the third morally differentiates between eliminative and opportunistic intentional agency.

The epistemic distinction

When assessing our ethical obligations to avoid killing other animals in the production of food, it is important to distinguish between; (a) farmers killing animals for the production of food; (b) consumers purchasing farmers' products that have involved the killing of animals, and; (c) consuming farmers' products that have involved the killing of animals.[10] Whereas situation (a) involves the farmer engaging in the *direct* infliction of harm, situations (b) and (c) involve *indirectly* participating in harm through encouragement or collaboration (through supply and demand), or at the least, a tacit complicity or condoning of a harm-inflicting practice (the act of animal product consumption can be understood at the least as a form of acquiesce to such practices). Sometimes it can be as wrongful, or even more wrongful, to be the indirect cause of harm. Think about somebody hiring a hitman to commit murder. The hirer in this instance is surely *at least* as morally responsible for the murder as the hitman. One cannot argue, either in law or morality, that one's culpability is diminished simply by virtue of the wrongful act not being done by one's hands.

However, a relevant consideration to be made in assessing the wrongfulness of indirect causers of harm is the degree to which they are aware of the harm that they are contributing towards, or the degree to which they are aware of the likelihood that their conduct may contribute to harm. It might be thought that, all else equal, the greater the epistemic certainty one has that one's actions will indirectly lead to harm, the more morally culpable one will be for the harm they contribute to. Where there are two courses of action, one which involves a harm as a certainty and another in which there is a risk that harm may occur, one ought, all else equal, to choose the later course of action.

When one supports animal agriculture, one supports an industry that requires the killing of animals. In the context of plant agriculture, the extent to which animals are killed depends on the methods of farming deployed. This sort of information is not standardly transparent to consumers, so when it comes to the purchasing of plant foods, there is epistemic uncertainty about the extent to which field animals will be killed by farmers. When one is choosing between different courses of action that may result in harm one ought to take account of, *inter alia*, the gravity of that potential harm weighted by the probability that it will arise. Focusing just on the harm of intentional killing, it is clear that this is a certain outcome of animal agriculture. In plant agriculture, the risk of contributing to intentional animal deaths will vary according to many factors, including the methods of farming used, the natural environment the crops are grown in, available technologies, the regulatory regime that governs agricultural operations, the norms and values of the farming community etc. As there is epistemic certainty that animal agriculture (as well as hunting and fishing) involves the killing of animals then one ought to purchase products where there is less certainty.

Turning to the gravity of the harm – again, just focusing on intentional killing – there is good reason to think that meat products have the higher death toll. First, it should be noted that, in addition to the animals that are killed for their flesh, free living animals are also intentionally killed to protect animal agriculture. For example, in the United Kingdom badgers are routinely "culled" in response to the alleged threat of disease they pose to cattle and in the United States large carnivores such as grey wolves are trapped and killed to prevent the perceived threat they pose to farmed animals.

Second, it is worth remembering that most farmed animals are themselves fed harvested crops, including many so-called "grass-fed" cattle who are fed grains and hay indoors over the winter months. As such, in most instances of meat purchase, the problem of the intentional killing of animals to protect crops actually multiplies rather than diminishes.

Third, there is empirical analysis that suggests that the animal death toll per calorie of food produced is higher *even in pasture-fed cattle* than in crop harvesting.[11] In other words, the total number of animal deaths – both intentional and accidental – is lower in plant agriculture than in the farming of pasture-raised cattle.

All of these comparative features of animal and plant farming add up to robust epistemic reasons for consumers to choose plant products over animal products, even if the avoidance of supporting the intentional killing of animals is their only concern. So even if we assume, *arguendo*, that a farmer acts equally wrongfully when she intentionally kills farmed animals and field animals, it does not follow from this that consumers of these products also act equally wrongfully.

The contingent/necessary distinction

A second distinction, related to the first, is that, whereas farmed animal deaths are a *necessary* feature of animal agriculture, field animal deaths are only a *contingent* feature of plant farming. In other words, whereas animal agriculture *requires* intentionally killing animals to produce its products, plant agriculture does not. The moral import of this distinction is twofold. On the one hand, it simply seems more wrongful to participate, even indirectly, in practices that *necessarily* involve harm than ones that do not. Think about the following example: A consumer is buying a table. She has two options: (1) purchase it from Slave PLC – a company that operates openly through the use of slave labour or; (2) purchase it from Nice PLC – a multinational company that standardly uses wage labour but some of its subcontractors in the past have been exposed as engaging in slave labour. Perhaps the reason why it seems worse to support Slave PLC in comparison to Nice PLC is largely attributable to the epistemic reasons outlined above.

In addition to that, it seems as if there are also relevant consequentialist considerations for preferring to indirectly participate in contingently wrongful practices as opposed to necessarily wrongful practices. Slavery is a necessary constitutive feature of Slave PLC's operations – if it did not use slavery it would not be Slave PLC in the relevant sense. By contrast, Nice PLC would still be the same brand if it eliminated slavery from its supply chain, indeed, it's brand image would be enhanced as a result. Nice PLC is therefore *reformable* in a way that Slave PLC is not. Likewise, while plant agriculture can be reformed, animal agriculture cannot, at least not to the same degree. It is possible, conceivably at least, to reduce the number of animals killed by plant farming practices significantly. Animal agriculture cannot reduce the number of animals that are killed to produce the equivalent amount of meat in the same way.[12] (It is true that selective breeding practices can produce more meat/milk/eggs per animal, but this often involves a trade off with animal welfare, and so raises a different set of ethical problems.)

Whilst it is conceivable that animal deaths – *intentionally* inflicted animal deaths at least – could be reduced to negligible levels in plant agriculture, this is not the case with regard to animal agriculture. Moreover, the prospects for reform of plant agriculture in a more animal-friendly direction seem far more likely to occur in a society that doesn't unnecessarily kill and exploit animals for their body parts and secretions. After all, if increasing numbers of people desisted from participating in and supporting the later sort of practices out of moral respect for animals, this would likely engender a shift in methods of crop production that would seek to minimise free living animal deaths, both intended and unintended. These seem to supply practical reasons to prioritise the boycott of animal products over plant products if one's objective is to reduce unnecessary animal deaths.

The eliminative/opportunistic distinction

There is now a growing body of literature on the ethics of self-defense around the distinction between eliminative and opportunistic harming.[13] I believe that this distinction is also of relevance to the present discussion. Eliminative harming involves the infliction of harm against a threat so as to neutralize that threat. Opportunistic harming, by contrast, involves inflicting harm on a non-threat (a "by-stander") in order to benefit from the presence of their body. In instances of eliminative harming the person inflicting the harm does not benefit from the individual who is having harm inflicted upon them: The eliminative harmer is merely eliminating the threat that she poses.

In order to make concrete the relevance of the eliminative/manipulative distinction let us consider two cases. First, consider:

> *Bad Acid Trip*: Anton is spiked with LSD without his knowledge (let us suppose that there is nothing that he could have done to avert this situation) and experiences a bad trip. He hallucinates that Bryony is a giant snake attacking him, so he charges at her with a knife. Fortunately for Bryony, she has a gun, which she can shoot Anton with in order to protect herself.

Even though Anton is entirely morally innocent, most have the intuition that it is permissible for Bryony to kill him in self-defence in this case. However, compare:

> *Bad Acid Trip II*: The facts are the same as *Bad Acid Trip* except this time Bryony does not have a gun. The only way she can protect herself from Anton's attack is to use a bystander, Carys, as a human shield, pulling her body in front of herself to block Anton's attack. This would kill Carys but allow Bryony to escape to safety.

In *Bad Acid Trip II* it seems impermissible for Bryony to use Carys as a human shield. Yet both instances involve Bryony intentionally killing a morally innocent individual to protect herself. It is possible that the distinction between eliminative and opportunistic harming explains the different intuitions about these cases: In *Bad Acid Trip* Bryony is eliminating the threat that Anton is posing to her whereas in *Bad Acid Trip II* she is benefiting from the existence of Carys' body. The moral distinction might be explained in the following way: Opportunistic harming entails using your victim as a *mere* means to your ends because you treat them as a tool in order to do something that you could not do in their absence. By contrast, eliminative harming does not treat your victim as a mere means; rather you simply remove the threat that they pose. As Quong puts it, the distinction between eliminative and opportunistic agency corresponds to a distinction between killing someone as a means to survival and killing someone as a *mere* means to survival.[14]

If the eliminative/opportunistic distinction provides a sound moral basis for the intuitions in the *Bad Acid Trip* cases then it may also apply to the distinction between the intentional killing of field and farmed animals: In the former instance the animal is killed to eliminate the threat they pose to the crops/human health, whereas in the latter case they are killed opportunistically. The farmed animals are not posing any threat to any human interest: It is their bodies that confer a benefit to their killers in the form of flesh used for consumption. The opportunistic killing inflicted upon these animals can be argued to be more difficult to justify, all else equal, than the eliminative killing inflicted upon field animals.

The *Bad Acid Trip* cases are intended to make vivid the distinction between eliminative and opportunistic intentional agency. However, there are important and obvious dis-analogies between the *Bad Acid Trip* cases and the killing of animals in farming practices. Animals in the field do not pose the sort of immediate, lethal threat that the attacker in the *Bad Acid Trip* cases does. In fact, it might often, or even usually, be the case that farmers use lethal force against field animals not to protect their lives, or even their livelihoods, but rather to safeguard their profit margins. The use of lethal force to protect profits clearly raises more difficult issues of proportionality than the *Bad Acid Trip* cases do. Moreover, the *Bad Acid Trip* cases are premised upon the assumption that Bryony has no option but to use lethal force to defend herself. In the context of field animal deaths, there may well be alternative non-lethal methods that can be deployed by farmers to protect their crops, thus there is also the question of whether the eliminative force deployed is necessary.

The extent to which the lethal defensive force deployed by farmers to protect their crops is necessary and proportionate to the threat that field animals pose raises complex empirical and normative questions that I cannot address here. I will assume that a great deal of the intentional deaths that farmers inflict on farmed animals fail to satisfy both the requirements of necessity and proportionality, thereby rendering them morally unjustifiable. Nevertheless, even in such instances, the unjustifiable eliminative killing that crop farmers engage in is still, other things equal, less morally wrongful than the unjustifiable opportunistic killing that is required to produce animal products. This is because the latter type of action has the additional wrongful element of treating a non-threatening individual merely as a means to your ends. Even excessive force against field animals posing a threat to one's crops is easier to defend than killing farmed animals posing no threats to anyone. The fact that the opportunistic killing practised in animal agriculture is more *pro tanto* wrongful than the eliminative killing in plant agriculture provides consumers a further reason to purchase plant products over animal products.

Some philosophers may dispute whether treating animals merely as a means to human ends raises any distinctive ethical problems. This is because it is often assumed, following Immanuel Kant, that the duty to treat individuals with respect (which is to, not to treat them merely as a means to ones ends) only applies to *persons*, i.e. to rational, autonomous, self-aware moral agents who are capable of determining their own ends. According to this line of thought, it is the status of persons as end-setters that makes treating them as a mere means to your ends rather than ends in themselves morally objectionable. Animals, so it is assumed, lack this autonomous control over their ends and therefore using them as mere means is not wrong *per se*. As Tadros puts it:

> As humans we are capable of determining what ends to set for ourselves. We do so by determining what we will value amongst the range of things that are valuable. Our right to set ends for ourselves can be understood as derived from our independence from each other – the autonomous control that we have over our own lives. Non-human animals lack this status because they lack the relevant capacity – to determine what is valuable and to pick out particular things to valuable.[15]

Thus, it is the capacity to determine ends that grounds the moral right not to be treated as a mere means to another's end. Call this the *end-setter argument*. Even if we assume that animals lack any capacity to determine their own ends, there are at least two reasons to doubt the end-setter argument. First, this argument leads to highly counter-intuitive conclusions about permissible treatment of other humans. There are certain severely cognitively impaired human beings who lack the capacity for end-setting but it seems wrongful to treat them as a mere means to the ends of others. Given this, it seems impossible to reconcile our deeply held moral convictions that it is wrongful to treat any human as mere instruments with the end-setter argument.[16]

A second, positive reason to question the end-setter argument comes from Korsgaard. Korsgaard, drawing upon a Kantian metaphysical framework, argues that the realm of intrinsic values is epistemically inaccessible, even to rational beings. As such we have to build up our system of values up from what we do know: That there are certain things that are good or bad *for us*. From that starting point we can take those things to be good or bad absolutely – and in doing so take ourselves to matter absolutely as ends in ourselves. Yet "ourselves" here cannot refer just to ourselves *qua* rational beings. For there are many things we take to be valuable – our desire to live, love, experience pleasure and avoid pain – that we value not by virtue of our rational nature, but rather by our animal nature. Korsgaard argues that we value ourselves as "ends in ourselves" not just as rational beings, but as beings for whom things can be good or bad. Because animals – sentient animals at least – are also creatures for whom things can be good or bad, it follows that we must also treat animals as ends in themselves. Since they are ends in themselves we cannot treat them as mere means to our ends.[17]

As the opportunistic killing of farmed animals involves treating them as mere means to our ends, then, according to Korsgaard's persuasive argument, it follows that it is more *pro-tanto* wrongful than the eliminative killing of field animals which does not involve treating them as a mere means to our ends. Just to repeat, this does not make the unnecessary and disproportionate killing of field animals morally permissible, it merely makes such acts less wrongful in one sense, than the opportunistic killing of farmed animals.

Conclusion

This chapter has sketched out an argument that there are a number of important ethical distinctions between the killing of field animals and the killing of farmed animals that makes purchasing the products of plant agriculture generally easier to justify than purchasing the products of animal agriculture.

With that being said, it must be stressed that none of the preceding argument should be interpreted as justifying all current agricultural practices with regard to "pest" animals. We still owe duties to free living animals, including duties to avoid killing them as far as practicable and possible. Lethal defensive measures should be deployed only as a last resort and non-lethal, non-harmful alternatives should be pursued wherever possible. Lethal defensive force against field animals is only justifiable where the requirements of necessity and proportionality have been met.

Notes

1 I would like to thank Bob Fischer and Andy Lamey for their helpful feedback on earlier drafts of this chapter. Any errors are my own.
2 Academy of Nutrition and Dietetics, "Position of the Academy of Nutrition and Dietetics: Vegetarian Diets," *Journal of the Academy of Nutrition and Dietetics* 116 no. 12 (2016): 1970.
3 It is not only meat products that require the death of animals. Both the egg and dairy industries also standardly kill male offspring shortly after birth because they are surplus to the requirements of these industries. Both cows and hens are also standardly killed once their egg and milk "production" wanes to below profitable levels.
4 Stephen Davis, "The Least Harm Principle May Require That Humans Consume a Diet Containing Large Herbivores, Not a Vegan Diet," *Journal of Agricultural and Environmental Ethics* 16 (2003): 387–394; Mike Archer, "Ordering the Vegetarian Meal? There's More Animal Blood on Your Hands," The Conversation, December 15, 2011, https://theconversation.com/ordering-the-vegetarian-meal-theres-more-animal-blood-on-your-hands-4659.
5 Gaverick Matheny, "Least Harm: A Defense of Vegetarianism from Steven Davis's Omnivorous Proposal," *Journal of Agricultural and Environmental Ethics* 16 (2003): 505–511; Andrew Lamey, "Food Fight! Davis versus Regan on the Ethics of Eating Beef," *Journal of Social Philosophy* 38 no. 2 (2007): 331–348; Patrick Moriarty, "Vegetarians Cause Environmental Damage, but Meat Eaters Aren't Off the Hook," The Conversation, April 17, 2012, https://theconversation.com/vegetarians-cause-environmental-damage-but-meat-eaters-arent-off-the-hook-6090.
6 Matheny, "Least Harm," 508.
7 Henning Steinfeld et al., *Livestock's Long Shadow: Environmental Issues and Options* (Rome, Italy: United Nations Food and Agriculture Organisation, 2006).
8 Timothy Pachirat, *Every Twelve Seconds Industrialized Slaughter and the Politics of Sight* (New Haven, CT: Yale University Press, 2011).
9 Sherry Colb, "Exclusion, the Doctrine of Double Effect, and Animal Deaths," Dorf on Law, November 25, 2009, www.dorfonlaw.org/2009/11/exclusion-doctrine-of-double-effect-and.html; Alasdair Cochrane considers, and rejects the appeal to double-effect to distinguish field animal deaths see, Alasdair Cochrane, *Animal Rights without Liberation* (New York, NY: Colombia University Press, 2012), 96–98.
10 I am indebted to Bob Fischer for pointing out this distinction to me.
11 Matheny, "Least Harm," 506–507.
12 Though it should be noted that the future possibility of lab meat somewhat complicates this argument.

13 See e.g. Warren S. Quinn, "Actions, Intentions, and Consequences: The Doctrine of Double Effect," *Philosophy and Public Affairs* 18 (1989): 344; Helen Frowe, "Equating Innocent Threats and Bystanders," *Journal of Applied Philosophy* 25 (2008): 277; Jonathon Quong, "Killing in Self-Defence," *Ethics* 119 no. 3 (2009): 507; Victor Tadros, *The Ends of Harm: The Moral Foundations of the Criminal Law* (Oxford, England: Oxford University Press), 242.
14 Quong, "Killing in Self-Defence," 507.
15 Tadros, *The Ends of Harm*, 127.
16 McMahan, who also holds that only persons are entitled to what he calls "the morality of respect", is prepared to accept this consequence. McMahan acknowledges that on his account it would be permissible, all else equal, to kill a new born orphaned human infant to distribute her organs to save three other infants. McMahan himself admits that this outcome causes him "significant misgivings and considerable unease". Jeff McMahan, *The Ethics of Killing: Problems at the Margins of Life* (Oxford, Englans: Oxford University Press 2002), 360. It seems to me that such conclusions cannot withstand the process of reflective equilibrium and require the revision of "two tier" accounts of moral status such as McMahan's. For a recent critique of such theories see Rainer Ebert, "Mental-Threshold Egalitarianism: How Not to Ground Full Moral Status," *Social Theory and Practice* (2018), doi: 10.5840/soctheorpract201812330.
17 Christine Korsgaard, "A Kantian Case for Animal Rights," in *The Ethics of Killing Animals,* eds. Robert Garner and Tatjana Visak (Oxford, England: Oxford University Press, 2016), 163–167.

Bibliography

Academy of Nutrition and Dietetics. "Position of the Academy of Nutrition and Dietetics: Vegetarian Diets." *Journal of the Academy of Nutrition and Dietetics* 116, 12(2016): 1970.

American Dietetic Association. "Position of the American Dietetic Association: Vegetarian Diets." *Journal of the American Dietetic Association* 116, 12(2016): 1970–1980.

Archer, M. "Ordering the Vegetarian Meal? There's More Animal Blood on Your Hands." The Conversation. December 15, 2011. https://theconversation.com/ordering-the-vegetarian-meal-theres-more-animal-blood-on-your-hands-4659.

Cochrane, A. *Animal Rights without Liberation.* New York, NY: Columbia University Press, 2012.

Colb, S. F. "Exclusion, the Doctrine of Double Effect, and Animal Deaths." Dorf on Law. November 25, 2009. www.dorfonlaw.org/2009/11/exclusion-doctrine-of-double-effect-and.html.

Davis, S. "The Least Harm Principle May Require That Humans Consume a Diet Containing Large Herbivores, Not a Vegan Diet." *Journal of Agricultural and Environmental Ethics* 16(2003): 387–394.

Ebert, R. "Mental-Threshold Egalitarianism: How Not to Ground Full Moral Status." *Social Theory and Practice* (2018). doi: 10.5840/soctheorpract201812330.

Frowe, H. "Equating Innocent Threats and Bystanders." *Journal of Applied Philosophy* 25 (2008): 277.

Korsgaard, C. A. "Kantian Case for Animal Rights." In R. Garner R and T. Visak eds. *The Ethics of Killing Animals.* Oxford, England: Oxford University Press, 2015.

Lamey, A. "Food Fight! Davis versus Regan on the Ethics of Eating Beef." *Journal of Social Philosophy* 38, 2(2007): 331–348.

Matheny, G. "Least Harm: A Defense of Vegetarianism from Steven Davis's Omnivorous Proposal." *Journal of Agricultural and Environmental Ethics* 16(2003): 505–511.

McMahan, J. *The Ethics of Killing: Problems at the Margins of Life.* Oxford, England: Oxford University Press, 2002.

Moriarty, P. "Vegetarians Cause Environmental Damage, but Meat Eaters Aren't Off the Hook." The Conversation. April 17, 2012. https://theconversation.com/vegetarians-cause-environmental-damage-but-meat-eaters-arent-off-the-hook-6090.

Pachirat, T. *Every Twelve Seconds: Industrialized Slaughter and the Politics of Sight*. New Haven, CT: Yale University Press, 2011.

Quong, J. "Killing in Self-Defence." *Ethics* 119, 3(2009): 507–537.

Quinn, W. "Actions, Intentions, and Consequences: The Doctrine of Double Effect." *Philosophy and Public Affairs* 18(1989): 344–351.

Steinfeld, H., P. Gerber, T. Wassenaar, V. Castel, M. Rosales and C. de Haan. *Livestock's Long Shadow: Environmental Issues and Options*. Rome, Italy: Food and Agriculture Organization of the United Nations, 2006.

Tadros, V. *The Ends of Harm: The Moral Foundations of the Criminal Law*. Oxford, England: Oxford University Press, 2011.

3.6

HOW VISUAL CULTURE CAN PROMOTE ETHICAL DIETARY CHOICES

Hadas Marcus

Peering through glass walls

Most animal advocates today are familiar with the oft-quoted line by Sir Paul McCartney and his late wife Linda that "if slaughterhouses had glass walls, everyone would be a vegetarian." And indeed, McCartney narrated a gruesome online video entitled *Glass Walls* (2009) to explicitly demonstrate this idea. As we endure the harrowing experience of watching this PETA film featuring undercover footage of slaughter, we might ask ourselves how successful this and other documentaries with such shocking content are in disseminating messages meant to arouse people to embrace veganism or vegetarianism. Might these films even be alarming and desensitizing viewers by inundating them with scenes of brutality? Could less inflammatory, alternative forms of media also cultivate empathy toward farmed animals? In other words, are those figurative glass walls really necessary to inspire a paradigm shift toward more ethical dietary choices?

As the title of her book *Regarding the Pain of Others* (2003) implies, cultural analyst Susan Sontag pondered whether the act of witnessing past atrocities (mainly war and genocide) through the ever-evolving medium of photography is motivated by our moral conscience or by morbid curiosity. Sontag claimed that to watch such bloodbaths was permissible only if the viewer could do something tangible to alleviate suffering (such as the surgeon in the military hospital) or if it initiated change – anything else was voyeuristic pleasure.[1] Film scholar Anat Pick explores the "war on animals" in a book chapter entitled "Animal Rights, Organized Violence, and the Politics of Sight."[2] She points out how Sontag's theory is applicable when chronicling the horrors of factory farming that relates to "the regulation and codification of vision on which the representation of atrocities committed against animals depends." In keeping with Sontag's line of thought, we can also ask: What benefit does the general public derive from watching actual film footage of unspeakable acts taking place on a daily basis in the intentionally concealed world of the slaughterhouse? Does screening these atrocities serve the animal welfare movement by enabling viewers to contribute anything of lasting value, propel them to make significant lifestyle changes, or educate others as to the need to mitigate nonhuman suffering? In documentaries shot in slaughterhouses, the camera typically captures macabre moments that create angst and revulsion. Is it possible that viewers are able to withstand them by detaching themselves emotionally from situations that are not occurring before their very eyes?

To examine these issues, this chapter will review selected visual works from various historical periods intended to evoke a broad range of emotional responses, including grief, anger, shock, guilt, and sometimes even ironic laughter. It will also draw upon a theoretical framework demonstrating the link between visual culture and critical animal studies, such as the worthy contributions made by Steve Baker and Jonathan Burt. This chapter will briefly consider some examples to probe the relationship between visual representations of animals and the role they might play as a catalyst for positive change that, ideally, would result in a critical mass of people willing to commit to veganism, vegetarianism, or even minor steps (e.g., Meatless Monday), for ethical and environmental reasons.

To look or not to look

In his celebrated essay "Why Look at Animals?" (1980), John Berger condemns the exploitation of animals as marginalized objects of spectatorship to satisfy our desire for amusement, particularly in zoos, and claims this visual indulgence has led to their gradual "disappearance" and continual suffering.[3] Responding to Berger's claim, Anat Pick asks the inverse question ("Why *not* look at animals?"), and her forthcoming book, *Vegan Cinema: Looking, Eating, Letting Be*, draws from Simone Weil's theory of vulnerability, and the tension between looking and eating, while promoting the nonviolent gaze to endorse veganism.[4]

Whereas *ecocriticism* (literature and other cultural media dealing with nature and environmental themes) is an ever-expanding field, many fictional works tend to formulaically reduce animals to purely anthropomorphized, cute, symbolic, or allegorical status, creating what Steve Baker terms the "cultural denial of the animal."[5] Furthermore, Jonathan Burt points out the need "to move away from emphasis on the textual, metaphorical animal, which reduces the animal to a mere icon, to achieve a more integrated view of the effects of the presence of the animals and the power of its imagery in human history."[6] Undoubtedly, classic works of literature such as "The First Step" (1909) by Leo Tolstoy, an ardent vegetarian, and Upton Sinclair's *The Jungle* (1906), that depict the harrowing conditions for both animals and workers alike, were seminal in exposing the harsh realities of abattoirs and the meatpacking industry. Unfortunately, such literary masterpieces reach a very limited audience today, as few people read such lengthy and outdated descriptions, and turn instead to the internet, television, or cinema for entertainment and information.

Recent books, such as Jonathan Safran Foer's *Eating Animals* (2009), have given enormous momentum to the animal welfare and vegan/vegetarian movement and belong on the shelf together with the groundbreaking contributions of Peter Singer and Tom Regan. However, we live in a rapid-paced, digital world where people are less inclined to read stories and novels and are bombarded with easily accessible visual images, clips, and documentaries that often take sovereignty over the printed page.

Visual culture communicating blatant animal injustice has received widespread public attention in recent years, especially through the unflagging dedication of animal rights/welfare organizations and fiercely committed individuals who often put themselves at extreme risk to obtain undercover blood-and-guts footage of torture and slaughter. Nonetheless, despite their highly commendable and courageous efforts to speak for the voiceless, caution should be exercised with categorical depictions of animal cruelty, as it can lead to a sense of futility and hopelessness. To illustrate, the compelling award-winning documentaries *Earthlings* (Monson, 2005), *The Cove* (Psihoyos, 2009), and *From Farm to Fridge* (Iovino, 2011) – which in many ways mirror earlier, slower-moving films such as Georges Franju's poetic and haunting *Le Sang de Betes* (1949) and Frederick Wiseman's uncompromising look at institutionalized America in *Meat* (1976) – can be traumatizing to the audience. For some viewers this may motivate them to embrace major lifestyle changes, while for others, such sickening images of mass butchery of innocent animals can actually create a boomerang effect. Films that are excessively graphic and offer no glimmer of hope may, paradoxically, generate the undesired attitude of indifference or paralysis. A study concerning the public's viewing of animal abuse suggested that:

> Images of animal cruelty will have less of a long-term detrimental impact on viewers if footage is interspersed with animals being humanely treated and discussion of ways in which viewers can participate to achieve this aim, empowering the audience rather than leaving them with overwhelming feelings of hopelessness.[7]

Although it has undoubtedly persuaded countless numbers of people to adopt veganism, the poignant film *Earthlings* also engenders despair. Its impassioned narration by vegan actor Joaquin Phoenix condemns speciesism as it draws historical parallels between animal mistreatment, discrimination and violent behavior. It has been described as a horror film demanding a certain spectatorial discipline:

> Viewers of *Earthlings* are meant to relinquish power and submit to a horrifying spectacle that rivets them with a simultaneous need to look and to look away. The intended payoff for the *Earthlings* viewer is the production of political awareness and engagement whose urgency is correlative to the affective intensity of the experience of spectatorship.[8]

Similarly, Adam Lowenstein claims that Georges Franju's *Le Sang des Bêtes* (1949, *Blood of the Beasts*), belongs to the genre of "shock horror," which is "the employment of graphic, visceral shock to access the historical substrate of traumatic experience."[9] The film, which takes place at an abattoir for horses, "opens with a dreamlike display of the outmoded ... a kaleidoscope of wildly contrasting Surrealist images that underline the impossibility of a soothingly familiar world to comfort us before we descend into the nightmare of the slaughterhouse."[10] As for Frederick Wiseman's *Meat*, spine-chilling austerity and detachment from all human warmth and emotion is summed up as the reduced scale of the animals in extreme long shot graphically foreshadows their treatment as raw materials for "animal fabrication" … *Meat*, even while it shows us the seeming terror in the eyes of the cattle as they are being herded, scrupulously avoids subjective camera techniques.[11]

Wiseman himself felt so distant from the subjects of his camera that he laughed about eating steak for dinner while making the documentary, stating that it probably came from one of the beef cattle he had filmed earlier in the day.[12] This strengthens the hypothesis that extensive observation of atrocities may harden the viewer and diminish the intended message.

One way to stimulate positive feelings of identification is in the simple beauty of animal portraiture – offering an intimate view of the expression in the subject's eyes and the contours of his face – rather than upsetting images of cows and pigs being chained, beaten, and murdered. This becomes clear when one looks at the stunning photography of Isa Leshko[13] and other animal artists who emphasize the unique character and quiet dignity emanating from their subjects. To support this notion, photographer Joe Zammit-Lucia and sociologist Linda Kalof discourage visual shock for the resolute purpose of animal rights and conservation measures, arguing that advocates, in their well-meaning campaigns, often employ grotesque images that deliberately "engender support through a combination of outrage and guilt." By creating a "divisive dichotomy – and a distance – between the Human and the Animal: The Human as the callous aggressor; the Animal as the helpless victim … The animal is portrayed as … a casualty of an undesirable human disposition and reprehensible human activity."[14] They further claim that the ultimate effect may be – instead of increased empathy with individual sentient beings – feelings of numbness, nausea and detachment from a persecuted mass of faceless animals. The authors conclude that while exposing animal suffering is a legitimate part of investigative journalism, and documenting animal abuse is vital to animal-rights organizations, it does not imply the necessity of a "visual monoculture of grisly imagery" and its "widespread dissemination as educational." They recommend investigating "what the long-term effects of these shock advocacy images will be on the cultural relationship between the human and the animal – particularly now that exposure to acts of animal cruelty has moved beyond the still image to ubiquitously available graphic videos."[15]

Let us now turn to less disturbing perspectives – those that may seem incongruous for portraying the deplorable fate of farmed animals. While it would be highly inappropriate to sugar-coat the grave issues of animal affliction, a less blatant approach might offer an effective mode for generating a paradigm shift toward plant-based dietary choices.

Banksy's squealing truck

Many city dwellers never witness the appalling plight of farmed animals during their short, miserable existence on industrial farms. If these conditions were more visible, it is plausible that some people might be persuaded to reconsider their consumer habits. How would they react to newborn calves bawling for their distraught mothers, chickens crammed into such tiny spaces that their bodies are maimed, or sows trapped in claustrophobic gestation crates? When do urbanites ever behold a truckload of farmed animals en route to the slaughterhouse, eyes bulging in panic as they gaze through the metal slats of their final moving prison? To that end, Banksy's travelling exhibit in New York City, *Sirens of the Lambs* (2013) was an astonishing feat. The anonymous British graffiti artist sent a creative art installation to various international sites, featuring a meat delivery truck loaded with 60 cuddly-looking stuffed animals squealing for help while being driven to a supposed abattoir. The very ludicrousness of it elicited uncomfortable laughter from surprised onlookers as the animatronic sheep, cows, pigs, and chickens cried out in fear while being hauled off to hypothetical slaughter. By metaphorically re-enacting the spectacle of frightened, flesh-and-blood animals on the last day of their wretched lives, *Sirens of the Lambs* delivered a powerful message without subjecting its "audience" of bewildered bystanders to the upsetting experience of candid violence and animal cadavers. One cannot help but ask if this softer representation might not raise more empathy and willing support.

Classic paintings of meat

Prehistoric cave paintings featured hunting scenes of large mammals, and animal images in art have been universal and ubiquitous for countless centuries. While it may seem like an odd subject for fine art, the portrayal of meat or dead animals can have significant aesthetic, moral, and social value. Some examples are Aertsen's *Meat Stall* (1551), Rembrandt's *The Flayed Ox* (1655), and Soutine's *Carcass of Beef* (1925); all three are suffused with strong religious overtones.

Flemish artist Pieter Aertsen's monumental panel *Meat Stall* (1551) is an enigmatic piece, which has been endlessly interpreted by art historians, scholars, and casual museum-goers alike. Even enthusiastic meat-eaters might find the 21 square foot painting of butchered animal flesh morose and upsetting. Amongst the conglomeration of glistening body parts in the forefront and miniature humans in the background, viewers of *Meat Stall* cannot ignore the reproachful gaze of the partly skinned, decapitated ox head lying on a table, his doleful eye staring out as if to accuse his executioners. As Charlotte Houghton aptly describes:

> Great bloody hunks of it press outward from the picture plane … a pig's body, cleaved in half and gutted; a marbled haunch; a lung hung up by a ragged windpipe … Aertsen portrayed the properties of the animal flesh itself in exquisite sensory detail.[16]

In *Meat Stall*, a miniature biblical scene from Matthew is barely discernable as Mary sits astride a donkey and offers a starving beggar some bread, sharing her family's meager sustenance as they flee to Egypt. This spiritual act of compassion and sacrifice is juxtaposed with the lavish display of animal corpses and copious mounds of meat rendered with painterly precision in the foreground, portraying another side of human nature: Our gluttonous carnal appetites.

In *The Flayed Ox* (1655) by Rembrandt, an enormous bovine carcass hangs from a crossbeam by its truncated hind legs, spread apart so that the bone and muscle of the inner body are fully visible. In contrast to the crowded images in *Meat Stall*, Rembrandt's *The Flayed Ox* emanates a mood of sober tranquility in which the viewer is confronted with the singular, startling image of dead flesh. The split body, rendered with broad brushstrokes and muted colors, is hung in a manner reminiscent of a crucifixion, and it similarly dominates the entire canvas.[17]

Parisian artist Chaim Soutine was inspired by Rembrandt to paint *Carcass of Beef* (1925) and other works featuring slaughtered animals such as geese, rabbits, and chickens as the subjects for his still-life paintings. Avigdor Poseq describes these paintings as erotic metaphors, stating that "the allusive character of these images, especially their oblique reference to sexual experiences, reflects deeply ingrained taboos, which seem to have conditioned Soutine's emotional attitudes."[18] Frequently going to the Louvre to contemplate *The Flayed Ox*, Soutine's obsession with animal carcasses was more impressionistic and visceral, though no less disturbing.

Contemporary protest art

Based on firsthand experiences in abattoirs since her childhood in Surrey, England, political artist Sue Coe portrays bold, nightmarish scenes of immeasurable sorrow in haunting black and white images. Her works, which she refers to as either graphic or reportage, are often mixed with rage and sarcasm as echoed in her book titles – *Dead Meat, Sheep of Fools*, and *Cruel: Bearing Witness to Animal Exploitation* – all of which probe the thorny question of farmed animal torment and abuse. As she explores the complex moral issues of social injustice, animal rights, and the ethics of slaughter, many of Coe's pieces depict the meatpacking industry, which she portrays as a genocidal massacre of nonhuman species. The artist describes what motivated her as a child:

> I grew up in the time of huge media attention to animal cruelty, and lived a block away from a slaughterhouse and next door to a hog farm. The animals lived in tin sheds, and I could see what happened to them, and could not ignore it … Human beings, even those who are aware, still want to see themselves as the center of attention in a painting holding the viewer's gaze. It's an uphill fight to make non-human animals the star of their own reality that have their own emotions and own lives.[19]

Compared to the intrepid, monochromatic works by Coe with their harsh and accusatory tone, the delicate paintings of Sunaura Taylor, an animal activist living in California, echo her own vulnerability and the stigmatization she faces at being severely disabled[20] through images of downed cows too weak to stand and a mountain of male chicks about to be ground up while still alive. Keenly aware of the discrimination shown to nonhuman species, her anger is fueled by the fact that animals have always been unfairly measured according to "a human yardstick."[21] Taylor draws an analogy between the harsh prejudice and cruel treatment shown to disabled humans and the exploitation of animals used for food and purposely bred for factory farming in a way that mutilates their bodies through genetic manipulation. As she states:

> The animals people eat are largely manufactured to be disabled. Animals are bred to have too much muscle for their bodies to hold, cows and chickens develop broken bones and osteoporosis from the overproduction of milk and eggs. Very often the very thing animals are bred for is, or leads to, disability.[22]

Perhaps the most unusual installation of animal rights activism exhibited in recent years is that of Miru Kim, who illegally entered industrial farms housing swine in order to photograph her own curvaceous nude female figure in semi-erotic poses as she snuggled up against the startlingly similar skin and abdomens of live pigs. Regarding Kim's series of photographs, "The Pig That Therefore I Am" (alluding to Jacques Derrida's famous essay often referred to in posthumanist theory), Elizabeth Cherry points out how Kim's photography sheds light upon animal practices by "making the invisible visible," and thus "plays a significant role in viewers' initial awareness of and potential mobilization into the animal rights movement."[23]

The lighter side of farmed animal welfare

Sadly, most humans relish the taste, convenience, and presumed "healthiness" of meat, eggs, fish, and dairy products. In Western society, the elimination or reduced consumption of these foods is construed as denying some of life's greatest pleasures. This desire for self-gratification often goes hand-in-hand with an attitude of resentment and hostility toward vegans, who may be blamed for creating havoc and proselytizing to an unwilling audience at every opportunity. Indeed, outspoken vegans and animal-rights activists who display a sanctimonious or militant approach are often met with antagonism, derision and dismissal. The same holds true when excessive animal suffering is displayed in the media, especially in documentaries and advocacy campaigns perceived as an attempt to "brainwash" viewers into changing their eating habits, but in fact can provoke defensive or apathetic reactions.

Therefore, films geared for general and young audiences that skillfully promote ethical dietary choices, sometimes turning to humor to suffuse a tragic situation with comic elements, are a praiseworthy endeavor. A few examples of this lighter side to vegan/vegetarian advocacy can be found in *Chicken Run, Babe*, and *The Meatrix*. These highly anthropomorphic, amusing films about animal welfare and the consumption of farmed animals have been wildly successful.

The British claymation comedy *Chicken Run* (2000), by Aardman Studios, features factory-farmed poultry in a parody of the classic war film *The Great Escape*.[24] While many scenes in *Chicken Run* may seem hilarious, such as when two protagonists narrowly flee an ominous mechanized assembly line making chicken pies, one cannot ignore its darker, underlying themes. The film became a box-office hit despite its slapstick approach to factory farming as well as its thinly veiled allusions to the Holocaust – with Nazi-like farmers whose chickens bear intolerable conditions behind barbed wire in barracks reminiscent of Auschwitz.

Babe (1995), another smash hit film, depicts the ups-and-downs of a mixed menagerie of talking farmed animals whose main protagonist, a brilliant pig, becomes the pride and joy of his carer, Farmer Hoggett (James Cromwell), by learning to herd sheep with flawless accuracy. The sturdy emotional bond that is formed between the farmer and his precious pig ultimately spares the latter from the butcher's knife, and the film created a movement that Nathan Nobis coined as "Babe Vegetarians." In fact, being the main human protagonist in *Babe* so deeply influenced Cromwell to regard farmed animals as sentient and intelligent beings, that he became a vegan activist, later narrating *Farm to Fridge* (2011) from Mercy for Animals undercover investigations.

A slightly different genre, *The Meatrix* (2003) is an animated cult video that highlights the potential of humor to deliver strong ethical and environmental messages to a wide public through media technology. Rather than Keanu Reeves, the handsome, virile main protagonist of *The Matrix*, it features Leo, a squat animated pig with a moral conscience who is curious about how farmed animals are raised in contemporary society. The truth is revealed to him by a bull wearing a black coat and sunglasses named Moopheus, who accompanies Leo into the ugly world of industrial farming, thus demonstrating the cruelty and environmental ills it encompasses. Passionately, Moopheus calls out for solidarity, admonishing "We are going to spread the word. But it's you, the consumer who has the real power! Don't support the factory farming machine! There is a world of alternatives!"[25]

Humorous visual works such as political cartoons can also deliver an ethical message in gentler ways. For example, animal-rights activist Dan Piraro, author of the syndicated series known as *Bizarro!*, addresses the ethics of eating meat in numerous single-panel cartoons promoting veganism.[26] In many of these, animals are highly anthropomorphized; talking cows, chickens, turkeys and pigs appear in absurd circumstances, often engaged in conversation with humans or with each other to accentuate how unjust and detrimental it is to consume meat and other animal products. One cartoon depicts a bluish-skinned, obese pig with a glum expression, wearing a baseball cap backwards and a T-shirt that reads "Meat Kills (*Both* of Us)," as he holds a protest sign asking: "Are Humans Carnivores?" To address this question, Piraro wrote a long, sarcastic essay comparing the anatomical (dentition, length of intestines) and behavioral (attraction to rotting meat, ability to hunt) traits of carnivores, omnivores, and herbivores to prove that humans are naturally herbivorous. He converted this piece into a vegan video narrated by another anthropomorphized pig who beckons: "Gimme a break and barbeque a veggie burger next time."[27]

Conclusion: a subtler approach

To foster empowerment and avoid hopelessness, it is vital to place greater emphasis on the aesthetic value and emotional impact of positive rather than negative visual representation of animals. Thus, we can avoid the potential psychological harm generated by watching explicit suffering, which frequently entails a ghastly scene of a writhing animal hoisted up by one leg before its throat is slit. Used appropriately, visual images of animals are capable of delivering realistic and moving messages regarding their welfare, saturated with multiple levels of meaning expressed in many creative forms. Visual culture functions as a powerful and constructive vehicle for advocating for the many billions of defenseless animals killed annually for human consumption while endorsing alternative, more ethical dietary choices. Clearly, some forms of animal representation encourage the adoption of plant-based diets and stir public awareness of the inhumane methods of raising farmed animals, especially on industrialized farms. However, it is worth noting that the predilection of even the most adept filmmakers to overuse vivid and extremely violent images of animals in agonizing pain and distress sometimes has an opposite effect. Pointing the finger of culpability at humans for abusing and killing animals by showing repeated, heart-wrenching images of slaughter may actually be counterproductive to the desired goal of instilling compassion toward our fellow species.

Notes

1 Susan Sontag, *Regarding the Pain of Others* (London, England: Penguin Books, 2004), 37–38.
2 Anat Pick, "Animal Rights, Organized Violence and the Politics of Sight," in *The Routledge Companion to Cinema and Politics*, ed. Yannis Tzioumakis and Claire Molloy (London, England: Routledge, 2016), 91–102.
3 John Berger, "Why Look at Animals?" in *About Looking* (New York, NY: Pantheon, 1980), 1–28.
4 "Anat Pick," Queen Mary, University of London, accessed August 14, 2016, http://filmstudies.sllf.qmul.ac.uk/filmstudies/people/pick.html.
5 Steve Baker, *Picturing the Beast: Animals, Identity, and Representation* (Manchester, England: Manchester University Press, 1993), 216.
6 Jonathan Burt, "The Illumination of the Animal Kingdom: The Role of Light and Electricity in Animal Representation Society and Animals," *Journal of Human–Animal Studies* 9 no. 3 (2001): 203.
7 Catherine Tiplady et al., "Ethical Issues Concerning the Public Viewing of Media Broadcasts of Animal Cruelty," *Journal of Agricultural and Environmental Ethics* 28 (2015): 644.
8 Jason Middleton, "Documentary Horror: The Transmodal Power of Indexical Violence," *Journal of Visual Culture* 14 no. 3 (2015): 288.
9 Adam Lowenstein, "Films Without a Face: Shock Horror in the Cinema of Georges Franju." *Cinema Journal*, 37 no. 4 (1998): 37.
10 Lowenstein, "Films Without a Face," 41.
11 Barry Grant, "Blood of the Beasts," in *Voyages of Discovery: The Cinema of Frederick Wiseman* (Urbana, IL: University of Illinois Press, 1992), 119–120.
12 I Grant, "Blood of the Beasts."
13 Isa Leshko's photography can be viewed at http://isaleshko.com/press/.
14 Joe Zammit-Lucia and Linda Kalof, "From Animal Rights and Shock Advocacy to Kinship with Animals: Lessons from the Visual Culture of Endangered Species," *Antennae*, 12 (2012): 100.
15 Zammit-Lucia and Kalof, "From Animal Rights and Shock Advocacy," 106.
16 Charlotte Houghton, "This Was Tomorrow: Pieter Aertsen's Meat Stall as Contemporary Art," *The Art Bulletin* 86 no. 2 (2004): 281–282.

17 Stephen F. Eisenman, *The Cry of Nature: Art and the Making of Animal Rights* (London, England: Reaktion, 2013), 74–76.
18 Avigdor Poseq, "Soutine's Dead Birds as Metaphors of Sexuality," *Konsthistorisk Tidskrift*, 66 no. 4 (1997): 258.
19 Giovanni Aloi, "In Conversation with Sue Coe," *Antennae – The Journal of Nature in Visual Culture* 5 (2008): 55–56.
20 Taylor has a congenital joint disorder known as arthrogryposis, and is bound to a wheelchair.
21 Erica Grossman, "An Interview with Sunaura Taylor," *Critical Animal Studies Journal* 12 no. 2 (2014): 15.
22 Grossman, "An Interview with Sunaura Taylor," 14.
23 Elizabeth Cherry, "'The Pig That Therefore I Am': Visual Art and Animal Activism," *Humanity and Society* 40 (2016): 64–85.
24 Mark Glancy, *Hollywood and the Americanization of Britain: From the 1920s to the Present* (London, England: I. B. Tauris, 2014), 273.
25 Louis Fox and Jonah Sachs, *The Meatrix*, Flash animation, directed by Louis Fox (Oakland, CA: Free Range Studios, 2003).
26 Dan Piraro, "Are Humans Carnivores?" KPAO by Dave Cortright, October 15, 2011, www.kpao.org/2011/10/are-humans-carnivores-by-dan-piraro-bizarro-cartoonist-and-vegan.html.
27 "Vegan Video," YouTube, April 24, 2006, www.youtube.com/watch?v=05zhL1YUd8Q.

Acknowledgement

To Jessica, the clever and affectionate goat of my childhood, whose long, serene life should be the inherent right of all farmed animals today.

Bibliography

Aloi, Giovanni. "In Conversation with Sue Coe." *Antennae – The Journal of Nature in Visual Culture* 5(Winter 2008): 54–59.

Baker, Steve. *Picturing the Beast: Animals, Identity, and Representation*. Manchester, England: Manchester University Press, 1993.

Berger, John. "Why Look at Animals?" in *About Looking*. New York, NY: Pantheon, 1980.

Burt, Jonathan. "The Illumination of the Animal Kingdom: The Role of Light and Electricity in Animal Representation." *Society and Animals* 9, 3(2001): 203–228.

Burt, Jonathan. *Animals in Film*, Vol. 1. London, England: Reaktion Books, 2002.

Cherry, Elizabeth. "'The Pig That Therefore I Am': Visual Art and Animal Activism." *Humanity and Society* 40(February 2016): 64–85.

Derrida, Jacques. "The Animal That Therefore I Am (More to Follow)." *Critical Inquiry* 28, 2(2002): 369–418.

Eisenman, Stephen F. *The Cry of Nature: Art and the Making of Animal Rights*. London, England: Reaktion, 2013.

Foer, Jonathan Safran. *Eating Animals*. New York, NY: Little, Brown and Company, 2009.

Fox, Louis and Jonah Sachs. *The Meatrix*. Flash Animation. Directed by Louis Fox. Oakland, CA: Free Range Studios, 2003.

Glancy, Mark. *Hollywood and the Americanization of Britain: From the 1920s to the Present*. London, England: I. B. Tauris, 2014.

Grossman, Erica. "An Interview with Sunaura Taylor." *Critical Animal Studies Journal* 12, 2 (2014): 9–16.

Grant, Barry Keith. "Blood of the Beasts" in *Voyages of Discovery: The Cinema of Frederick Wiseman*. Urbana, IL: University of Illinois Press, 1992, 104–132.

Houghton, Charlotte. "This Was Tomorrow: Pieter Aertsen's 'Meat Stall' as Contemporary Art." *The Art Bulletin* 86, 2(2004): 277–300.

Kalof, Linda and Amy Fitzgerald, eds. *The Animals Reader: The Essential Classic and Contemporary Writings*. Oxford, England: Berg Publishers, 2007.

Lowenstein, Adam. "Films without a Face: Shock Horror in the Cinema of Georges Franju." *Cinema Journal* 37, 4(1998): 37–58.

Middleton, Jason. "Documentary Horror: The Transmodal Power of Indexical Violence." *Journal of Visual Culture* 14, 3(2015): 285–292.

Nobis, Nathan. "The 'Babe' Vegetarians: Bioethics, Animal Minds and Moral Methodology." In *Bioethics at the Movies*, edited by Sandra Shapshay. Baltimore, MD: Johns Hopkins University Press, 2009.

Pick, Anat. "Why Not Look at Animals?" *Necsus*2015, Accessed August 9, 2016. www.necsusejms.org/portfolio/spring-2015_animals/.

Pick, Anat. "Animal Rights, Organized Violence and the Politics of Sight." In *The Routledge Companion to Cinema and Politics*, edited by Yannis Tzioumakis, and Claire Molloy. London, England: Routledge, 2016.

Piraro, Dan. "Are Humans Carnivores?" KPAO by Dave Cortright. October 15, 2011. www.kpao.org/2011/10/are-humans-carnivores-by-dan-piraro-bizarro-cartoonist-and-vegan.html.

Poseq, Avigdor. "Soutine's Dead Birds as Metaphors of Sexuality." *Konsthistorisk Tidskrift* 66, 4(1997): 251–260.

Regan, Tom. *The Case for Animal Rights*. Berkeley, CA: University of California Press, 1983.

Sinclair, Upton. *The Jungle*. New York, NY: Airmont Publishing Company, Inc., 1965.

Singer, Peter. *Animal Liberation*, Second edition. London, England: Random House, 1990.

Sontag, Susan. *Regarding the Pain of Others*. London, England: Penguin Books, 2004.

Tiplady, Catherine, Deborah-Anne Walsh, and Clive Phillips. "Ethical Issues Concerning the Public Viewing of Media Broadcasts of Animal Cruelty." *Journal of Agricultural and Environmental Ethics* 28(2015): 635–645.

Tolstoy, Leo. "The First Step." *The New Review* (1892): 23–41.

Queen Mary, University of London. "Anat Pick." Accessed August 14, 2016. http://filmstudies.sllf.qmul.ac.uk/filmstudies/people/pick.html.

YouTube. "Vegan Video." April 24, 2006. www.youtube.com/watch?v=05zhL1YUd8Q.

Zammit-Lucia, Joe and Linda Kalof. "From Animal Rights and Shock Advocacy to Kinship with Animals: Lessons from the Visual Culture of Endangered Species." *Antennae* 12 (2012): 98–111.

3.7

LEADERSHIP, PARTNERSHIP AND CHAMPIONSHIP AS DRIVERS FOR ANIMAL ETHICS IN THE WESTERN FOOD INDUSTRY

Monique R. E. Janssens and Floryt van Wesel

Introduction: why study animal ethics in food companies?

Today, concepts like Corporate Social Responsibility and Triple P (People, Planet, Profit)[1] are well known in all industries and offer companies opportunities to define their own responsibilities toward people, nature, animals, and sustainability of the planet and its ecological and social systems. Business ethics offers extra food for thought, not only for philosophers, but also for decision makers in companies. However, business ethics has ignored the ongoing debate in animal ethics for too long.[2, 3] Furthermore, Janssens and Kaptein found that some multinationals express responsibilities towards animals, whereas others do not.[4] Although keeping animals and being part of the animal-derived food industry are both factors that correlate positively with expressed commitment to animals, this does not account for all differences.

In the Business Benchmark on Farm Animal Welfare, Amos and Sullivan identify ethical and business reasons for companies to set animal welfare as a corporate issue. Business arguments are: Legislation, market opportunities, and stakeholder expectations, especially from clients and customers.[5] Though their study offers many interesting insights, even more knowledge is needed about the drivers and challenges companies and their corporate responsibility managers (RMs) encounter with regard to animal welfare.

Based on the above, our research question is: *What are the drivers and challenges for large Western-Europe-based food companies to take responsibility for animals?* Answering this question will not only explain differences between companies; at the same time, it will offer companies and their stakeholders tools for change. We chose the term "drivers" for stimulatory factors because it is a common term in managerial literature,[6] and the term "challenges" for obstructive factors, for reasons of positive framing: A challenge is surmountable. To explore the drivers and challenges RMs encounter, we conducted a qualitative study among nine Western-Europe-based companies producing or selling meat, poultry, fish, or dairy products. By conducting and analyzing interviews, we have identified several drivers and challenges and the way they interact.

This chapter provides a connection between theory and practice. We identify drivers and challenges for incorporating animal ethics into corporate responsibility in food companies. Our findings can be of interest for other sectors too, since all large companies hire caterers for their company restaurants, or have impacts on animals through transport, development and construction.

This chapter is organized according to the qualitative study reporting guidelines by Corbin and Strauss.[7] In this section we outlined the significance of the issue of animal ethics as a theme in corporate responsibility and presented our research question. In section 2 we explore the theoretical background. In addition, we explain how this study contributes to animal and business ethics theory and practice. In section 3 we present our research method. In section 4 we offer our findings, which we discuss in section 5.

Theoretical background

In this section previous research is presented and related concepts are explained, as a starting point of the research and as an analytical lens.

Status quo of the capitalist system

People have bred and slaughtered animals for ages. In the second half of the twentieth century an industrial system emerged that systematically breeds as many animals as possible at the lowest possible cost, and sells them at the best possible price. McMullen argues that the place of animals in our economy is determined by: Low-cost technology, competition-driven efficiency and accountability, and specialization.[8] It is very hard for individuals to change the system. This is why RMs may encounter challenges related to the system. Rollin adds that a positive correlation between individual animal welfare and individual animal productivity from traditional agriculture, was probably extrapolated to a correlation between overall animal welfare and system productivity in industrial agriculture.[9]

External pressure

Schultz and Wehmeier see competition, regulative norms, professional norms, and public pressure as triggers for institutionalization of CSR (Corporate Social Responsibility).[10] One of the regulative norms could be legislation about reporting,[11] as could non-coercive regulations, such as political pressure or the international CSR guideline ISO 26000.[12] Castka and Balzarova predicted in 2007 that multinationals would adopt this standard to legitimize their policies.[13] It is conceivable that these kinds of external pressure will also influence the way a company takes responsibility for animals.

Managerial decision making

It can also be expected that individual (managerial) decision making plays a role. While arguing how factors from both within and outside the company influence managerial decision making about corporate responsibility measures, Hoffman and Bazerman also identified external regulatory pressure and internal reward systems, amongst others, as influential. In addition, they argue that institutional resistance can result from resource constraints, fear of the unknown, threats to political interests, and habitual distrust.[14]

Responsibility awareness

Constantinescu and Kaptein found that under particular conditions, responsibility awareness of corporations and of individuals within these corporations can be mutually enhancing. The most important condition for that effect to emerge is for corporate structure and culture to be disconnected: Then there is room for doing something better or worse than corporate regulations dictate. However, if people in the company act according to culture more than to structure, this strengthens the corporate ethical culture.[15]

Collective responsibility and leadership

Isaacs explores moral responsibilities in the collective context of a company. She sees companies as having collective responsibilities because they have agency. Professional actions can be seen as contributions to a larger collective act. Decision-makers bear extra responsibility, because outside their role their actions and decisions would not have the same impact. Isaacs concludes that individual agency is not absorbed by collective agency, but on the contrary, adds power. At the same time, an individual agent in the company has to deal with policies, structures, interests, attitudes, practices, and cultures. In addition, Isaacs identifies potential collective agents: Groups that act upon certain moral issues. In fact, she sees a responsibility to form them. This responsibility can be diminished by vagueness about what can be done or by a status quo of wrongful social practice. These are all factors that can be expected to be perceived as challenges. In addition, we learn from Isaacs that when decision-makers are aware of their power, they can take individual responsibility and stand up as leaders.[16]

Mostovicz et al. find that leadership as a personal quality plays an important role in corporate ethical behavior.[17] Angus-Leppan et al. follow Matten and Moon[18] in making a distinction between implicit and explicit CSR (the former being collective and embedded in the relations of the political system, and the latter being more a voluntary and often strategic choice of the company) and explore different forms of leadership. A combination of leadership types can lead to conflicts or to healthy friction.[19]

Organizational culture

Ingenbleek et al. studied four European "criteria formulating organizations" and found elements within organizational culture that influence the development of ethical standards for animal welfare. They found that standards are uplifted by strong positive shared ethical values and are lowered by negative shared values. In addition, norm diversity leads to lower standards in the short term, but more upgrading standards in the longer term. Strong shared values lead to more symbolic artifacts which stimulate compliance and innovative behavior, which leads to higher standards.[20] These influences could also be present in food companies.

More institutional determinants

Basu and Palazzo made a study of internal institutional determinants of CSR. They distinguish between different kinds of dimensions of sensemaking in CSR: 1) cognitive dimensions (identity orientation and legitimacy), 2) linguistic dimensions (justification and transparency), and 3) conative dimensions (posture, consistency, and commitment).[21]

Van Tulder et al. identified more than 70 different managerial tipping points for transition to sustainability in organizations, moving from an inactive position, through a reactive, and an active, to a proactive one. Influencers that take companies from tipping point to tipping point include perception, involvement and satisfaction of employees, purchasing policy, chain management, organizational communication, internal coherence between departments, openness to stakeholders, the need to set priorities and appointing a manager with excellent communication skills.[22]

Mauser identified intra-organizational drivers for the environmental performance of businesses from a case study of the Dutch dairy sector. She finds four drivers for environmental performance: Policy and strategy, communication, organization structure, and management commitment; to achieve optimal environmental performance, the right balance between these four has to be found.[23]

Visser holds five principles of Transformative CSR (or: CSR 2.0) responsible for its potential success: Creativity, scalability, responsiveness (being proactive, in dialogue with stakeholders), glocality (tailor-made adaptations to global solutions) and circularity (renewability and no waste).[24]

Resuming

In this section we have seen that influencers of CSR have been identified by many studies, but none for animal ethics have been identified for the food industry. This makes us alert to finding similar elements and open to finding new ones. We use the definition of animal ethics from a previous study in the field of animal ethics and business ethics. Janssens and Kaptein define it as being about the moral status of animals, not as a species but as individuals with interests.[25]

CSR in this chapter is approached as all ethical responsibility of a company beyond what is legally required.

Method

The current study is explorative in nature. Our focus is on animal-derived food industry, as that is where large numbers of animals are bred, kept, cared for, transported and slaughtered. Our methodology is based on grounded theory as described by Boeije[26] and by Corbin and Strauss,[27] a research method that is easily applicable to organizational research.[28, 29, 30, 31, 32]

We studied nine large, internationally operating food companies based in Western Europe. Western Europe was chosen because it is where we ourselves are based (the Netherlands) and because there are signs that Europe is one of the areas where animal ethics in businesses is the most developed.[33] The companies were found by searching on the internet for the most important players in the animal-derived food industry and by asking interviewees for more names (snowball method). Seven companies were based in the Netherlands, one in the United Kingdom and one in Switzerland. They all operated internationally and had over 1,000 employees (up to 172,000). Some were very old (the first shop opening in 1887) and some relatively new (2007). They are producers, processors, wholesalers, and retailers.

For reasons of data triangulation, we studied the companies' responsibility reports, their websites. and interviews we conducted with the managers responsible for animal welfare policy (CSR managers, quality managers, et cetera, from here: RMs). We used the "semi-structured interview" method, which provides the richest source of data for theory building. Each interview was started with announcing the main topic (drivers and challenges for their company to take responsibility for animals) and an invitation to elaborate on it. At the end we checked our topic list to remind them about topics we missed. Guiding topics on this list were derived from the theory presented earlier. The interviews took 40 to 90 minutes each. Audio recordings were transcribed by one of the researchers or an assistant, and in the latter case checked by one of the researchers. Trust was created by offering the research plan beforehand by e-mail, promising to use the data anonymously and for research purposes only, and conducting the interview in person. One interview was done by telephone because of the geographical distance. Interviews were conducted by the first author, who has 25 years of experience in journalism and communications.

Data were handled in a cyclical process of interviewing, coding, identification, comparison, and analysis. Three of the interviews were open coded by both authors. Inter-coder reliability was high (around 90%) and differences in coding were agreed upon. Hereafter, the first author open coded the rest of the transcripts according to the adapted coding system. All additional text documents (websites and CSR reports) were also coded (with NVivo software, QSR International, Australia) by the first author. After coding the information from the ninth company, saturation was reached. Further analysis consisted of axial coding where the authors defined the codes, restructured them, and investigated relationships between the codes, and selective coding, where a theoretical model of the findings was created.

Findings

As a result of the axial coding we created a matrix of types of influencers (emotional, attitudinal, ethical, cultural, communicative, historical, geographical, practical, economical, and strategic ones) and different levels at which these influencers occurred (the RM, the company, several external organizations, the consumer, and the broader public). In selectively coding this matrix, we found three reoccurring challenges and three reoccurring drivers by which these challenges seemed manageable. These drivers we called leadership, partnership, and championship. How they are connected to the company and its surroundings is shown in Figure 3.7.1 Leadership of the company, a CEO, the board or at least the RM is crucial. Partnerships with partners inside and outside the production chain are very helpful, and so are Championships: Ways of celebrating moments of importance, for example when a goal is set or an achievement is reached.

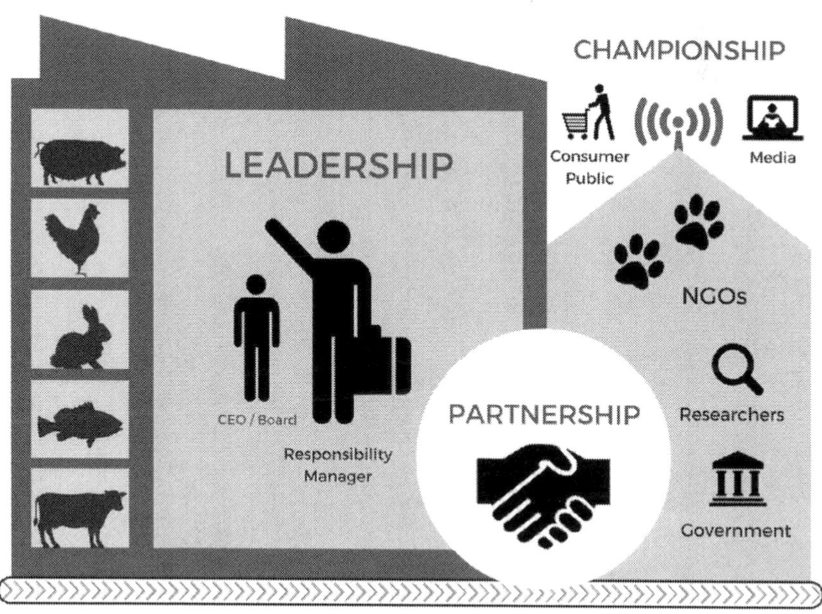

FIGURE 3.7.1 Leadership, partnership and championship as drivers for animal ethics in companies

Leadership overcomes immobility

A large collection of challenges can be grouped under the term "immobility," meaning: Having fixed ideas, being satisfied with the status quo, or not being prepared to move. We found fixed ideas about: The responsibility level of the company towards animals, what animals prefer, what animals are, the taste of meat, et cetera. We also found fixed ideas like "good care is enough," "animal welfare equals animal health," "animal welfare equals animal productivity," "animals are lower creatures," and "the farmer represents the animals."

RM1: The farmer is a stakeholder, and that includes the cows and the land and biodiversity and the birds and the flora and fauna. And the farmer represents them, he's our most important stakeholder, because he's our stockholder and our biggest milk supplier, and he knows the best way to do that. Immobility can take the form of seeing animal welfare as the exclusive concern of the RM, in feelings of powerlessness of the RM, in a reward system with a short-term sales focus, and in clinging to a traditional money-driven culture in which there is no room for compassion. External factors of immobility are the capitalist system, the public's prejudices, strictness of animal welfare label requirements, rigidity of NGOs, or entrenched values of stakeholders, originating from historical, geographical, national or cultural differences.

RM2: Ultimately, I think the commercial side is still winning … But after all, we're in a capitalist system, and money's important … Every morning, our Category Managers look at yesterday's margins.

RM6: [Some NGOs] actually want people to stop eating meat … If you want to improve things, then you have to make sure what you want to do is realistic, and not be too extreme with new standards. These different types of immobility can be overcome by leadership. RMs can play the devil's (animals') advocate or offer an ethical decision-making framework.

RM5: I … also wrestled for a long time over which decision-making framework to use. Ultimately I made it very simple: If I can't explain it on the Alexanderplatz in Berlin, in Soho in London or on the Dam Square in Amsterdam, we won't do it. It is helpful if the RMs are sensitive to trends and future issues, eager to learn, and open to new ideas, such as the idea that animals are individuals with their own needs, or that they can be stakeholders. Although RMs are aware that within the company their interest in animal welfare is above-average, they can have influence by setting clear standards. Their leadership stance is stronger if they encounter animals now and then, and if they are respectful and empathic towards animals, but also to people. In the following example the RM is connecting to farmers by translating animal suffering to children.

RM7: [I ASK THEM:]: Let's say thirty school kids take a bus trip, and three of them get out with a broken arm … is that normal? As is shown in Figure 3.7.1, a CEO, the board, or the company as a whole, can take leadership as well. The company can commit itself to an ethical approach to animals. It can communicate this stance, quantify animal welfare or create demand for products with higher animal welfare. In general, a proactive attitude is helpful, as well as real engagement, true involvement, and corporate pride in animal welfare.

RM8: If we're talking about people, animals and the environment, the principles of our CSR policy, but also how we see our formula, [animal welfare] is a part of that.

RM4: It's simply a fully accepted, and of course, an integrated element of our sustainable-agriculture approach. The Public Communication or Marketing department can make animal ethics part of trend identification or issue and reputation management, choose a "green" or distinguished brand, or be transparent and authentic. Websites play an important role in informing stakeholders on all levels, from ethical standards to product and label information.

RM3: I think that's also becoming more and more important … some authenticity, some honesty, openness about what you do. Organizational communication helps spread information and pride through the company. Extra optimizers are innovative ideas, short lines of communication, discussions with employees, and budget.

RM4: I've developed a kind of engagement platform with gamification … All employees are asked to participate … This year one of the challenges was about food … You could build a business case or a consumer behavior change case with our sustainable agriculture sourcing activities … [animal welfare] could be an element there. You could sign up to visit one of our pig farmers … Because our experience is: If you take people with you, then it comes alive. The leadership stance can be translated into goals, projects and programs, which are mentioned as very helpful. Another form of leadership is proactively preventing disruptive mistakes, for example in labeling. And when it is difficult to find the right suppliers (which can be quite a problem with farmed animals), leadership companies explore new markets and look for new solutions to practical problems. They invent new rewarding systems, spread risks over categories (for example: Sell vegetarian alternatives), conduct research on animal housing, or focus on value creation. They make animal welfare a business case, for example by building animal welfare into the cost model, choosing quality markets, or calculating the extra productivity of animals that live longer. The general line in these drivers is an open attitude and a willingness to anticipate and change.

Partnership overcomes dilemmas

A second cluster of factors is formed around the concept of dilemmas and finding balance, for example between strategies or competing values. Should we go for a low-price or a high-quality reputation? Invest in animal friendly production systems or in a reputation campaign? Take risks or play it safe? Offer free range or minimize emissions? Build a money-making culture or a responsibility culture? Take small steps or provoke radical change? Be paternalistic or not?

RM1: We're not going to tell farmers how to take care of their cows or dictate the size of [the stall] or what kind of bedding they have. Farmers can [determine] that much better than we can … It's almost patronizing to ask them to … It's so much part of their skills and their passion for farming, that it's almost a question of [laughs], how dare you, why would you think I wouldn't care about that? Operating internationally, companies encounter the dilemma of imposing national or European standards on suppliers, or accepting foreign standards. The materiality analysis, an assessment of the company's most important responsibility issues, plays an important role, but also creates dilemmas when many issues are identified as important. A dilemma can occur if stakeholders discriminate between animal species, or it can lie in the disputability of ethical claims about animals.

RM3: For example, how can you tell if an animal's in pain? That's always tricky for me. And everyone has an opinion about it. Partnership with other parties, as shown in Figure 3.7.1, is crucial for tackling these complex dilemmas and the numerous practical problems that come with it, like dealing with a specific traditional husbandry system, or aligning supply and demand, taking the life cycle of animals into account. Partnerships make it possible to exchange knowledge and do a well-informed materiality analysis in which choices can be made. Companies who show self-confidence about their animal welfare achievements are those who take part in extended partnerships, with mutual trust and understanding.

RM7: It's all about trust … You'd be surprised how little we write down. [That trust is there] because there's a fairly long-lasting relationship and [the retailer] has always presented themselves as a trustworthy partner. And we've tried to do the same. Partnerships with suppliers offer them some security and empower them to change. Farmers tend to appreciate the values naturalness, craftsmanship, respect for animals, and reciprocity between farmer and animal. Acknowledging these values in the partnership is helpful. Broad partnerships can also include governmental organizations.

RM1: Then the Ministry of Economic Affairs says: We actually want to look at what's possible. Could we do a project about it? Then we and the Agriculture and Horticulture Association become the mouthpiece, for the dairy farmers too. Some RMs get inspiration from international platforms.

RM2: For example, we work with the Consumer Goods Forum … That's an organization, 400 retailers and manufacturers are associated, and they also do a lot of research … Actually, it's how we try to signal all the trends, and work them into our policy. Partnerships with researchers and NGOs can be drivers too. RMs appreciate evidence for their decisions. Research can end discussions, show the way out of dilemmas and reveal animal preferences. The level of collaboration with NGOs ranges from participation in the materiality analysis to large-scale programs, involving labels, audits, research, certifications and awards. RMs mention as positive factors an open dialogue, and personal leadership of NGO directors.

Finally, companies can combine leadership and partnership by starting to work with new partners.

RM9: We then identified a third-party partner to work with us, to help us understand better the external landscape and to target our assets … We have been working with World Animal Protection. They've been helping us in terms of implementation, and also helping us formulate a vision … It was at that time the first partnership between an international food company and a global animal welfare NGO.

Championship overcomes slander

The third crucial element in Figure 3.7.1 is Championship, to overcome the third type of challenges, clustered around the concept of slander. Slander from animal-welfare organizations or influential individuals can be obstructive. As we will see later, a critical stance is accepted and is even experienced as an important driver, but prejudices, extreme demands, and strong opinions without knowledge have a negative effect on the willingness of the company to make a change.

RM8: Some NGOs take a harder stand than others … There's something going on now … that can have a pretty disruptive effect. While you're sticking your neck out as a retailer in so many ways, and you're really serious and committed, it can work against you at the point it gets unpleasant. And that absolutely has a huge impact on the people involved … Then it's not a positive energy … but more a negative energy. This negative influence of slander can be overcome by positive news in the traditional or social media, by which the general public can be reached as well as the consumer. Here NGOs play an important role too, because they can share an animal welfare item on a trustworthy base. The public are strongly influenced by the media, and at the same time express themselves in it. Media and NGOs can help companies celebrate their achievements through awards, reports, events, and news items. Publicly expressed ethical demands (transparency, integrity, accountability, traceability of products, naturalness of food, welfare) are extra drivers for companies to take action. Critical questions asked directly to the company (which today can be done publicly via social media) are supportive too.

RM2: People know more and more about it, and so they're asking more questions, and you have to have answers ready. Related to the concepts of slander and championship is the risk of reputation damage as a driver. Publicity on genuine positive achievements can be a buffer to future reputation damage.

Discussion, conclusions, and recommendations

We sketched a model of the central drivers for taking animals into account ethically. The most important drivers that help overcome challenges are: Leadership, partnership and championship. As shown in Figure 3.7.1, leadership from the company (or the decision makers in it) is important. It works best in combination with partnerships with governmental organizations, research institutions, individual researchers, and NGOs. Third, "championship" about this leadership and these partnerships is helpful: Showing and celebrating positive changes made by the company as leaders and partners. These moments of communication with the public or the consumer, through traditional or social media, eventually organized together with partners, have a positive impact too, and relate back to marketing decisions in the company through consumer behavior, which in its turn stimulates its leadership. See Figure 3.7.1 in which these connections are outlined.

In this study, several theoretical concepts from adjacent CSR fields were confirmed and illustrated for animal ethics in food companies. We saw how a money-driven culture[34] and a short-term sales focus[35] can be challenges. These and many others can be overcome by taking leadership,[36, 37, 38] initiating collectives,[39, 40, 41] and paying attention to the purchasing and supply chain as well as to organizational communication[42] and intra-organizational communication.[43] Finding balance in dilemmas seems to be difficult.[44] Market opportunities and stakeholder expectations appeared to be drivers,[45] just like competition, regulative norms, professional norms, and public pressure.[46] Identity, accountability and transparency issues play a role.[47] We also found traces of the mutually enhancing effect of responsibility awareness of corporations and of individuals within these corporations.[48] In our data political, legislative,[49, 50] and employee[51] influences were hardly mentioned. It is thinkable that the first two remained implicit and that the last one is stronger for CSR in general than for animal ethics, but employee motivation was no part of this study. Especially for animal ethics, we encountered as one of many fixed ideas that animal welfare would equal animal productivity.[52]

Though most of the individual challenges and drivers we found seem applicable to other CSR topics as well, the total of the model as we found it, is strongly related to what animals in the food industry are: Vulnerable living creatures, locked away from the public, not acknowledged as stakeholders themselves (whilst very much is at stake for them), unable to tell us in words what they experience. This combination of characteristics is not applicable to other stakeholders of CSR, like people, who have a voice, or the environment, which can be seen and experienced every day. Eventually, a comparison with children in child labor could be tenable. Those RMs who feel responsible for animal welfare, in surroundings where a different culture prevails, have a tough job in letting the voice of animals be heard. Hopefully, our model helps them and other parties to strengthen their case.

The small number of rather diverse companies in our study is one of its limitations. We tried to minimize this influence by asking interviewees for names of other RMs in the same industry, and by proceeding till saturation was reached.

We recommend further research be done into this topic by conducting quantitative studies into specific aspects of our study to quantify the weight of the influencers we found. Second, it would be helpful to explore the complex role of animal welfare NGOs, who on the one hand put off companies by naming and shaming, and on the other hand achieve big changes. In this light, a more thorough analysis of the role of communications would be useful.

Our recommendation to RMs and companies in the food industry would be that, on the basis of our model, they explore how they can make progress through leadership, partnership, and championship. This can be done exploring how the numerous examples we gave in this chapter apply to their own company.

Notes

1 John Elkington, *Cannibals with Forks: The Triple Bottom Line of 21st Century Business* (Oxford, England: Capstone Publishing, 1997).
2 Monique Janssens and Muel Kaptein, "The Ethical Responsibility of Companies Toward Animals: A Study of the Expressed Commitment of the Fortune Global 200," *Journal of Corporate Citizenship*, 63 (2016): 42–72.
3 Luc Van Liedekerke and Wim Dubbink, "Twenty Years of European Business Ethics: Past Developments and Future Concerns," *Journal of Business Ethics* 82 (October 2008): 273–280.
4 Janssens and Kaptein, "The Ethical Responsibility of Companies Toward Animals."
5 Nicky Amos and Rory Sullivan, *The Business Benchmark on Farm Animal Welfare, 2015 Report* (London, England: BBFAW, 2016).
6 See, for example, Anniek Mauser, *The Greening of Business: Environmental Management and Performance Evaluation: An Empirical Study of the Dutch Dairy Industry* (Delft, The Netherlands: Eburon Publishing, 2001).
7 Juliet Corbin and Anselm Strauss, *Basics of Qualitative Research, Techniques and Procedures for Developing Grounded Theory* (California, CA: Sage, 2015).
8 Steven McMullen, "Is Capitalism to Blame? Animal Lives in the Marketplace," *Journal of Animal Ethics*, 5 no. 2 (2015): 126–134.
9 Bernard Rollin, *Putting the Horse Before Descartes: My Life's Work on Behalf of Animals* (Philadelphia, PA: Temple University Press, 2011).
10 Frederike Schultz and Stefan Wehmeier, "Institutionalization of Corporate Social Responsibility Within Corporate Communications: Combining Institutional, Sense-making and Communication Perspectives," *Corporate Communications: An International Journal* 15 (2010): 9–29.
11 Olivier Delbard, "CSR Legislation in France and the European Regulatory Paradox: An Analysis of EU CSR Policy and Sustainability Reporting Practice," *Corporate Governance: The International Journal of Business in Society* 8 (2008): 397–405.
12 *International Standard ISO 26000, Guidance on Social Responsibility* (Geneva Switzerland: ISO, 2010).
13 Pavel Castka and Michaela Balzarova, "A Critical Look on Quality Through CSR Lenses: Key Challenges Stemming from the Development of ISO 26000," *International Journal of Quality and Reliability Management* 24 (2007): 738–752.
14 A. Hoffman and M. Bazerman, "Changing Practice on Sustainability: Understanding and Overcoming the Organizational and Psychological Barriers," in *Organizations and the Sustainability Mosaic: Crafting Long-term Ecological and Societal Solutions*, ed. B. Husted (Cheltenham, England: Edward Elgar Publishing, 2006), 84–105.
15 Mihaela Constantinescu and Muel Kaptein, "Mutually Enhancing Responsibility: A Theoretical Exploration of the Interaction Mechanisms Between Individual and Corporate Moral Responsibility," *Journal of Business Ethics* 129 (May 2015): 325–339.

16 Tracy Isaacs, *Moral Responsibility in Collective Contexts* (Oxford, England: Oxford University Press, 2011).
17 Isaac Mostovicz, Nada Kakabadse and Andrew Kakabadse, "CSR: The Role of Leadership in Driving Ethical Outcomes," *Corporate Governance: The International Journal of Business in Society* 9 (2009): 448–460.
18 Dirk Matten and Jeremy Moon, "'Implicit' and 'Explicit' CSR: A Conceptual Framework for a Comparative Understanding of Corporate Social Responsibility," *Academy of Management Review* 33 (2008): 404–424.
19 Tasmin Angus-Leppan, Louise Metcalf and Sue Benn, "Leadership Styles and CSR Practice: An Examination of Sensemaking," *Journal of Business Ethics* 93 (May 2010): 289–213.
20 Paul Ingenbleek, Menno Binnekamp and Silvia Goddijn, "Setting standards for CSR: A Comparative Case Study on Criteria-formulating Organizations," *Journal of Business Research* 60 (2007): 539–548.
21 Kunal Basu and Guido Palazzo, "Corporate Social Responsibility: A Process Model of Sensemaking," *The Academy of Management Review* 33 (January 2008): 122–136.
22 Rob van Tulder, Rob van Tilburg, Mara Francken and Andrea da Rosa, *Managing the Transition to a Sustainable Enterprise; Lessons from frontrunner companies* (New York, NY: Routledge, 2013).
23 Mauser, *The Greening of Business.*
24 Wayne Visser, *CSR 2.0: Transforming Corporate Sustainability and Responsibility* (Heidelberg, Germany: Springer, 2014).
25 Janssens and Kaptein, "The Ethical Responsibility of Companies Toward Animals."
26 Hennie Boeije, *Analyses in Qualitative Research* (London, England: SAGE, 2010).
27 Corbin and Strauss, *Basics of Qualitative Research.*
28 Joel Gehman, Linda Treviño and Raghu Garud, "Values Work: A Process Study of the Emergence and Performance of Organizational Values Practices," *Academy of Management Journal* 56 (February 2013): 84–112.
29 Patricia Martin and Barry Turner, "Grounded Theory and Organizational Research," *The Journal of Applied Behavioral Science* 22 (1986): 141–157.
30 Tammy McLean and Michael Behnam, "The Dangers of Decoupling: The Relationship between Compliance Programs, Legitimacy Perceptions, and Institutionalized Misconduct," *Academy of Management Journal* 53 (December 2010): 1499–1520.
31 Scott Sonenshein, "Emergence of Ethical Issues During Strategic Change Implementation," *Organization Science*, 20 (February 2009): 223–239.
32 Linda Treviño, Niki Den Nieuwenboer, Glen Kreiner and Derron G. Bishop, "Legitimating the Legitimate: A Grounded Theory Study of Legitimacy Work Among Ethics and Compliance Officers," *Organizational Behavior and Human Decision Processes* 123 (March 2014): 186–205.
33 Janssens and Kaptein, "The Ethical Responsibility of Companies Toward Animals."
34 McMullen, "Is Capitalism to Blame?"
35 Hoffman and Bazerman, "Changing Practice on Sustainability."
36 Tracy Isaacs, *Moral Responsibility in Collective Contexts.*
37 Mostovicz, Kakabadse and Kakabadse, "CSR."
38 Visser, *CSR 2.0.*
39 Mostovicz, Kakabadse and Kakabadse, "CSR."
40 Julia Rotter, P.-E. Airike and C. Mark-Herbert, "Category Management in Swedish Food Retail: Challenges in Ethical Sourcing," in *The Ethics of Consumption: The Citizen, the Market and the Law*, ed. Helena Röcklinsberg and Peter Sandin (Wageningen, The Netherlands: Wageningen Academic Publishers, 2013), 54–58.
41 Visser, *CSR 2.0.*
42 van Tulder, van Tilburg, Francken and da Rosa, *Managing the Transition to a Sustainable Enterprise.*
43 Mauser, *The Greening of Business.*
44 Mauser, *The Greening of Business.*

45 Amos and Sullivan, *The Business Benchmark on Farm Animal Welfare.*
46 Schultz and Wehmeier, "Institutionalization of Corporate Social Responsibility."
47 Basu and Palazzo, "Corporate Social Responsibility."
48 Constantinescu and Kaptein, "Mutually Enhancing Responsibility."
49 Amos and Sullivan, *The Business Benchmark on Farm Animal Welfare.*
50 Delbard, "CSR Legislation in France."
51 van Tulder, van Tilburg, Francken and da Rosa, *Managing the Transition to a Sustainable Enterprise.*
52 Rollin, *Putting the Horse Before Descartes.*

Bibliography

Amos, Nicky, and Rory Sullivan. *The Business Benchmark on Farm Animal Welfare, 2015 Report.* London, England: BBFAW, 2016.

Angus-Leppan, Tasmin, Louise Metcalf and Sue Benn. "Leadership Styles and CSR Practice: An Examination of Sensemaking." *Journal of Business Ethics* 93 (May 2010): 189–213.

Basu, Kunal, and Guido Palazzo. "Corporate Social Responsibility: A Process Model of Sensemaking." *The Academy of Management Review* 33 (January 2008): 122–136.

Boeije, Hennie. *Analyses in Qualitative Research.* London: SAGE, 2010.

Brambell, F. W. R. *Report of the Technical Committee to Enquire into the Welfare of Animals Kept Under Intensive Husbandry Systems.* London, England: Her Majesty's Stationery Office, 1965.

Castka, Pavel, and Michaela A. Balzarova. "A Critical Look on Quality Through CSR Lenses: Key Challenges Stemming from the Development of ISO 26000." *International Journal of Quality and Reliability Management* 24 (2007): 738–752.

Constantinescu, Mihaela, and Muel Kaptein. "Mutually Enhancing Responsibility: A Theoretical Exploration of the Interaction Mechanisms Between Individual and Corporate Moral Responsibility." *Journal of Business Ethics* 129 (May 2015): 325–339.

Corbin, Juliet M., and Anselm L. Strauss. *Basics of Qualitative Research, Techniques and Procedures for Developing Grounded Theory.* California, CA: Sage, 2015.

Delbard, Olivier. "CSR Legislation in France and the European Regulatory Paradox: An Analysis of EU CSR Policy and Sustainability Reporting Practice." *Corporate Governance: The International Journal of Business in Society* 8 (2008): 397–405.

Elkington, John. *Cannibals with Forks: The Triple Bottom Line of 21st Century Business.* Oxford, England: Capstone Publishing, 1997.

Gehman, Joel, Linda K. Treviño and Raghu Garud. "Values Work: A Process Study of the Emergence and Performance of Organizational Values Practices." *Academy of Management Journal* 56 (February 2013): 84–112.

Hoffman, A.J., and M.H. Bazerman. "Changing Practice on Sustainability: Understanding and Overcoming the Organizational and Psychological Barriers." In *Organizations and the Sustainability Mosaic: Crafting Long-term Ecological and Societal Solutions*, edited by B. Husted, 84–105. Cheltenham, England: Edward Elgar Publishing, 2006.

Ingenbleek, Paul, Menno Binnekamp and Silvia Goddijn. "Setting Standards for CSR: A Comparative Case Study on Criteria-formulating Organizations." *Journal of Business Research* 60 (2007): 539–548.

International Standard ISO 26000, Guidance on Social Responsibility. Geneva, Switzerland: ISO, 2010.

Isaacs, Tracy. *Moral Responsibility in Collective Contexts.* Oxford, England: Oxford University Press, 2011.

Janssens, Monique R.E., and Muel Kaptein. "The Ethical Responsibility of Companies Toward Animals: A Study of the Expressed Commitment of the Fortune Global 200." *Journal of Corporate Citizenship* 63 (2016): 42–72.

Martin, Patricia Y., and Barry A. Turner. "Grounded Theory and Organizational Research." *The Journal of Applied Behavioral Science* 22 (1986): 141–157.

Matten, Dirk, and Jeremy Moon. "'Implicit and 'Explicit' CSR: A Conceptual Framework for a Comparative Understanding of Corporate Social Responsibility." *Academy of Management Review* 33 (2008): 404–424.

Mauser, Anniek M. *The Greening of Business. Environmental Management and Performance Evaluation: An empirical Study of the Dutch Dairy Industry*. Delft, The Netherlands: Eburon Publishing, 2001.

McLean, Tammy L., and Michael Behnam. "The Dangers of Decoupling: The Relationship between Compliance Programs, Legitimacy Perceptions, and Institutionalized Misconduct." *Academy of Management Journal* 53 (December 2010): 1499–1520.

McMullen, Steven. "Is Capitalism to Blame? Animal Lives in the Marketplace." *Journal of Animal Ethics* 5, 2 (2015): 126–134.

Mostovicz, Isaac, Nada Kakabadse and Andrew Kakabadse. "CSR: The Role of Leadershipin Driving Ethical Outcomes." *Corporate Governance: The International Journal ofBusiness in Society* 9 (2009): 448–460.

Rollin, Bernard E. *Putting the Horse Before Descartes: My Life's Work on Behalf of Animals*. Philadelphia, PA: Temple University Press, 2011.

Rotter, Julia P., P.-E. Airike, and C. Mark-Herbert. "Category Management in Swedish Food Retail: Challenges in Ethical Sourcing." In *The Ethics of Consumption; The Citizen, the Market and the Law*, edited by Helena Röcklinsberg and Peter Sandin, 54–58. Wageningen, The Netherlands: Wageningen Academic Publishers, 2013.

Schultz, Frederike, and Stefan Wehmeier, "Institutionalization of Corporate Social Responsibility Within Corporate Communications: Combining Institutional, Sensemaking and Communication Perspectives." *Corporate Communications: An International Journal* 15 (2010): 9–29.

Sonenshein, Scott. "Emergence of Ethical Issues During Strategic Change Implementation." *Organization Science* 20 (February 2009): 223–239.

Treviño, Linda K., Niki A. Den Nieuwenboer, Glen E. Kreiner and Derron G. Bishop. "Legitimating the Legitimate: A Grounded Theory Study of Legitimacy Work Among Ethics and Compliance Officers." *Organizational Behavior and Human Decision Processes* 123 (March 2014): 186–205.

Van Liedekerke, Luc, and Wim Dubbink. "Twenty Years of European Business Ethics: Past Developments and Future Concerns." *Journal of Business Ethics* 82 (October 2008): 273–280.

Van Tulder, Rob, Rob van Tilburg, Mara Francken and Andrea da Rosa. *Managing the Transition to a Sustainable Enterprise; Lessons from frontrunner companies*. New York, NY: Routledge, 2013.

Visser, Wayne. *CSR 2.0: Transforming Corporate Sustainability and Responsibility*. Heidelberg, Germany: Springer, 2014.

INDEX